W9-AYB-049

R. D. Mindlin and Applied Mechanics

Raymond D. Mindlin

R. D. Mindlin and Applied Mechanics

A collection of studies in the development of Applied Mechanics

Dedicated to Professor Raymond D. Mindlin by his former students

Edited by

GEORGE HERRMANN

PERGAMON PRESS INC.

New York · Toronto · Oxford · Sydney

PERGAMON PRESS INC.
Maxwell House, Fairview Park, Elmsford, N.Y. 10523

PERGAMON OF CANADA LTD.
207 Queen's Quay West, Toronto 117, Ontario

PERGAMON PRESS LTD.
Headington Hill Hall, Oxford

PERGAMON PRESS (AUST.) PTY. LTD.
Rushcutters Bay, Sydney, N.S.W.

1974

Library of Congress Cataloging in Publication Data
Main entry under title:

R. D. Mindlin and applied mechanics.

CONTENTS: Herrmann, G. Preface.—List of publications of R. D. Mindlin (p. vii)—Drucker, D. C., Duffy, J., and Fox, E. A. Remarks of the past, present, and likely future of photoelasticity.—Goodman, L. E. Development of the three-dimensional theory of elasticity. [etc.]

1. Mechanics, Applied—Addresses, essays, lectures.
2. Mindlin, Raymond David, 1906– —Bibliography.
I. Mindlin, Raymond David, 1906– II. Herrmann, George, 1921– ed.
TA350.3.R2 1974 620.1 73-22346
ISBN 0-08-017710-7

BUS
R

CONTENTS

20.00

PREFACE

THIS VOLUME is dedicated to Professor Raymond D. Mindlin of Columbia University in recognition of his achievements as a teacher of numerous students and as a most influential scholar in Applied Mechanics.

The past several decades have witnessed rapid progress in Applied Mechanics, a branch of Applied Science and Engineering. Much of this progress was made in the area of Mechanics of Solids, and here many of the most significant advances bear the stamp and style of Mindlin's own contributions, which began to appear some forty years ago. This influence grew through the activities of Mindlin's students and others associated with him.

Since Mindlin's research work has appeared in many different publications, it seemed desirable to collect the results and to assemble them between the covers of a single volume. This could have been accomplished, for example, by publishing a representative collection of reprints. Another possible course of action, the one followed in the present volume, consisted of preparing accounts of the development of specific areas of Mechanics of Solids to which Mindlin has contributed most. Eight such areas were identified and are represented by the eight articles of this volume, all of which were written by authors who had been Mindlin's students. It fell upon the undersigned, who considers himself a *de facto* former student, to coordinate this project.

Each of the eight articles attempts to present, in a unified fashion, the substance of Mindlin's contributions, on the one hand, and the general development of a field, on the other hand. Thus each of the articles is an up-to-date and self-contained review and indicates the relevance and significance of Mindlin's own work. Hence, this volume should be regarded also as a tribute to R. D. Mindlin by his former students.

When one examines the evolution of Applied Mechanics during the last several decades, one is led to realize that the work of R. D. Mindlin possesses a distinctive style and character. It is not only a matter of what results are ultimately achieved, but also why and how. Above all, Mindlin's work is motivated by a concern for physical reality. His analytical studies always begin with an intense desire to explain and interpret, in mathematical terms, observed but poorly understood physical phenomena. It might suffice to mention, as examples, problems of elastic contact and of granular media related to the behavior of the carbon microphone and the problem areas of generalized elastic continua and anisotropic plates, related to the behavior of quartz crystals as frequency standards, filters, delay lines and transducers. His continued efforts to bridge the gap between continuum and lattice theories aided decisively in explaining certain observed anomalies. Mindlin's experimental studies in photoelasticity are characterized by attention and keen appreciation both of practical and theoretical considerations.

In planning the contents of the reviews, it became apparent that not all areas of Mindlin's work could be readily covered in this volume. Mention should be made in particular of Mindlin's contributions to the dynamics of package cushioning, which became a classical study and which has had much influence on the design of supports and containers for the protection of equipment from mechanical shock and vibration. Further, his work on the interaction of elastic shells with a surrounding fluid helped to set the stage for numerous later investigations on dynamic response of submerged structures.

In this Preface we have omitted accounts of Mindlin's dedicated activities in several professional societies and organizations, his work as a consultant to various Government agencies, and his many years of service and loyalty to Columbia University. In recognition of these, numerous awards and honors have been bestowed upon him, of which we mention here only the Presidential Medal for Merit, the Von Karman Medal of the American Society of Civil Engineers, and the Timoshenko Medal of the American Society of Mechanical Engineers. What his students surely remember best is what and how he taught them in the classroom and after lectures, as well as during discussions of research projects. It is the memory of those years which will live on in this volume.

GEORGE HERRMANN
Stanford University

List of Publications of R. D. Mindlin[†]

1. A Reflection Polariscope for Photoelastic Analysis, *Review of Scientific Instruments* **5,** 224–228 (1934).
2. Welded Joints Studied with New Type Polariscope, *Engineering News-Record,* 621–622 (1934).
3. Contribution au problème d'équilibre de l'élasticité d'un solide indéfini limité par un plan, *Comptes Rendus* **201,** 536 (1935).
4. Force at a Point in the Interior of a Semi-Infinite Solid, *Physics,* **7,** 195–202 (1936).
5. Note on the Galerkin and Papkovitch Stress Functions, *Bulletin of the American Mathematical Society* **42,** 373–376 (1936).
6. Pressure Distribution on Retaining Walls, *Section J—Earth Pressure, Proceedings of the International Conference on Soil Mechanics,* Cambridge, Mass., Vol. III (1937).
7. On the Distribution of Stress around a Pile, *Section E—Stress Distribution in Soils, Proceedings of the International Conference on Soil Mechanics,* Cambridge, Mass., Vol. III (1937).
8. Analysis of Doubly Refracting Materials with Circularly and Elliptically Polarized Light, *Journal of the Optical Society of America* **27,** 288–291 (1937).
9. Stress Systems in Circular Disk under Radial Forces, *Journal of Applied Mechanics* **4,** A115–118 (1937).
10. Distortion of the Photoelastic Fringe Pattern in an Optically Unbalanced Polariscope, *Journal of Applied Mechanics* **4,** A170–172 (1937).
11. The Equiangular Strain-Rosette, *Civil Engineering* **8,** 546–547 (1938).
12. Gravitational Stresses in Bi-polar Coordinates, *Proceedings of the Fifth International Congress of Applied Mechanics,* Cambridge, Mass., pp. 112–116 (1938).
13. Stresses in a Heavy Disk Suspended from an Eccentric Peg, *Journal of Applied Physics* **9,** 714–717 (1938).
14. Stresses in an Eccentrically Rotating Disk, *London, Edinburgh and Dublin Philosophical Magazine,* Ser. 7, **26,** 713–719 (1938).
15. Stress Distribution around a Tunnel, *Proc. Amer. Soc. C.E.* 619–642 (1939).
16. A Review of the Photoelastic Method of Stress Analysis, *Journal of Applied Physics* **10,** 222–241, 273–294 (1939).
17. An Extension of the Photoelastic Method of Stress Measurement of Plates in Bending, *Journal of Applied Mechanics* **8,** A187–189 (1941).
18. Stress Analysis by Three-Dimensional Photoelastic Methods (with D. C. Drucker), *Journal of Applied Physics* **11,** 724–732 (1940).
19. Optical Aspects of Three-Dimensional Photoelasticity, *Journal of the Franklin Institute* **233,** 349–363 (1942).
20. The Present Status of Three-Dimensional Photoelasticity, *Civil Engineering* **12,** 255–258 (1942).
21. Dynamics of Package Cushioning, *Bell System Technical Journal* **24,** 353–461 (1945).

† These publications are referred to in text by the letter M before a number in brackets, e.g., [M2].

22. Response of Damped Elastic Systems to a Pulse Acceleration, *6th International Congress of Applied Mechanics*, Paris, France, September 1946.
23. Analogy between Multiply-Connected Slabs and Slices, *Quarterly of Applied Mathematics* **4**, 279–290 (1946).
24. Activities of the Society for Experimental Stress Analysis, *International Technical Congress*, Paris, France, September 1946, and *Proceedings of the Society for Experimental Stress Analysis*, **4**(2), ix–xi (1947).
25. Stress Distribution around a Hole near the Edge of a Plate under Tension, *Proceedings of the Society for Experimental Stress Analysis* **5**(2), 56–68 (1948).
26. Response of Damped Elastic Systems to Transient Disturbances (with W. F. Stubner and H. L. Cooper), *Proceedings of the Society for Experimental Stress Analysis* **5**(2), 69–87 (1948).
27. The Optical Equations of Three-Dimensional Photoelasticity (with L. E. Goodman), *Journal of Applied Physics* **20**, 89–95 (1949). Also, *Proceedings of the Seventh International Congress for Applied Mechanics*, London, 1948, p. 353.
28. A Mathematical Theory of Photo-Viscoelasticity, *Journal of Applied Physics* **20**, 206–216 (1949).
29. Compliance of Elastic Bodies in Contact, *Journal of Applied Mechanics* **16**, 259–268 (1949).
30. Thermo-Elastic Stress around a Cylindrical Inclusion of Elliptic Cross-Section (with H. L. Cooper), *Journal of Applied Mechanics* **17**, 265–268 (1950).
31. Beam Vibrations with Time-Dependent Boundary Conditions (with L. E. Goodman), *Journal of Applied Mechanics* **17**, 377–380 (1950).
32. Analogies (with M. G. Salvadori), *Handbook of Experimental Stress Analysis*, Wiley, New York, 1950, pp. 700–827.
33. Nuclei of Strain in the Semi-Infinite Solid (with David H. Cheng), *Journal of Applied Physics* **21**, 926–930 (1950).
34. Thermo-Elastic Stress in the Semi-Infinite Solid (with David H. Cheng), *Journal of Applied Physics* **21**, 931–933 (1950).
35. Influence of Rotatory Inertia and Shear on Flexural Motions of Isotropic, Elastic Plates, *Journal of Applied Mechanics* **18**, 31–38 (1951).
36. Thickness-Shear and Flexural Vibrations of Crystal Plates, *Journal of Applied Physics* **22**, 316–323 (1951).
37. Development of the Mathematical Theory of Three-Dimensional Photoelasticity, *Applied Mechanics Reviews* **4**, 537–539 (1951).
38. Forced Thickness-Shear and Flexural Vibrations of Piezoelectric Crystal Plates, *Journal of Applied Physics* **23**, 83–88 (1952).
39. A One-Dimensional Theory of Compressional Waves in an Elastic Rod (with G. Herrmann), *Proceedings of the First U.S. National Congress of Applied Mechanics*, pp. 187–191 (1951).
40. Effects of an Oscillating Tangential Force on the Contact Surfaces of Elastic Spheres (with W. P. Mason, T. F. Osmer and H. Deresiewicz), *Proceedings of the First National Congress of Applied Mechanics*, pp. 203–208 (1951).
41. Response of an Elastic Cylindrical Shell to a Transverse, Step Shock-Wave (with H. H. Bleich), *Journal of Applied Mechanics*, **20**, 189–195 (1953).
42. Elastic Spheres in Contact under Varying Oblique Forces (with H. Deresiewicz), *Journal of Applied Mechanics*, **20**, 327–344 (1953).

43. Thickness-Shear and Flexural Vibrations of Contoured Crystal Plates (with M. Forray), *Journal of Applied Physics* **25,** 12–20 (1954).
44. Thickness-Shear Vibrations of Piezoelectric Crystal Plates with Incomplete Electrodes (with H. Deresiewicz), *Journal of Applied Physics* **25,** 21–24 (1954).
45. Suppression of Overtones of Thickness-Shear and Flexural Vibrations of Crystal Plates (with H. Deresiewicz), *Journal of Applied Physics* **25,** 25–27 (1954).
46. Force at a Point in the Interior of a Semi-Infinite Solid, *Proceedings of the First Midwestern Conference on Solid Mechanics,* 56–59 (1955).
47. Thickness-Shear and Flexural Vibrations of a Circular Disk (with H. Deresiewicz), *Journal of Applied Physics* **25,** 1329–1332 (1954).
48. Axially Symmetric Flexural Vibrations of a Circular Disk (with H. Deresiewicz), *Journal of Applied Mechanics* **22,** 86–88 (1955).
49. Mechanics of Granular Media, *Proceedings of the Second U.S. National Congress of Applied Mechanics,* 13–20 (1954).
50. Timoshenko's Shear Coefficient for Flexural Vibrations of Beams (with H. Deresiewicz), *Proceedings of the Second U.S. National Congress of Applied Mechanics,* pp. 175–178 (1954).
51. Mathematical Theory of Three-Dimensional Photoelasticity, Colloque de Photoélasticité et Photoplasticité, Union Internationale de Mécanique Théorétique et Appliquée, Bruxelles, 1954, p. 47.
52. Thickness-Shear and Flexural Vibrations of Rectangular Crystal Plates (with H. Deresiewicz), *Journal of Applied Physics* **26,** 1435–1442 (1955).
53. Flexural Vibrations of Rectangular Plates (with A. Schacknow and H. Deresiewicz), *Journal of Applied Mechanics* **23,** 430–436 (1956).
54. High-Frequency Extensional Vibrations of Plates (with T. R. Kane), *Journal of Applied Mechanics* **23,** 277–283 (1956).
55. Simple Modes of Vibration of Crystals, *Journal of Applied Physics* **27,** 1462–1466 (1956).
56. Waves on the Surface of a Crystal (with H. Deresiewicz), *Journal of Applied Physics* **28,** 669–671 (1957).
57. Velocities of Long Waves in Plates (with D. C. Gazis), *Journal of Applied Mechanics,* **24,** 541–546 (1957).
58. Stress-Strain Relations and Vibrations of a Granular Medium (with J. Duffy), *Journal of Applied Mechanics* **24,** 585–593 (1957).
59. Vibrations of a Monoclinic Crystal Plate (with E. G. Newman), *Journal of the Acoustical Society of America* **29,** 1206–1218 (1957).
60. Vibrations of an Infinite, Elastic Plate at Its Cut-Off Frequencies, *Proceedings of the Third U.S. National Congress of Applied Mechanics,* pp. 225–226 (1958).
61. Extensional Vibrations of Elastic Plates (with M. A. Medick), *Journal of Applied Mechanics* **26,** 561–569 (1959).
62. Axially Symmetric Waves in Elastic Rods (with H. D. McNiven), *Journal of Applied Mechanics* **27,** 145–151 (1960).
63. Vibrations and Waves in Elastic Bars of Rectangular Cross Section (with E. A. Fox), *Journal of Applied Mechanics* **27,** 152–158 (1960).
64. Dispersion of Flexural Waves in Elastic, Circular Cylinder (with Y. H. Pao), *Journal of Applied Mechanics* **27,** 513–520 (1960).
65. Extensional Vibrations and Waves in a Circular Disk and a Semi-Infinite Plate (with D. C. Gazis), *Journal of Applied Mechanics* **27,** 541–545 (1960).

66. Waves and Vibrations in Isotropic, Elastic Plates, *Structural Mechanics*, Pergamon Press, New York, 1960, pp. 199–232.
67. On the Equations of Extensional Motion of Crystal Plates (with H. L. Cooper), *Quart. Appl. Math.* **19,** 111–118 (1961).
68. On the Equations of Motion of Piezoelectric Crystals, *Problems of Continuum Mechanics*, Soc. for Ind. and Appl. Math., Philadelphia, 1961, pp. 282–290. Muskhelishvili Anniversary Volume.
69. High Frequency Vibrations of Crystal Plates, *Quart. Appl. Math.* **19,** 51–61 (1961).
70. Forced Vibrations of Piezoelectric Crystal Plates (with H. F. Tiersten), *Quart. Appl. Math.* **20,** 107–119 (1962).
71. Vibrations of an Infinite, Monoclinic Crystal Plate at High Frequencies and Long Wave Lengths (with R. K. Kaul), *Journal of the Acoustical Society of America* **34,** 1895–1901 (1962).
72. Frequency Spectrum of a Monoclinic Crystal Plate (with R. K. Kaul), *Journal of the Acoustical Society of America* **34,** 1902–1910 (1962).
73. Extensional Vibrations of Thin Quartz Disks (with A. G. Lubowe), *Journal of the Acoustical Society of America* **34,** 1911–1918 (1962).
74. Dispersion of Axially Symmetric Waves in Elastic Rods (with M. Onoe and H. D. McNiven), *Journal of Applied Mechanics* **29,** 729–734 (1962).
75. Strong Resonances of AT-Cut Quartz Plates (with D. C. Gazis), *Proc. Fourth U.S National Congress of Applied Mechanics*, pp. 305–310 (1962).
76. Effects of Couple-stresses in Linear Elasticity (with H. F. Tiersten), *Archive for Rational Mechanics and Analysis* **11,** 415–448 (1962).
77. Influence of Couple-stresses on Stress Concentrations, *Experimental Mechanics* **3,** 1–7 (1963).
78. Complex Representation of Displacements and Stresses in Plane Strain with Couple-stresses, *Proceedings of the International Symposium on Applications of the Theory of Functions in Continuum Mechanics*, Tbilisi, USSR, 1963, pp. 256–259.
79. Extensional Waves Along the Edge of a Plate (with J. J. McCoy), *Journal of Applied Mechanics* **30,** 75–78 (1963).
80. High Frequency Vibrations of Plated Crystal Plates, *Progress in Applied Mechanics*, Macmillan, New York, 1963, pp. 73–84. Prager Anniversary Volume.
81. Micro-structure in Linear Elasticity, *Archive for Rational Mechanics and Analysis* **16,** 51–78 (1964).
82. On the Equations of Elastic Materials with Micro-structure, *International Journal of Solids and Structures* **1,** 73–78 (1965).
83. Thickness-twist Vibrations of an Infinite, Monoclinic, Crystal Plate, *International Journal of Solids and Structures,* **1,** 141–145 (1965).
84. Stress Functions for a Cosserat Continuum, *International Journal of Solids and Structures* **1,** 265–271 (1965).
85. Second Gradient of Strain and Surface-tension in Linear Elasticity, *International Journal of Solids and Structures,* **1,** 417–438 (1965).
86. Thickness-shear and Flexural Vibrations of Partially Plated, Crystal Plates (with P. C. Y. Lee), *International Journal of Solids and Structures* **2,** 125–139 (1966).
87. Bechmann's Number for Harmonic Overtones of Thickness-Twist Vibrations of Rotated-Y-Cut Quartz Plates, *Journal of the Acoustical Society of America* **41,** 969–973 (1967).
88. Anharmonic, Thickness-twist Overtones of Thickness-shear and Flexural Vibrations of

Rectangular, AT-cut, Quartz Plates (with W. J. Spencer), *Journal of the Acoustical Society of America* **42,** 1268–1277 (1967).

89. On First Strain-gradient Theories in Linear Elasticity (with N. N. Eshel), *International Journal of Solids and Structures* **4,** 109–124 (1968).
90. Polarization Gradient in Elastic Dielectrics, *International Journal of Solids and Structures* **4,** 637–642 (1968).
91. Optimum Sizes and Shapes of Electrodes for Quartz Resonators, *Journal of the Acoustical Society of America* **43,** 1329–1331.
92. Face-shear Waves in Rotated-Y-cut Quartz Plates, *The Joseph Marin Memorial Volume,* University of Toronto Press, 1968, pp. 143–145.
93. Theories of Elastic Continua and Crystal Lattice Theories, *Mechanics of Generalized Continua,* Springer-Verlag, 1968, pp. 312–320.
94. Lattice and Continuum Theories of Simple Modes of Vibration in Cubic Crystal Plates and Bars *Symposium on Dynamics of Structural Solids,* American Society of Mechanical Engineers, 1968, pp. 2–9.
95. Note on Vibrations of Prisms and Polygonal Plates, *The L. I. Sedov Anniversary Volume,* USSR Academy of Sciences and Society for Industrial and Applied Mathematics, Philadelphia, 1969.
96. Crystal Lattice Theory of Torsion of a Rectangular Bar of Simple Cubic Structure, The B. Seth Anniversary Volume, *Indian Journal of Mechanics and Mathematics,* 1969.
97. Continuum and Lattice Theories of Influence of Electro-Mechanical Coupling on Capacitance of Thin Dielectric Films, *International Journal of Solids and Structures* **5,** 1197–1208 (1969).
98. Lattice Theory of Shear Modes of Vibration and Torsional Equilibrium of Simple-Cubic Crystal Plates and Bars, *International Journal of Solids and Structures* **6,** 725–738 (1970).
99. *Polarization Gradient in Elastic Dielectric,* International Centre for Mechanical Sciences, Udine, Italy, 1970.
100. Thickness-Twist Vibrations of a Quartz Strip, *International Journal of Solids and Structures* **7**(1), 1–4 (1971).
101. Forced Electromechanical Vibrations of a Sodium Chloride Plate. (Published as Item 105 below.)
102. Acoustical and Optical Activity in Alpha Quartz (with R. A. Toupin) *International Journal of Solids and Structures* **7**(9), 1219–1228 (1971).
103. A Continuum Theory of a Diatomic, Elastic Dielectric, *International Journal of Solids and Structures* **8**(3), 369–384 (1972).
104. Coupled Elastic and Electromagnetic Fields in a Diatomic, Dielectric Continuum *International Journal of Solids and Structures* **8**(4), 401–408 (1972).
105. Electromechanical Oscillations of Centro-Symmetric Crystal Plates with Cubic Lattice, *Applied Mathematics and Mechanics* (*PMM*) **35** (3), 446–450 (1971).
106. Galerkin Stress Functions for Non-Local Theories of Elasticity, *Galerkin Anniversary Volume,* edited by A. Y. Ishlinski, Moscow, USSR (1972).
107. High Frequency Vibrations of Piezoelectric Crystal Plates, *International Journal of Solids and Structures* **8,** 895–906 (1972).
108. Equations of High Frequency Vibrations of Thermopiezoelectric Crystal Plates, *International Journal of Solids and Structures,* forthcoming.
109. Elasticity, Piezoelectricity and Crystal Lattice Dynamics, *Journal of Elasticity* **2,** 217 (1972).

110. On the Electrostatic Potential of a Point Charge in a Dielectric Solid, *International Journal of Solids and Structures* **9**, 233–236 (1973).
111. Electromagnetic Radiation from a Vibrating Quartz Plate, *International Journal of Solids and Structures* **9**, 697–702 (1973).
112. Coupled Piezoelectric Vibrations of Quartz Plates, *International Journal of Solids and Structures*, **4**, 453–460 (1974).

Remarks on the Past, Present, and Likely Future of Photoelasticity

D. C. DRUCKER, J. DUFFY, and E. A. FOX

University of Illinois (Urbana), Brown University, and Rensselaer Polytechnic Institute

Abstract—The *Applied Mechanics Reviews* summary paper by Mindlin† is taken as the point of departure for a review of the past and present status of photoelasticity and a projection of the future. Some later developments are described and supplemental remarks made on earlier work. Recent advances in optical techniques and in optical analysis are seen as opening the way to many more new and valuable experimental and theoretical studies. Far more sophisticated modeling of materials and the more incisive use of photoelasticity as a tool of materials science and engineering appear as likely consequences.

1 Introduction

The optical techniques of experimental mechanics have enormous appeal beyond their intrinsic intellectual content and their utility in practice. Over the past fifty years photoelasticity, holography, and moiré methods provided one of the most useful experimental methods to research worker and practicing engineer alike. High as this utility has been, the continued dominance today of optically oriented papers in the technical literature probably is more a result of esthetic satisfaction and of the host of fascinating observations which challenge the researcher to provide explanations.

In the last few years finite element methods have taken over many of the utilitarian functions which photoelasticity used to serve. Simultaneously, however, fundamental advances in our understanding of materials, and the desire for further advances, have opened new areas for photoelastic research. As a result one can foresee for the coming years profound changes in the directions of research and the areas of application of photoelasticity.

There will be greater understanding and use made of the meaning of the birefringence which is produced in a wide variety of materials. Attention will be directed to large deformations with both large strain and large rotation in the visco-elastic-plastic domain. Birefringence of material, as a material property, will be given far more attention in living as well as in inanimate fluids and solids. Crystals and polycrystals will be studied more for their own sake.

† Reprinted in full as an Appendix to this paper.

Somewhat as in the early days of purely linear elastic studies in which the photoelastic technique was checked by simple known solutions and the technique used to obtain new solutions, practical problems will be solved with complex material behavior and the birefringence patterns obtained will be employed to help develop usable constitutive relations. With greater understanding, optically sensitive materials will be tailored to model complex prototype materials and so lead rather than follow the mathematical modeling. Fiber reinforced plastics are an obvious candidate, but all problems involving failure of viscoelastic materials exceed our present capacity to model mathematically beyond a crude overall first approximation. Finite element techniques or matrix representation for machines, structures, and continua will be combined naturally with matrix techniques of optical analysis to learn far more from birefringence readings and patterns than can be determined now. Many of the tools now are at hand.

It is remarkable how much of the present high state of development shines out in the very brief summary article by R. D. Mindlin included as an appendix to this paper. Written in 1951 it was updated through 1963 but otherwise revised only a little. There is every reason to suppose that the trends will continue which Mindlin did so much to establish: clear understanding of modeling, careful examination of the basic and applied aspects of optics, fundamental understanding of mechanics and materials, an appreciation of the practical along with the theoretical, and the importance of doing first-class experimental work. In the move we foresee to far greater complexity of material properties, shape, loading, and deformation in the static and dynamic range, there will be many opportunities for error. Careful attention must be devoted to all aspects of each study from the delineation of the problem, through the experimental work and its interpretation, to the solution.

2 A Supplement to Mindlin's Applied Mechanics Review's Summary†

The discovery that transparent materials become birefringent when placed under load in polarized light occurred early in the nineteenth century and is due to Brewster [1]. However, photoelasticity did not become a practical method of stress analysis until much later. This was due in part to the poor optical sensitivity of the materials then available and also to a lack of understanding of the underlying causes of the photoelastic effect. Over the years many investigators contributed to the understanding of the fringe patterns and to their interpretation for purposes of experimental stress analysis. Among these the earliest is Neumann who performed both analytic and experimental studies [2]. He established theorems which made it possible to evaluate the difference in the principal strains and also their directions. Maxwell also con-

† See Appendix.

tributed to the field, as did Wertheim through his experiments in simple tension and compression [3, 4]. The first practical application of photoelasticity, however, did not occur until almost one hundred years after its discovery. In 1913 Mesnager constructed a complete glass model of a bridge in order to check the stress calculations [5]. Vertical loads were applied by springs and even an axial load could be imposed. A Babinet compensator was used in this investigation.

Coker and Filon's treatise on photoelasticity appeared in 1931 [6]. It combined the available knowledge in elasticity and optics; added much practical information and made many original contributions. The result was a landmark in the history of the subject. It was now possible for laboratories to be established throughout the world which could make further advances as well as employ the available techniques in design or analysis. Their treatise, furthermore, came at a time when new materials were appearing and when improvements could be made in the polariscopes. Coker, himself, had introduced "Celluloid" as a photoelastic material in 1906 and Filon started using "Bakelite" in 1922. "Polaroid" sheets were to appear shortly after the publication of this treatise so that the overall result was a very great increase in interest in photoelasticity throughout the world during the 1930s. The field attracted many workers in solid mechanics because the technique promised solutions to boundary-value problems of elastostatics which were then not accessible to analysis. The early work dealt with photoelastic investigations of plane stress for which the existing theory seemed adequate, the check of theory being the photoelastic analysis of problems with known analytic solutions.

Mindlin, however, was concerned equally with the validity of the experimental procedures. He provided, for instance, a detailed analysis of the function of the quarter wave plates in the polariscope. In the ordinary monochromatic polariscope two quarter wave plates are generally used, one to either side of the specimen, to produce a circular polariscope and so eliminate the isoclinics from the observed fringe pattern (Fig. 1). When the axis of the

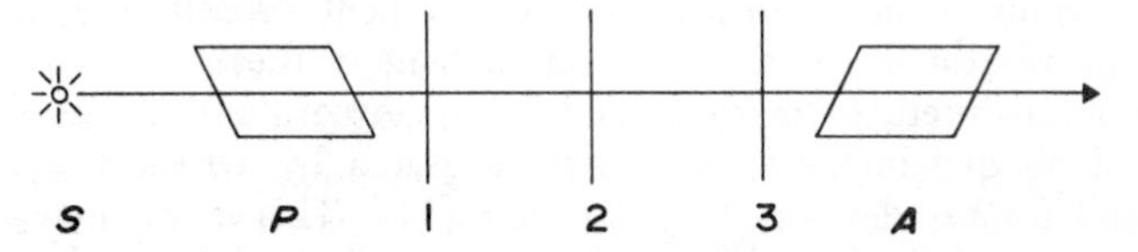

Fig. 1. Schematic diagram of photoelastic polariscope. (S = monochromatic light source, P = polarizer, 1 and 3 = retardation plates, 2 = model, A = analyzer.) (Source: From R. D. Mindlin "Distortion of the Photoelastic Fringe Pattern in an Optically Unbalanced Polariscope" [M10]. Reproduced by permission of *Journal of Applied Mechanics*.)

first quarter wave plate is at 45° to the polarizer, the light through the model is circularly polarized. With the second quarter wave plate and the analyzer in the crossed position, the observed fringe pattern for a specimen provides a measure of the principal stress difference at each point of the body. The question arises, however, as to the influence of any error in the relative retardations of each of the quarter wave plates. As a practical matter it is easy to obtain a plate with a uniform relative retardation, but it is difficult and expensive to make this relative retardation precisely equal to one quarter of the wavelength of the light used. Mindlin presented an analysis of circularly and elliptically polarized light going through a doubly refracting material and showed that if the two retardation plates are crossed and identical, the isochromatic-fringe pattern is unaffected by inaccuracies in the plates (although the isoclinics are not completely removed) [M8 and M10].† However, if the retardation plates are not identical, the fringe pattern is distorted. Mindlin provided all the necessary equations for calculating the error and showed, by an example as well as analytically, that the error is negligible for small errors in the retardation plates. Photoelasticians then could employ available quarter wave plates properly with confidence particularly if the plates were cut from the same sheet.

One of the major problems in the early days of photoelasticity was the interpretation of fringe patterns in terms of individual values of stress. Coker and Filon covered most of these methods extensively and they will not be repeated here. Many investigators devised ingenious methods for the determination of the individual stress components. These range from analytic methods based on equilibrium, to measurements of the sum of the principal stresses as suggested by Mesnager [7] and performed either by lateral extensometers of which an early example is due to Coker himself [8] or by optical methods as in Favre's interferometer [9]. Other methods available include use of the membrane analogy. In practice, however, most of the experimental methods are found wanting, generally because they involve delicate measurements to be made with a high degree of precision. As a result the methods used in practice are mainly the analytic methods mentioned above and the method of oblique incidence The analytic methods are described by Frocht in his classic volumes on photoelasticity [10]. Frocht himself had been an important contributor in the development of some of these methods, in particular the shear-difference method, and he made constant use of them. The method of oblique incidence was first suggested by Drucker and Mindlin [M18] and further developed by Drucker [11]. This is a full-field method requiring as specialized equipment only the ability to rotate the specimen. Its accuracy has been shown to be quite high for double oblique incidence.

† Numbers preceded by the letter M in brackets refer to the Publications of R. D. Mindlin listed after the Preface to this volume.

It has also the advantage that it can be applied very easily to analyze "frozen stress" patterns.

The technique of freezing and slicing was developed toward the end of the 1930s. It provided a means of evaluating internal stresses experimentally and, to this day, remains the most effective experimental technique available for this purpose. Although a number of investigators (Maxwell [12], Filon and Harris [13], Tuzi [14], and Solakian [15]) had been aware of the existence of "frozen stresses" and had attempted to explain their significance as well as to use them in stress analysis, it was not until the work of Oppel in 1936 [16] that the technique was well understood. Hetényi [17], Kuske [18], Hiltscher [19], and Frocht and Guernsey [20] provided explanations of the frozen stress phenomenon and presented a number of methods to evaluate the complete stress state of the body. They also performed experiments employing the technique. Nevertheless there remained questions as to the relationship between the interference pattern observed in the slices and the state of stress in the prototype. These questions are of two types. The first deals with the mode of propagation of light in an anisotropic heterogeneous material, and the second deals with the troublesome question implied in the Introduction of the relationship between the stress field in a viscoelastic model and the stress field in an elastic prototype. Mindlin made major contributions toward the resolution of these two questions. Without these contributions the photoelastic fraternity could never have been quite sure of the validity of their results. Mindlin and his students [M17, M18, M27] showed that the equations used by photoelasticians in relating fringe patterns to stress fields, usually derived in a somewhat *ad hoc* fashion, were approximations within electromagnetic theory valid for typical applications. He also proved that the practice of transferring the results of tests on viscoelastic models to elastic prototypes could be safely done if the problem satisfied a set of necessary and sufficient conditions that he explicitly set forth [M28]. A variety of very useful and interesting applications of the "frozen stress" method have been made in many countries in connection with the analysis and design of pressure vessels. Among others, the work of Prigorovsky, Durelli, Leven, Sampson, Taylor, Hiltsher, Fessler, and Mönch [21–23] has proved helpful in the development of design codes.

There are, of course, other photoelastic methods besides "frozen stress" for finding internal stresses. For particular states of stress, e.g., radial symmetry, the internal stress can be calculated by passing light through the body, measuring the total relative retardation, and combining the result with elasticity theory. This approach was first suggested by Poritsky [24]. Mindlin's review, which is included as an appendix to this article, contains a description of progress in this area of the subject.

Another method available for the analysis of stress in three-dimensional bodies utilizes the scattered light phenomenon in plastics discovered in 1939 by Weller [25]. This scattering effect may be employed either as polarizer or as analyzer. In either case it affords a technique for the measurement of internal stresses without the need for freezing and slicing. Description of the method with advantages and disadvantages has been presented by Drucker [26]. Jessop following Refs. [M12] and [26] separated the principal stresses in a two-dimensional problem [27]. Now that a variety of lasers are available at moderate cost, scattered light techniques can be employed for dynamic as well as static problems. The complementary if not simultaneous development of theory and the technique of application enabled significant progress to occur while solving real problems. Over the course of the years so many technical advances have been made that only a practicing photoelastician can appreciate their worth. They involve all parts of photoelasticity from light sources, to the polariscope, to loading techniques, materials, photography, reduction of data, etc. It is impossible in a review of this sort to describe the range of the problems faced and overcome, such as the difficulties involved in time-edge effects or in machining specimens, or in casting special birefringent materials. The reader, therefore, is referred to a paper by Leven [23], books and articles on stress analysis, as for instance Hetenyi [28], Dove and Adams [29], Dally and Riley [30], and Hendry [31], and to current and past issues of *Experimental Mechanics*, the Proceedings of the Society for Experimental Stress Analysis. An extensive reference list through 1959 is contained in the book by H. Wolf [32]. Two techniques, however, do warrant additional attention because of their widespread application. These are the methods of measuring fractional fringe orders and the reflection polariscope.

The Babinet compensator has been in use for over one-hundred years. Along with the Soleil–Babinet compensator it depends for its effect on wedge-shaped prisms of naturally refracting materials moved by calibrated screws in and out of the field. The Coker compensator, on the other hand, uses a previously calibrated birefringent material with a convenient loading method. There have been numerous variations in individual laboratories. A different compensating method, due to Tardy [33], involves rotation of the analyzer. Today, in some polariscopes, this rotation is effected automatically. All these methods of compensation are described very clearly in the article by Dolan and Murray [34]. Interest has remained high during the intervening years [35].

Another method of measuring fractional fringe orders is Post's well-known technique of fringe multiplication [36] by means of multiple reflections. It was used by Mylonas and Drucker [37] to measure the stress in twisted tape and by Day, Kobayashi, and Larson [38] and by Bohler and Schumann [39] to improve the resolution of birefringent coatings.

One of the consequences of the considerable use of birefringent coatings is the return of the reflection polariscope as one of the most important tools of photoelasticity. The optical arrangement is shown in Fig. 2. As Mindlin states [M1], "the chief characteristic of the optical arrangement is that the light, after passing through the usual arrangement of polarizing prism, retar-

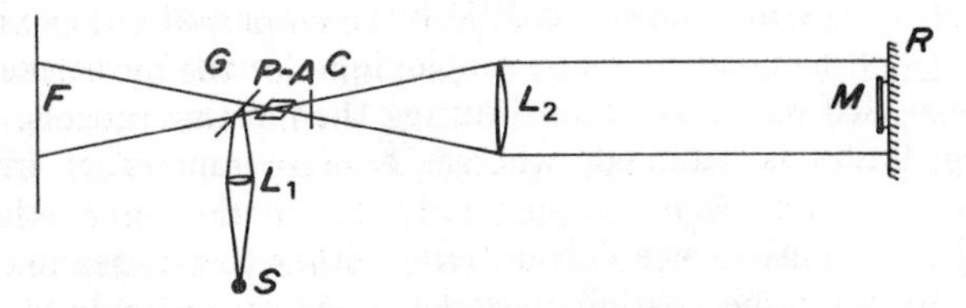

Fig. 2. Optical arrangement of the reflection polariscope. (Source: From R. D. Mindlin "A Reflection Polariscope for Photoelastic Analysis" [M1]. Reproduced by permission of *Review of Scientific Instruments.*)

dation plate and model, is reflected directly back through the same system." This idea has been used extensively by many investigators. The reflection polariscope is easy to install, requires less room than the conventional polariscope and is convenient to use. The results are more distinct fringes and twice as many of them. The instrument can be used also whenever a mirror is imbedded within a photoelastic model or whenever a birefringent coating is used. As such, it is the basis of the birefringent coating technique, of so-called photoelastic strain gages, etc.

Mindlin presented the idea of the reflection polariscope in his characteristic fashion with examples showing applications to a number of practical problems, and an analytic treatment of the method. Details are carefully worked out. A beam of light comes in at right angles to the collimation axis of the instrument with a clear glass plate to reflect the light (Fig. 2). Some later investigators in setting up their reflection polariscope reversed the position of light source and image. This resulted in multiple reflections and a large loss in light intensity, both of which could have been avoided [M17] by a more careful reading of Mindlin's original paper [M1].

In photoelasticity, the idea of using a mirrored surface which is integral with a model or prototype came to Mindlin's mind at about the same time as to Mesnager's [40]. These two investigators, however, made somewhat different use of the mirror. Mesnager suggested finding the strains at the surface of a metal, or other opaque body, by polishing the outer surface of the metal to a mirror finish and measuring the relative retardation induced in a birefringent coating cemented to the mirrored surface when the body is loaded. No model of the prototype is needed and deformations are not limited to elastic ones. However, Mesnager did not have either appropriately sensitive birefringent materials or adequate adhesives. These difficulties account for the almost twenty year delay from Oppel's experimental demonstration [41] in 1937 of the value of Mesnager's concept to its real use. The method has since been used successfully to study a number of problems, from Dantu's exploratory study of the deformation of concrete [42] as well as the more recent work in the same area by Khesin and Sakharov [43], and by Swamy [44], to the fracture in tension of plates of various materials by Dixon and Visser [45].

In Mindlin's version, followed in 1941 by Goodier and Lee [46] and Papirno and Becker [47], a model is made employing a birefringent material, but a mirror is embedded within the model during the forming process. A fairly high relative retardation is attained, whereas a high number of fringes can be attained with a birefringent coating only by employing a relatively thick coating and as a consequence encountering either thickness effects [48] or a reinforcing effect by the coating itself [49]. As an example of the method, Mindlin studied the stresses in welded joints [M2] doing all the work himself, from setting up the polariscope, to machining and assembling the specimen, performing the photography and interpreting results.

3 Dynamic Photoelasticity

Dynamic photoelasticity is a very active field today and promises to remain so for some time to come. A more accurate name than dynamic photoelasticity would be perhaps birefringent methods applied to dynamic problems in mechanics. The reason for this is that the field includes dynamic photoviscoelasticity, the study of plastic deformation in metals by means of birefringent coatings, as well as the study of fracture phenomena. An excellent review of the field up to 1962 is Goldsmith's article in *Experimental Techniques in Shock and Vibration* [50].

Dynamic work in photoelasticity requires either a high-speed camera and a strong light source or a light source of very short duration combined with a means of moving the image on the film. In the latter case there are advantages if the light source can give repeated flashes. Progress in the field therefore was limited by the rate at which either of these instruments was developed. In spite of this, however, Tuzi [51] presented a series of papers starting in 1928 on dynamic photoelasticity, while in 1932 Frocht made use of a high-speed movie camera to record moving fringes [52]. Since even a high-speed camera in those days took a maximum of only sixty-four frames per second, Frocht distinguished individual fringes by making use of a low-modulus plastic. Nonetheless, the patterns photographed represent the quasi-static stress state at various loads rather than the stress waves. An advance occurred in 1946 when Senior and Wells used a flash source of one microsecond duration and a material of medium modulus to photograph stress waves [53]. Following this early work a growing number of studies has been made employing various specimen materials, cameras and light sources. Durelli, along with Dally and Riley, have obtained remarkable sets of consecutive photographs showing a pulse traveling in a bar or in a plate and giving the details of the state of stress at various distances and after reflections. They employ principally low-modulus materials so that dispersion occurs both for reasons of geometry and internal damping [54–56]. They extended their work to the study of dynamic stress concentration due to stress waves incident on holes of various geometries. References to most of this work appear in Durelli and Riley's text [57]. Similar work has been done by Khesin, Kostin, and Zateev [58].

P. D. Flynn has used both low-modulus and high-modulus materials in his work [59]. He and Frocht showed how to separate principal stresses under dynamic conditions [60] and Flynn constructed a special polariscope to perform simultaneously oblique incidence and normal incidence studies in dynamic photoelasticity [61]. This is an excellent approach to the problem of resolving individual stress components; a problem which is generally much more complicated for dynamic than for static conditions. Frocht and Flynn have also made great use of streak photography [62], which had been used as early as 1935 by Tuzi and Nisida [63].

The use of an intermittent source of light in dynamic photoelasticity has not progressed as rapidly as might be expected following the early work of Senior and Wells. One difficulty is the need for a repeated flashing light of sufficient intensity and of a duration of a microsecond or less to produce high-quality photographs; another has to do with the complicated optics or mechanism to place each image at a different place on the film. Single pictures of "birefringent strips" have been taken by Duffy and Lee [64] and by Frasier and Robinson [65].

Dynamic photoviscoelasticity is also an active field. Work in this area has been performed by Daniel [66] and by Khesin, Kostin, Roshdestvenski, and Shpyakin [67]. A most promising solution to this problem involves the use of laser light sources which can be flashed repeatedly for very brief intervals of time. The earliest work combining lasers and photoelasticity was undertaken by Taylor, Bowman, North, and Swinson [68], but Khesin, Shavoronok, Babyshkina, and Khe [69] have also made contributions. These were static studies and great strides in themselves. Since then, repeated flashing lasers have been employed in dynamic problems [70].

4 Advances in Related Optical Techniques and Analysis

The great and increasing attention paid by photoelasticians during the past decade to interferometric [39], moiré [71, 72], and holographic [73–76] techniques separately and in combination [77–80], in part represents the continuing forward thrust toward purely optical procedures of determination of the components of stress. A key paper by Nisida and Saito [81] discusses the meaning and peculiarities of the optical patterns obtained. In part, it is a matching of the trend in theory and in analysis to take into account many aspects of inelastic behavior [28].

When the present use of birefringent coatings became feasible upon the development of the epoxies and other polymeric materials suitable for coatings and cements, the first application was to the plastic behavior of an aluminum specimen [82]. Dantu [42] employed a coating to bring out the role of the aggregate and the mortar in actual concrete specimens whose behavior in detail is far from linear elastic at working stresses.

Moiré procedures [83, 78, 72] had their adherents and persistent workers, following Ligtenberg's [84] and Mesmer's [85] lead in the early days of very limited possible accuracy. In large measure this was because measurements could be made on actual parts. One immediate appeal of holographic techniques, once they were made feasible experimental tools by the development of laser light sources, was again that machines, structures, devices, and parts could be examined directly [86].

An extensive review and analysis of viscoelastic behavior and associated birefringence by Van Geen and other Belgian authors [87, 88] builds upon and advances the earlier work by Dill [89], Mindlin [M28], Toupin [90], and others. However, despite the understanding achieved, especially for general linear viscoelastic materials, it is plastic rather than viscoelastic behavior which has attracted the major share of attention in "photoelastic" studies of the inelastic range. The method for the study of inelastic behavior which has developed most over the years is to strain birefringent polymers well into the range where their inelastic response dominates their elastic [28, 91–94]. Similarities between the stress-strain curve in simple tension for polymeric materials and for metals are cited in justification of photoplasticity as a natural extension of photoelasticity. Birefringence in the regions of the model which remain linear elastic (or as elastic as in classical photoelasticity) is then interpretable in the usual way in terms of stress in the model. As suggested by Hiltscher [95], no problem then exists in finding model stresses outside of the inelastic domain. The stress in the corresponding region of a metal prototype, however, may not bear any relation to the stress in a viscoelastic model.

This interesting difficulty need not divert us here. The important point is that real as well as idealized material behavior has become an important concern of photoelasticians. Optical response in terms of birefringence, rather than interpreted in terms of stress, brings attention back to interferometric methods and many other techniques of physical optics and optical analysis. Some of these methods of analysis are very old, such as the Poincaré sphere [96]; others are newer but long established, such as Jones' calculus [97]; some concepts of physical optics, the associated instrumentation and types of observations are at a high level of sophistication, as in the work of Nelson and Lax [98, 99].

5 A View of the Future

As mentioned in the introductory remarks, ideas and understanding now can flow effectively in both directions between the optical response of solids or fluids and the properties of materials. The greater appreciation of the power of optical analysis and of the physical basis for optical effects will lead to an increasing variety of new types of measurements for the determination of fields of stress and strain. This enhanced ability to interpret new and standard optical measurements also will provide new means of obtaining fundamental knowledge of the basic properties of transparent materials. High initial birefringence, inhomogeneity of birefringence and of other physical properties no longer pose so great a barrier to photoelastic analysis.

Studies of single crystals [99] and polycrystals can now progress far beyond the pioneering work of Nye [100] and of Goodman and Sutherland [101] on silver chloride crystals. Random arrays of randomly oriented birefringent crystals are meaningful objects for study, both organic and inorganic, living as well as inanimate, fluid as well as solid, for very strong and very weak initial or added birefringence. Grain boundary behavior and properties at free surfaces and in the interior can be explored. Thin films as well as bulk parts are now amenable prospects for research and application when the strain induced birefringence is small in comparison with the initial birefringence. Interface questions, such as the relative degree of mechanical and chemical bonding between fibers and matrix in a composite, can be studied directly. Surface layer investigations can be aided by birefringent coatings and strips.

The time is ripe for this reversal of the traditional role of photoelasticity, to turn photoelastic techniques far more incisively than in the past toward elucidation of the fundamental behavior of materials. Scattering, absorption, optical rotation, and dispersion can provide a wealth of information about material composition and structure when coupled with the highly developed techniques of modern photoelasticians. However, there is a limitation to the scale of the study when visible light is employed. Wavelengths in the neighborhood of 5000 Å or 0.5 μm can only explore details of surfaces or volumes whose linear dimensions range upward from the micron range. Those fascinating and often governing details at the dislocation level of scale and below, which provide the major share of the understanding of the behavior of crystals and polycrystals, are invisible in the optical microscope. Except for the average surface or volume effects which are observable, as for interfaces or as in neutron or gama-ray irradiation [102, 103], wavelengths shorter by a factor of 10 or 100 are needed. These smaller dimensions are still large by comparison with the atomic spacings of a few Angstrom units. However, animate or inanimate molecular chains are often very long and hence can be made visible in a sense quite different from the visibility of interatomic forces through their effect on the velocity, scattering, rotation, and absorption of light. Concepts familiar to the photoelastician should prove extremely helpful to the materials scientist and vice versa.

Interesting and important as this basic materials science interaction is, it is likely that far more attention will be devoted to topics more closely allied to present-day photoelasticity and to continuum mechanics. As simple analogs, photoelasticity and allied optical techniques have become and will continue to be less and less needed. The increasing power and accessibility of the computer, along with the development of suitable mathematical procedures to handle singularities properly and to deal with discrete presentations of continua, will relegate linear analogs of all types to educational tools rather than practical means of solving real world problems. Educational tools for the practitioners of engineering as well as for the students in universities and schools do play a significant role. However, even the delightful visual display in color or black and white, which made photoelasticity so attractive, will be simulated more and more faithfully by computer displays in two and three dimensions. Bit by bit, those elastic problems solvable by computer with finite element or other techniques will pass entirely from the photoelastic domain. Yet photoelasticity as an analog tool will remain viable for a long time to come in problems of complex geometry, especially for dynamic or wave propagation studies. The tailoring of materials to match the requirements of an actual or idealized prototype will call for elaboration on the work of Durelli and others [104] on variable modulus materials or on the use of a birefringent substitute for the more usual granular materials [105].

Just as analog experiments tend to fade in value with time as applied mathematics takes over, the realization grows that the most elegant of mathematical solutions or complex computer solutions are based on drastic simplifications of the real behavior of materials on the continuum scale. The more carefully one looks at the details of behavior at all levels down to the submicroscopic, the more likely it is that there will be significant departures from linear-elastic behavior in the most conservative and conventional of designs for static loading and room temperature. When operating conditions include dynamic loading or elevated temperatures or neutron irradiation or any of the host of current design conditions, accurate modeling of the material under the conditions of use, either mathematically or by substituting another material, is an almost unattainable goal.

Simplified models are extremely valuable in design, but their validity can only be ascertained by comparison with model or full-scale tests on the actual material under actual working conditions. This difficulty cannot be overcome by the apparent infinite complexity of what are still oversimplified models, although of course the greater the understanding of the behavior of materials and the sensitivity of solutions to the form and nature of constitutive relations, the better the simplified model one can construct for the given environmental and loading conditions. Consequently, greater and greater attention will be devoted to the comparison between the mathematical or computer solution to boundary-value problems employing more and more complicated but idealized elastic-viscous-plastic constitutive relations and the birefringent patterns obtained with real elastic-viscous-plastic transparent model materials. Such comparisons are still at a very primitive stage, unable to handle well even such elementary cases as homogeneous states of time-varying uniaxial tension or simple shear. A beginning has been made for viscoelastic material and the 1973 IUTAM Symposium on the Photoelastic Effect and its Applications should report further progress in this "rheo-optic" area [106].

The closely related direction in which effort is certain to increase, is to utilize photoelastic procedures to make measurements on the inelastic behavior of models or actual parts made of real materials of construction. Birefringent coatings or strips of the conventional type or in the scored or strip form proposed by O'Regan [107] perform a function which also cannot be replaced by a computer. Use of the actual material for the model or prototype avoids the impossibility of exact modeling of the material. Use of the birefringent coating or strip in the linear elastic or viscoelastic range [108] avoids the most serious problems of interpretation of the measurements at the surface of the model. Once a set of such measurements has been made and understood, the ability to develop and critically test suitable mathematical models of material behavior is greatly enhanced.

An infinite host of possibly useful directions for research and practical application are now open in this merging of photoelasticity and the inelastic behavior of materials. To minimize bringing a raft of imaginary problems to formal but useless solutions, the twin objectives of fundamental understanding and the development of useful techniques should be kept firmly in mind as each direction is explored. If Mindlin's incisive advanced level but pragmatic view is followed, with its clear tie to the real world and its place in the mainstream of knowledge, all will be well.

References

1. D. Brewster, On the communication of the structure of doubly refracting crystals of glass, muriate of soda, fluor spar, and other substances by mechanical compression and dilatation, *Phil. Trans. Roy. Soc.*, London, 156 (1816).
2. F. E. Neumann, Über die Gesetze der Doppelbrechung des Lichtes in comprimierten und ungleichförmig erwärmten unkrystallinischen Körpern, *Abh. Dtsch. Kön. Akad. Wiss. Berlin*, Part II, 1 (1841).
3. J. C. Maxwell, On the equilibrium of elastic solids, *Trans. Roy. Soc. Edin.* **20,** 87 (1853).
4. G. Wertheim, Mémoire sur la double réfraction temporairement produite dans les corps isotropes, et sur la relation entre l'élasticité mécanique et l'élasticité optique, *Annales de Chimie et de Physique Série III* **XL,** 156 *et seq.* (1854).
5. A. Mesnager, Détermination complète sur un modèle réduit des tensions qui se produiront dans un ouvrage, *Annales des Ponts et Chaussées* **16,** 133 (1913).
6. E. G. Coker and L. N. G. Filon, *A Treatise on Photoelasticity*, Cambridge University Press, 1931.
7. A. Mesnager, Contribution à l'étude de la déformation élastique des solides, *Annales des Ponts et Chaussées* **14,** 128 (1901).

8. E. G. Coker, A laboratory apparatus for measuring the lateral strains in tension and compression members, with some applications to the measurement of the elastic constants of metals, *Proc. Roy. Soc. Edin.* **25** (1904–1905).
9. H. Favre, Sur une nouvelle méthode optique de détermination des tensions intérieures, *Revue d'optique* **8,** 193, 241, 289 (1929).
10. M. M. Frocht, *Photoelasticity*, Wiley, New York, 1941, 1948, Vols. I and II.
11. D. C. Drucker, The method of oblique incidence in photoelasticity, *Proceedings of the Society for Experimental Stress Analysis* **8** (1951).
12. J. C. Maxwell, On the equilibrium of elastic solids, *Trans. Roy. Soc. Edin.* **20,** 87 (1853).
13. L. N. G. Filon and F. C. Harris, On the di-phasic nature of glass as shown by photoelastic observations, *Proc. Roy. Soc.* **A103,** 561 (1923).
14. Z. Tuzi, Photo-elastic study of stress in a heat-treated column, *Sci. Pap. Inst. Phys.-Chem. Res., Tokyo* **7,** 104 (1927).
15. A. G. Solakian, A new photoelastic material, *Mech. Eng.* **57,** 767 (1935).
16. G. Oppel, Polarisationsoptische Untersuchung räumlicher Spannungs—und Dehnungszustände, *Forschung auf dem Gebiete des Ingenieurwesens* **7,** 240 (1936).
17. M. Hetényi, The fundamentals of three-dimensional photoelasticity, *J. appl. Mech., Trans. ASME* **5,** A149 (1938).
18. A. Kuske, Das Kunstharz Phenolformaldehyd in der Spannungsoptik, *Forschung auf dem Gebiete des Ingenieurwesens* **9,** 139 (1938).
19. R. Hiltscher, Polarisationsoptische Untersuchung des räumlichen Spannungszustandes im konvergenten Licht, *Forschung auf dem Gebiete des Ingenieurwesens* **9,** 91 (1938).
20. M. M. Frocht and R. Guernsey, Jr., Studies in three-dimensional photoelasticity—The application of the shear difference method to the general space problem, *Proc. 1st U.S. Nat'l Congress Applied Mech.*, 301 (June 1951).
21. N. I. Prigorovsky, *Thermal Stress*, Academy Sciences USSR, Moscow, 1972.
22. H. K. Aben and N. I. Prigorovsky, *Proceedings of the Seventh All-Union Conference on Photoelasticity*, Talinn, Esthonian Academy of Science, 1971.
23. M. M. Leven, Epoxy resins for photoelastic use, *Photoelasticity, Proc. Intl. Symposium at IIT*, Chicago, 1961; Pergamon Press, 1963, p. 145.
24. H. Poritsky, Analysis of thermal stresses in sealed cylinders and the effect of viscous flow during anneal, *Physics* **5,** 406 (1934).
25. R. Weller, A new method for photoelasticity in three dimensions, *J. appl. Phys.* **10,** 266 (April 1939).
26. D. C. Drucker, Three-dimensional photoelasticity, *Handbook of Experimental Stress Analysis*, Wiley, New York, 1950, p. 950.
27. H. T. Jessop, The scattered light method of exploration of stresses in two- and three-dimensional models, *Br. J. appl. Phys.* **2,** 249 (1951).
28. M. Hetényi, Photoelasticity and photoplasticity, *Structural Mechanics*, edited by J. N. Goodier and N. J. Hoff, Pergamon Press, 1960, p. 483.
29. R. C. Dove and P. H. Adams, *Experimental Stress Analysis and Motion Measurement*, Merrill, Columbus, Ohio, 1964.
30. J. W. Dally and W. F. Riley, *Experimental Stress Analysis*, McGraw-Hill, New York, 1965.
31. A. W. Hendry, *Elements of Experimental Stress Analysis*, Pergamon Press, Oxford, 1964.
32. H. Wolf, *Spannungsoptik*, Springer-Verlag, Berlin, 1961.

33. H. L. Tardy, Méthode pratique d'éxamen et de mesure de la biréfringence des verres d'optique, *Revue d'optique* **8,** 59 (1929).
34. T. J. Dolan and W. M. Murray, Photoelasticity: I. Fundamentals and two-dimensional applications, *Handbook of Experimental Stress Analysis,* edited by M. Hetényi, Wiley, New York, 1950, p. 828.
35. S. M. Sathikh and G. W. Biggs, On the accuracy of goniometric compensation methods in photoelastic fringe-order measurements, *Experimental Mechanics* **12,** 47 (1972).
36. D. Post, Isochromatic fringe sharpening and fringe multiplication in photoelasticity, *Proceedings of the Society for Experimental Stress Analysis* **12,** 143 (1954).
37. C. Mylonas and D. C. Drucker, Twisting stresses in tape, *Proceedings of the Society for Experimental Stress Analysis* **18,** 23 (1960).
38. E. E. Day, A. S. Kobayashi, and C. N. Larson, Fringe multiplication and thickness effects in birefringent coating, *Experimental Mechanics* **2,** 115 (1962).
39. P. Bohler and W. Schumann, On a multiplication technique applied to very thin photoelastic coatings, *Experimental Mechanics* **11,** 289 (1971).
40. A. Mesnager, Sur la détermination optique des tensions intérieures dans les solides à trois dimensions, *C.R., Académie des Sciences, Paris* **190,** 91 (1930).
41. G. Oppel, Das polarisationsoptische Schichtverfahren zur Messung der Oberflächenspannung am beanspruchten Bauteil ohne Modell, *Z.V.D.I.* **81,** 803 (1937).
42. P. Dantu, Étude des contraintes dans les milieux hétérogènes. Application au béton, *Annales de l'Institut technique du Bâtiment et des Travaux Publics,* France **121,** 53 (1958).
43. G. L. Khesin and V. N. Sakharov, Methods of strain measurements on the surface of concrete and reinforced concrete by means of photoelastic coatings, *4th International Conference of Experimental Stress Analysis,* Cambridge, April 1970.
44. R. N. Swamy, Study of the micro-mechanical behavior of concrete using reflective photoelasticity, *Matériaux et Constructions* **4,** 357 (1971).
45. J. R. Dixon and W. Visser, An investigation of the elastic-plastic strain distribution around cracks in various sheet materials, *Proc. Intern'l Sym. Photoelasticity,* edited by M. M. Frocht, 231 (1961).
46. J. N. Goodier and G. H. Lee, An extension of the photoelastic method of stress measurement to plates in transverse bending, *J. appl. Mech., Trans. ASME* **63,** A27 (1941).
47. R. Papirno and H. Becker, A bonded polariscope for three-dimensional photoviscoelastic studies, *Experimental Mechanics* **6,** 609 (1966).
48. J. Duffy, Effects of the thickness of birefringent coatings, *Experimental Mechanics* **1,** 74 (1961).
49. F. Zandman, S. S. Redner, and E. I. Riegner, Reinforcing effect of birefringent coatings, *Experimental Mechanics* **2,** 55 (1962).
50. W. Goldsmith, Dynamic photoelasticity, *Experimental Techniques in Shock and Vibration,* edited by W. J. Worley, A.S.M.E., 1962, p. 25.
51. Z. Tuzi, Photographic and kinematographic study of photoelasticity, *Scientific Papers,* Inst. Phys. Chem. Res., Japan, **8,** 247 (1928).
52. M. M. Frocht, Kinematography in photoelasticity, *Trans. ASME* **54,** 83 (1932).
53. D. A. Senior and A. A. Wells, A photo-elastic study of stress waves, *Philosophical Magazine* **37,** 463 (1946).
54. J. W. Dally, A. J. Durelli, and W. F. Riley, Photoelastic study of stress wave propagation in large plates, *Proceedings of the Society for Experimental Stress Analysis* **17,** 33 (1960).

55. A. J. Durelli and W. F. Riley, Experiments for the determination of transient stress and strain distributions in two-dimensional problems, *J. appl. Mech.* **24,** 69 (1957).
56. J. W. Dally, W. F. Riley, and A. J. Durelli, A photoelastic approach to transient stress problems employing low-modulus materials, *J. appl. Mech.* **26,** 613 (1959).
57. A. J. Durelli and W. F. Riley, *Introduction to Photomechanics*, Prentice-Hall Series in Solid and Structural Mechanics, 1965.
58. G. L. Khesin, I. K. Kostin and V. B. Zateev, Investigations of the stress concentration around holes in thin plates affected by pressure wave, *Collection of Papers no. 73, The Modeling of Problems by the Use of the Optical Polarization Methods*, edited by G. L. Khesin, Moscow, 1970, p. 33.
59. P. D. Flynn, J. T. Gilbert, and A. A. Roll, Some recent developments in dynamic photoelasticity, *SPIEJ* **2,** 128 (1964).
60. P. D. Flynn and M. M. Frocht, On the photoelastic separation of principal stresses under dynamic conditions by oblique incidence, *J. appl. Mech.* **28,** 144 (1961).
61. P. D. Flynn, A dual-beam polariscope for oblique incidence, *Experimental Mechanics* **4,** 182 (1964).
62. M. M. Frocht, P. D. Flynn, and D. Landsberg, Dynamic photoelasticity by means of streak photography, *Proceedings SESA* **14,** 81 (1957).
63. Z. Tuzi, and M. Nisida, Photoelastic study of stresses due to impact, *Philosophical Magazine* **21,** 448 (1936).
64. J. Duffy, and T. C. Lee, The measurement of surface strains by means of bonded birefringent strips, *Experimental Mechanics* **1,** 109 (1961).
65. J. T. Frasier, D. N. Robinson, The measurement of dynamic strains using birefringent strips, *J. appl. Mech.* **33,** 173 (1966).
66. I. M. Daniel, Viscoelastic wave interaction with a cylindrical cavity, *J. Engr. Mech. Div. Proc. ASCE* **92,** 25 (1966).
67. G. L. Khesin, I. K. Kostin, K. N. Roshdestvenski, and V. N. Shpyakin, Technique and results of calibrating investigations of epoxy birefringent polymers under impulsive loadings, *Collection of Papers no. 73, The Modeling of Problems by the Use of the Optical Polarization Methods*, edited by G. L. Khesin, Moscow, 1970, p. 8.
68. C. E. Taylor, C. E. Bowman, W. P. North, and W. F. Swinson, Applications of lasers to photoelasticity, *Experimental Mechanics* **6,** 289 (1966).
69. G. L. Khesin, I. V. Shavoronok, S. N. Babyshkina, and V. I. Khe, Use of lasers to the investigations of dynamic processes by the photoelasticity method, *Collection of Papers no. 73, The Modeling of Problems by the Use of the Optical Polarization Methods*, edited by G. L. Khesin, Moscow, 1970, p. 74.
70. R. E. Rowland, C. E. Taylor and I. M. Daniel, A multiple-pulse ruby-laser system for dynamic photomechanics: applications to transmitted- and scattered-light photoelasticity, *Experimental Mechanics* **9,** 385 (1969).
71. P. S. Theocaris, *Moiré Fringes in Strain Analysis*, Pergamon Press, 1969.
72. A. J. Durelli, N. A. Chechenes, and J. A. Clark, Developments in the Optical Spatial Filtering of Superposed Crossed Gratings, *Experimental Mechanics* **12,** 496 (1972).
73. D. C. Holloway and R. H. Johnson, Advancements in holographic photoelasticity, *Experimental Mechanics* **11,** 57 (1971).
74. J. D. Hovanesian, V. Brcic, and R. L. Powell, A new experimental stress-optic method: Stress-Holo-Interferometry, *Experimental Mechanics* **8,** 363 (1968).

75. M. E. Fourney, Application of holography to photoelasticity, *Experimental Mechanics* **8,** 33 (1968).
76. R. O'Regan and T. D. Dudderar, A new holographic interferometer for stress analysis, *Experimental Mechanics* **11,** 241 (1971).
77. D. Post, Holography and interferometry in photoelasticity, *Experimental Mechanics* **12,** 113 (1972).
78. D. C. Holloway, W. F. Ronson, and C. E. Taylor, A neoteric interferometer for use in holographic photoelasticity, *Experimental Mechanics* **12,** 461 (1972).
79. C. A. Sciammarella and G. Quintanilla, Techniques for the determination of absolute retardation in photoelasticity, *Experimental Mechanics* **12,** 57 (1972).
80. R. J. Sanford and A. J. Durelli, Interpretation of fringes in stress-holo-interferometry, *Experimental Mechanics* **11,** 161 (1971).
81. M. Nisida and H. Saito, A new interferometric method of two-dimensional stress analysis, *Experimental Mechanics* **4,** 366 (1964).
82. J. D'Agostino, D. C. Drucker, C. K. Liu, and C. Mylonas, An analysis of plastic behavior of metals with bonded birefringent plastic, *Proceedings of the Society for Experimental Stress Analysis* **XII,** 115 (1955).
83. C. A. Sciammarella and A. J. Durelli, Moiré fringes as a means of analyzing strains, *J. Engr. Mech. Div. Proc. ASCE* **87,** 54 (1961).
84. F. K. Ligtenberg, The Moiré Method. A new experimental method for the determination of moments in small slab models, *Proceedings SESA* **12,** 83 (1955).
85. G. Mesmer, The interference screen method for isopachic patterns (Moiré Method), *Proceedings SESA* **13,** 21 (1956).
86. T. D. Dudderar and H. J. Gorman, The determination of mode I stress-intensity factors by holographic interferometry, *Experimental Mechanics* **13,** 145 (1973).
87. A. Noblet, G. Sylin, M. Vandaele-Possche, and R. Van Geen, La biréfringence mécanique differée de sign négatif et sa représentation analogique, *Bulletin de la Classe des Sciences de l'Académie Royale de Belgique,* 5e série, **L,** 534 (1964).
88. P. Boulanger, G. Mayne, A. Hermanne, J. Kestens, and R. VanGeen, L'effet photo-élastique dans le cadre de la mécanique rationelle des milieux continus, *Revue de l'industrie minérale-mines,* Special number, 3 (June 15, 1971).
89. E. H. Dill and C. Fowlkes, Photoviscoelasticity, *Report 65-1 Department of Aeronautics and Astronautics, College of Engineering, University of Washington* (1965).
90. R. A. Toupin, A dynamical theory of elastic dielectrics, *Int. J. Engng. Sci.* **1,** 101 (1963).
91. D. H. Morris and W. F. Riley, A photomechanics material for elastic plastic stress analysis, *Experimental Mechanics* **12,** 448 (1972).
92. H. F. Brinson, Rate and time dependent yield behavior of a ductile polymer, *Deformation and Fracture of High Polymers,* edited by H. Kausch *et al.*, Plenum Press, New York, 1973.
93. B. Fried, Some observations on photoelastic materials stressed beyond the elastic limit, *Proceedings SESA* **8,** 143 (1951).
94. K. Ito, Studies on photoelasto-plastic mechanics, *Journal Scientific Research Institute,* Tokyo **50,** 19 (1956).
95. R. Hiltscher, Theorie und Andwendung der Spannungsoptik im elastoplastischen Gebiet, *Z.V.D.I.* **97,** 49 (1955).
96. H. Poincaré, *La théorie mathématique de la lumière,* Gauthier-Villars, Paris, 1892.

97. R. C. Jones, New calculus for the treatment of optical systems, *J. Opt. Soc. Amer.* **31**, 488 (1941).
98. D. F. Nelson and M. Lax, Theory of the photoelastic interaction, *Physical Review B* **3**, 2778 (1971).
99. D. F. Nelson, P. D. Lazay, and M. Lax, Brillouin scattering in anisotropic media: calcite, *Physical Review B* (October 15, 1972).
100. J. F. Nye, Plastic deformation of silver chloride, *Proc. Roy. Soc.* **A198,** 190 (1949).
101. L. E. Goodman and J. G. Sutherland, Elasto-plastic stress-optical effect in silver chloride single crystals, *J. appl. Phys.* **24,** 577 (1953).
102. J. F. Gross-Peterson, The gamma-ray-irradiation method applied to three-dimensional thermal photoelasticity, *Experimental Mechanics* **12,** 414 (1972).
103. C. Mylonas and R. Truell, Radiation effects from (n, α) reactions in boron glass, and energy of the reacting neutrons, *J. appl. Phys.* **29,** 1252 (1958).
104. A. J. Durelli, H. Miller, and F. Jones, A photoelastic material with variable modulus of elasticity, *Experimental Mechanics* **12,** 381 (1972).
105. A. Drescher and G. de Josselin de Long, Photoelastic verification of a mechanical model for the flow of a granular material, *J. Mech. & Phys. Solids* **20,** 337 (1972).
106. J. Kestens, in *Proceedings of the IUTAM Symposium on Photoelasticity Effect and Its Applications,* Belgium (to appear).
107. R. O'Regan, New method for determining strain on the surface of a body with photoelastic coatings, *Experimental Mechanics* **5,** 241 (1965).
108. P. S. Theocaris and C. Mylonas, Viscoelastic effects in birefringent coatings, *J. appl. Mech.* **28,** 601 (1961).

APPENDIX

Development of the Mathematical Theory of Three-Dimensional Photoelasticity†

R. D. Mindlin

The First Ninety Years

In the first systematic treatment of the laws relating strain and optical fields, Neumann [1],‡ in 1841, gave not only the complete, linear relation between the small strain and index of refraction tensors, but also approximate, ordinary differential equations governing the almost rectilinear propagation of a ray of light in the

This contribution is a revised version of that published originally in the October 1951 issue of AMR.

† Reprinted with permission from *Applied Mechanics Surveys.* Edited by H. N. Abramson, H. Liebowitz, J. M. Crowley, and S. Juhasz, Spartan Books (1966).

‡ Numbers in brackets refer to the references at the end of this appendix.

anisotropic and heterogeneous optical medium resulting from the application of a heterogeneous strain. Although this was a great stride toward the establishment of a theory on which an experimental method of three-dimensional stress analysis could be based, many questions were left unanswered—especially: what optical measurements should be made and how should they be interpreted.

Neumann's extraordinary accomplishment remained unused and unimproved for many years; despite a pregnant hint from Maxwell's [2] experiment with isinglass, in 1850, which failed to stimulate an early development of the freezing method. The first reappearance of Neumann's results is found in the great treatise by Coker and Filon [3] ninety years later. Here Neumann's strain-optical relations are extended to finite strain and his differential equations are rederived and utilized in a discussion of optical effects accompanying the quadratic variation of stress through the thickness of a plate in a state of plane stress.

Occupying a unique position, between the old and new regimes in photoelasticity, is an interesting paper by Poritzky [4] in 1934. Although dealing with a two-dimensional problem of thermal stresses in long coaxial cylinders, Poritzky gave a correct interpretation of the integrated birefringence along a chord of an equatorial section—an essentially three-dimensional optical-mathematical problem which became the subject of later investigations.

Resurgence

A revival of activity in three-dimensional photoelasticity was stimulated, during the period 1936–1940, by reports and interpretations of experiments with the freezing and slicing technique by Oppel [5], Hetényi [6] and Kuske [7]; studies of stress-optical relations and measurements with convergent light by Hiltscher [8]; and the introduction of the scattered-light technique by Weller [10] and Menges [11]. From the point of view of the mathematical theory, Hiltscher's paper is important for containing the discovery that the two-dimensional state of stress can be resolved completely by optical measurements involving only relative, rather than absolute, retardations; while the paper by Menges reports the first use of Neumann's differential equations in the analysis of experimental data.

Laws of Propagation of Light

Whereas Neumann's differential equations are based on purely kinematical considerations, involving approximations of hidden character and magnitude, Maxwell's electromagnetic theory, combined with Neumann's strain-optical or Maxwell's stress-optical relations, supplies the general field equations governing the propagation of light in the anisotropic, heterogeneous medium encountered in three-dimensional photoelasticity. These equations, being a system of partial differential equations with variable coefficients, are formidable in appearance; but they have been solved for two cases of special interest in the interpretation of photoelastic observations, namely, when the magnitudes only and the orientations only, of the principal axes of the section of the index ellipsoid normal to the ray, vary linearly along the path of the ray. The first of these exact solutions, by Mindlin [13], led to the conclusion that variation in magnitude alone has no significant influence for index variations at rates ordinarily encountered. The second solution, by Drucker and Mindlin [12] revealed some curious and useful effects of rotation of polarizing axes on the light vector and also led to the establishment of a *rotation-retardation ratio* as a criterion for assessing the influence of rotation on the relation between relative phase retardation and secondary principal stress difference. They found this influence to be small.

When terms of negligible magnitude are dropped from the two exact solutions, they may be shown to be identical with the corresponding solutions of Neumann's equations. Thus, approximations to exact solutions of the general electromagnetic equations are identical with exact solutions of Neumann's approximate equations in the two cases for which direct comparisons can be made. That this will always be the case was shown by Mindlin and Goodman [27] who executed the passage between the equations themselves; showing, in the process, the nature and order of magnitude of the approximations required and establishing the generality of the rotation-retardation ratio employed by Drucker and Mindlin. Similar results were obtained subsequently by O'Rourke [34] who applied the Sommerfeld-Runge method for deriving laws of geometrical optics from the electromagnetic equations.

Resolution of Principal Stresses

Measurements involving relative, rather than absolute, phase retardations at interior points were shown by Hiltscher [8] to yield the three principal stress differences in the general case and the principal stresses themselves in two-dimensional states of stress. He applied point by point measurements with convergent polarized light to both cases [8, 19], as also did Kuske [16]. Drucker and Mindlin extended this idea to observations on whole plates or slices of a frozen model by suggesting passage of light in an appropriate number of directions either in the plane of the plate or slice or at an angle to it. These preliminary studies culminated in the development by Drucker [18, 31] of the *method of oblique incidence*.

Many of the methods for resolving principal stresses in two dimensions have counterparts in three dimensions. Examples are: measurements of change of thickness of a reheated frozen slice by Kuske [7], graphical integration of the stress-equations of equilibrium along stress trajectories (in a plane of symmetry) by Jessop [24] and in rectangular coordinates by Frocht and Guernsey [35].

Methods of resolving principal stresses by means of combinations of graphical integration and oblique incidence and scattered light observations have been given by Jessop [36], Frocht and Guernsey [37], Frocht and Srinath [40] and Cheng [45].

Flexure of Plates

In the problem of flexure of plates, the essential difficulty is that of the optical antisymmetry with respect to the middle plane—resulting in cancellation of the relative phase retardation. Sandwich and layer devices have been employed to eliminate the antisymmetry and reduce the optical situation to an essentially two-dimensional one. An alternative procedure, suggested by Mindlin [13], is to freeze an initial tension in the plate so that the subsequent application of bending produces an asymmetric, rather than an antisymmetric, index field. Mindlin proposed to measure the resulting birefringence by using the phenomenon of rotation of the light vector previously discovered by Drucker and Mindlin [12]; but Drucker [17] found that greater precision and simplicity would be gained by observing the fringe pattern in the ordinary transmission polariscope. Drucker's measurement of stress-concentration around a hole and subsequent comparison [20] of his results with the predictions of E. Reissner's theory of flexure constitute both an interesting sequence of events and a landmark in the accomplishments of three-dimensional photoelasticity.

Torsion

One of the simplest three-dimensional states of stress occurs in a cylindrical bar under torsion. This case is an excellent one for the beginner to study in order to become acquainted with some of the complexities of three-dimensional photoelasticity. The optical properties of a birefringent bar in torsion are described by Mindlin [9] and the identity of scattered-light fringe-patterns with the contours of the membrane analogy is demonstrated by Drucker and Frocht [22] and Saleme [23].

Integrated Phase Retardation

A very intriguing idea in three-dimensional photoelasticity is the possibility of measuring only the integrated relative phase retardation, after complete passage of light through the model, and finding the stresses at interior points by using auxiliary mathematical theory. This would dispense with the tedious freezing-slicing procedure and the difficult scattered-light technique. Considerable progress has been made along these lines since the early work of Poritzky [4]. An ingenious study by Read [28] of stresses in glass bulbs accomplishes this result in a special case of much interest in industry. Indications that the complete problem has some possibility of solution are contained in a remarkable series of papers by O'Rourke and Saenz [23, 33, 34]. Considering the cases of light passed along chords of equatorial sections of symmetrically strained cylinders and spheres, they showed that the gross retardation can be expressed in terms of the interior stresses by means of an Abel integral equation. Since this equation can be solved, they find simple quadrature formulas for the interior stresses.

In the case of the cylinder, O'Rourke and Saenz assumed that the axial stress is equal to the sum of the radial and circumferential stresses at each point. Drucker and Woodward [38] showed how this restriction can be removed if an additional measurement of integrated retardation is made along chords of an oblique plane. They also indicated how the method can be extended to axially symmetric stress in any solid of revolution; but they pointed out an essential difficulty in the case of general three-dimensional states of stress.

Photoviscoelasticity

Most experiments in photoelasticity are actually performed with viscoelastic models so that questions arise regarding the interpretation of the optical measurements in terms of stresses in an elastic body. In a paper by Mindlin [25] there are given general, three-dimensional relations connecting stress, strain, time, temperature and birefringence. These constitutive relations are based on a model of springs and dashpots in which birefringence is assigned only to the springs—as suggested by the experiments of Maxwell [2], Oppel [5], Hetényi [6] and Kuske [7]. The constitutive equations are combined with equations of equilibrium and compatibility to form a field theory which is used, in the special case of incompressible materials, to find conditions under which the birefringence in viscoelastic materials may be interpreted directly in terms of the stress in an elastic body. It is found that this interpretation is valid except if there are moving loads. Another conclusion reached is that the freezing method in three-dimensional photoelasticity is not restricted to models made of cross-linked polymers. Both conclusions have been confirmed in numerous experiments [41].

Mindlin's equations of the photoviscoelastic field have been revised and extended by Read [30] who included compressible media, inertia terms and any type of boundary condition. Both papers are limited to small strain and linear viscoelasity.

Birefringent Coatings

The modern use of birefringent coatings to reveal surface strains in opaque materials was initiated by D'Agostino, Drucker, Liu and Mylonas [39]. The connection of this development with the mathematical theory of three-dimensional photoelasticity stems from the fact that surface-curvatures and surface-gradients of strain induce three-dimensional effects and viscoelastic deformations in the coatings. The former have been investigated by Duffy [42] and by Lee, Mylonas and Duffy [43]; while the latter have been studied by Theocaris and Mylonas [44].

References

1. F. E. Neumann, "Über die Gesetze der Doppelbrechung des Lichtes in comprimierten und ungleichförmig erwärmten unkrystallinischen Körpern," *Abh. d. Kon. Akad. d. Wiss.*, Berlin, v. 2, 1841, p. 1.
2. J. C. Maxwell, "On the equilibrium of elastic solids," *Trans. Roy. Soc. Edin.*, v. 20, 1850, p. 87.
3. E. G. Coker and L. N. G. Filon, *A Treatise on Photo-Elasticity*, Cambridge Univ. Press, 1931.
4. H. Poritsky, "Analysis of thermal stresses in sealed cylinders and the effect of viscous flow during anneal," *Physics*, v. 5, 1934, p. 406.
5. G. Oppel, "Polarisationsoptische Untersuchung räumlicher Spannungs—und Dehnungszustände," *Forsch. Geb. Ing.-Wes.*, v. 7, 1936, p. 240.
6. M. Hetényi, "The fundamentals of three-dimensional photoelasticity," *J. Appl. Mech.* v. 5, 1938, p. 149.
7. A. Kuske, "Das Kunstharz Phenolformaldehyd in der Spannungsoptik," *Forsch. Geb. Ing.-Wes.*, v. 9, 1938, p. 139.

8. R. Hiltscher, "Polarisationsoptische Untersuchung des räumlichen Spannungszustandes im konvergenten Licht," *Forsch. Geb. Ing.-Wes.*, v. 9, 1938, p. 91.
9. R. D. Mindlin, "A review of the photoelastic method of stress analysis," *J. Appl. Phys.*, v. 10, 1939, pp. 222 and 273.
10. R. Weller, "A new method for photoelasticity in three dimensions," *J. Appl. Phys.*, v. 10, 1939, p. 266; *Nat. Adv. Comm. Aero.*, Tech. Note 737, 1939.
11. H. J. Menges, "Die experimentelle Ermittlung räumlicher Spannungszustande an durchsichtgen Modellen mit Hilfe des Tyndalleffects," *Z. Angew. Math. Mech.*, v. 20, 1940 p. 210.
12. D. C. Drucker and R. D. Mindlin, "Stress analysis by three-dimensional photoelastic methods," *J. Appl. Phys.*, v. 11, 1940, p. 724.
13. R. D. Mindlin, "An extension of the photoelastic method of stress measurement to plates in transverse bending," *J. Appl. Mech.*, v. 8, 1941, p. 187 (discussion of paper by J. N. Goodier and G. H. Lee, p. 27).
14. R. D. Mindlin, "Optical aspects of three-dimensional photoelasticity," *J. Franklin Inst.*, v. 233, 1942, p. 349.
15. R. D. Mindlin, "The present status of three-dimensional photoelasticity," *Civ. Engng.*, *Am. Soc. Civ. Engnrs.*, v. 12, 1942, p. 255.
16. A. Kuske, "Vereinfachte Auswerteverfahren räumlicher Spannungsoptischer Versuche," *Z. Ver. Dtsch. Ing.*, v. 86, 1942, p. 541.
17. D. C. Drucker, "The photoelastic analysis of transverse bending of plates in the standard transmission polariscope," *J. Appl. Mech.*, v. 9, 1942, p. 161.
18. D. C. Drucker, "Photoelastic separation of principal stresses by oblique incidence," *J. Appl. Mech.*, v. 10, 1943, p. 156.
19. R. Hiltscher, "Vollständige Bestimmung des ebenen Spannungszustandes nach dem Achsenbildvefahren," *Forsch. Geb. Ing.-Wes.*, v. 15, 1944, p. 12.
20. D. C. Drucker, "The effect of transverse shear deformation on the bending of elastic plates," *J. Appl. Mech.*, v. 13, 1946, p. 250 (discussion of paper by E. Reissner, v. 12, p. 69).
21. C. B. Norris and A. W. Voss, "An improved photoelastic method for determining plane stress," *Nat. Adv. Comm. Aero.*, Tech. Note 1410, 1948.
22. D. C. Drucker and M. M. Frocht, "Equivalence of photoelastic scattering patterns and membrane contours for torsion," *Proc. Soc. Exp. Stress Anal.*, v. 5, no. 2, 1948, p. 34.
23. E. M. Saleme, "Three-dimensional photoelastic analysis by scattered light," *Proc. Soc. Exp. Stress Anal.*, v. 5, no. 2, 1948, p. 49.
24. H. T. Jessop, "The determination of the separate stresses in three-dimensional stress investigations by the frozen stress method," *J. Sci. Instrum.*, v. 26, 1949, pp. 27–31.
25. R. D. Mindlin, "A mathematical theory of photoviscoelasticity," *J. Appl. Phys.*, v. 20, 1949, p. 206.
26. D. C. Drucker, "Three-dimensional photoelasticity," Chapter 17, Part 2, *Handbook of Experimental Stress Analysis*, M. Hetényi, ed., John Wiley and Sons, New York, 1950.
27. R. D. Mindlin and L. E. Goodman, "The optical equations of three-dimensional photoelasticity," *J. Appl. Phys.*, v. 20, 1949, p. 89.
28. W. T. Read, Jr., "An optical method for measuring the stress in glass bulbs," *J. Appl. Phys.*, v. 21, 1950, p. 250.
29. H. T. Jessop and M. K. Wells, "The determination of the principal stress differences at a point in a three-dimensional photoelastic model," *British J. Appl. Phys.*, v. 1, 1950, p. 184.

30. W. T. Read, Jr., "Stress analysis for compressible viscoelastic materials," *J. Appl. Phys.*, v. 21, 1950, p. 671.
31. D. C. Drucker, "The method of oblique incidence in photoelasticity," *Proc. Soc. Exp. Stress Anal.*, v. 8, no. 1, 1950, p. 51.
32. R. C. O'Rourke and A. W. Saenz, "Quenching stresses in transparent isotropic media and the photoelastic method," *Q. Appl. Math.*, v. 8, 1950, p. 303.
33. A. W. Saenz, "Determination of residual stress of quenching origin in solid and concentric hollow cylinders from interferometric observations," *J. Appl. Phys.*, v. 21, 1950, p. 962.
34. R. C. O'Rourke, "Three-dimensional photoelasticity," *J. Appl. Phys.*, v. 22, 1951, p. 872.
35. M. M. Frocht and R. Guernsey, Jr., "Studies in three-dimensional photoelasticity," *Proc. First U.S. Natl. Cong. Appl. Mech.*, Am. Soc. Mech. Engrs., New York, 1951, p. 301.
36. H. T. Jessop, "The scattered light method of exploration of stresses in two- and three-dimensional models," *British J. Appl. Phys.*, v. 2, 1951, p. 249.
37. M. M. Frocht and R. Guernsey, Jr., "A special investigation to develop a general method for three-dimensional photoelastic stress analysis," *Nat. Ad. Comm. Aero.*, Tech. Note 2822, 1952.
38. D. C. Drucker and W. B. Woodward, "Interpretation of photoelastic transmission patterns for a three-dimensional model," *J. Appl. Phys.*, v. 25, 1954, p. 510.
39. J. D'Agostino, D. C. Drucker, C. K. Liu and C. Mylonas, "An analysis of plastic behavior of metals with bonded birefringent plastic," *Proc. Soc. Exptl. Stress Anal.*, v. 12, no. 2, 1955, p. 115.
40. M. M. Frocht and L. S. Srinath, "A non-destructive method for three-dimensional photoelasticity," *Proc. Third U.S. Natl. Cong. Appl. Mech.*, Am. Soc. Mech. Engrs., 1958, p. 329.
41. M. Hetényi, "Photoelasticity and photoplasticity," in *Structural Mechanics*, J. N. Goodier and N. J. Hoff (eds.), Pergamon Press, 1960, p. 483.
42. J. Duffy, "Effects of the thickness of birefringent coatings," *Exptl. Mech.*, v. 1, 1961, p. 74.
43. T. C. Lee, C. Mylonas and J. Duffy, "Thickness effects in birefringent coatings with radial symmetry," *Exptl. Mech.*, v. 1, 1961, p. 134.
44. P. S. Theocaris and C. Mylonas, "Viscoelastic effects in birefringent coatings," *J. Appl. Mech.*, v. 28, 1961, p. 601.
45. Y. F. Cheng, "Some new techniques for scattered light photoelasticity," *Exptl. Mech.*, v. 3, 1963, p. 275.

Development of the Three-Dimensional Theory of Elasticity

L. E. GOODMAN

University of Minnesota, Minneapolis, Minnesota

1 Introduction

The theory of elasticity has for its objective the determination of the state of stress induced by load in bodies composed of an ideal elastic material. Such a determination is a prerequisite for progress in the study of a wide variety of technological and scientific questions, ranging from structural engineering to the assessment of geophysical hypotheses. It is not surprising, therefore, that the subject developed relatively early in the history of continuum mechanics. By the close of the nineteenth century it had assumed a form recognizable today. This development at the hands of Cauchy, Navier, de St. Venant, F. Neumann, Kirchhoff, and Kelvin, among many distinguished names, has been described in the treatise of Todhunter and Pearson [1] and has been comprehensively reviewed by Truesdell [2]. It is understandable also that a brief account can no longer do justice to the entire theory of elasticity. Fortunately, a number of recent works develop the subject at length. Among these the books by Green and Zerna [3], Lur'e [4], Sokolnikoff [5], and Novozhilov [6] have been influential. Excellent accounts by Trefftz [7], Sneddon, and Berry [8], Truesdell and Toupin [9], and Gurtin [10] appear in successive editions of the *Handbuch der Physik*. The treatises of Love [11], Timoshenko and Goodier [12], Biot [13], and Solomon [14] contain detailed descriptions of many particular problems.

Attention is here confined to a description of leading developments in the linear three-dimensional theory of continuum elastostatics. This excludes such important areas as finite deformations and structural stability, as well as the architectonic synthesis of two-dimensional elastostatics associated with the names of Kolosov and Muskhelishvili†. The limitation is intended to make it possible to trace, in brief space, several unifying lines of twentieth-century activity in three-dimensional elasticity: the development of general methods of analysis, interaction with other field theories, and embedment in an axiomatically rigorous field theory of continuum mechanics.

† These topics are developed by Green and Adkins [15], Truesdell and Noll [16], Ziegler [17], and England [18], *inter alia*.

2 Notation

The summation convention of Cartesian tensor notation is employed. An alphabetical suffix appearing twice in any term is to be given all three possible values and the three terms so obtained are to be added. A suffix that appears only once in each term may have any of the three possible values (the same in each term of course). When we write "a vector u_i" we mean a tensor of the first order whose components are u_1, u_2, u_3. Similarly, when we write "a tensor e_{ij}" we mean a second-order tensor having nine components, e_{11}, e_{12}, e_{13}, e_{21}, e_{22}, e_{23}, ...

σ_{ij} = Cauchy stress tensor
x_i = Cartesian coordinate
b_i = Body-force vector per unit volume
u_i = Displacement vector
e_{ij} = Strain tensor
δ_{ij} = Substitution tensor, if $i = j$ the component is $+1$; if $i \neq j$ the component is 0
e_{pks} = Alternating tensor; if any two of p, k, s are equal the corresponding component is 0; if b, k, s are all unequal and in cyclic order the component is $+1$; if the order is not cyclic the component is -1
λ, μ = Lamé's moduli of elasticity
E, ν = Young's modulus of elasticity; Poisson's ratio
β = Region occupied by loaded body
$\partial\beta$ = Bounding surface of β
$\partial\beta_1, \partial\beta_2$ = Disjoint subsets of β; $\partial\beta_1 \cup \partial\beta_2 = \partial\beta$
n_i = Outward unit normal vector to $\partial\beta$
f_i, g_i = Functions specified on $\partial\beta$; $i = 1, 2, 3$
$a_1, \ldots$ = Constants
$P(x_1)$ = Scalar point function
$H_i(x_j)$ = Vector point function
$\psi_i(x_j)$ = Vector point function appearing in the fundamental potential solution of the displacement equations of equilibrium. Class C^2
$\varphi(x_i)$ = Scalar point function appearing in the fundamental potential solution. Class C^3
F_i = Galerkin vector point function. Class C^4
$r = (x_1^2 + x_2^2)^{1/2}$
$R = (x_1^2 + x_2^2 + x_3^2)^{1/2} = (x_m x_m)^{1/2}$
ξ_i = Cartesian coordinates of source point
$G(\xi_i, x_i)$ = Green's function for Laplace's equation
P_i = Concentrated force in the direction x_i
c = Constant denoting position on x_3-axis; e.g., $(0, 0, c)$.
$x_3' = x_3 + it$; $i = \sqrt{-1}$; t is a nonce-variable here but in Section 6 it denotes time

$R' = [r^2 + (x_3 + it)^2]^{1/2} = [r^2 + (x_3')^2]^{1/2}$
$n_i(t)$ = Stress function associated with boundary traction or displacement in the x_i-direction
a = Radius of circle centered on the origin and in the plane $x_3 = 0$
$s_i(r)$ = Traction boundary condition associated with stress component σ_{3i}
$u_i(r)$ = Displacement boundary condition
$\delta(a - t)$ = Dirac delta function of argument $a - t$
p_0 = Fluid pressure
K = Modulus of cohesion in Barenblatt theory of equilibrium cracks
F_1, F_2 = Harmonic stress functions of class C^4
T = Temperature
α = Linear coefficient of thermal expansion
κ = Thermal conductivity
ρ = Density
c_v = Specific heat measured at constant strain
T_i = Applied surface traction components
S = Surface region
ϵ = Radius of S

3 Reduction to Potential Theory

In the linear theory of elasticity a static elastic state is defined as a displacement vector u_i, a strain tensor e_{ij} and a stress tensor σ_{ij} that satisfy the field equations

$$\frac{\partial \sigma_{ij}}{\partial x_j} + b_i = 0, \qquad \sigma_{ij} = \sigma_{ji} \tag{1}$$

$$\sigma_{ij} = c_{ijkl} e_{kl} \tag{2}$$

$$e_{ij} = \frac{1}{2}\left(\frac{\partial u_i}{\partial x_j} + \frac{\partial u_j}{\partial x_i}\right) \tag{3}$$

These equations are to hold throughout a body β with boundary $\partial\beta$. Displacements are assumed to be so small that all the field quantities may be referred to a fixed Cartesian system embedded in either the loaded or the unloaded configuration of the body. The reference state in classical elastostatics is one in which stress, strain, and displacement vanish concurrently.

Equation (1), due to Cauchy, embodies Newton's law of motion for the statical case. Equation (2), Cauchy's generalization of Hooke's law, defines the ideal elastic material. If the material is also homogeneous and isotropic

$$c_{ijkl} = \lambda \delta_{ij} \delta_{kl} + \mu(\delta_{ik} \delta_{jl} + \delta_{il} \delta_{jk}) \tag{4}$$

whence

$$\sigma_{ij} = \lambda e_{mm} \delta_{ij} + 2\mu e_{ij} \tag{5}$$

In what follows, except where the contrary is expressly stated, Eq. (5) will be assumed to hold and λ and μ taken as constants. They are related to Young's modulus of elasticity, E, and to Poisson's ratio, ν,

$$\lambda = \frac{E\nu}{(1+\nu)(1-2\nu)}, \qquad \mu = \frac{E}{2(1+\nu)} \tag{6}$$

The stress field σ_{ij} that satisfies Eq. (1) generates, through Eq. (2), a strain field e_{ij}. In order that a single-valued displacement field be then obtainable from Eq. (3) it is necessary that e_{ij} satisfy the geometrical compatibility condition of de St. Venant:

$$\epsilon_{pks} \frac{\partial}{\partial x_k}\left(\frac{\partial e_{sj}}{\partial x_i} - \frac{\partial e_{si}}{\partial x_j}\right) = 0 \tag{7}$$

If β is simply connected, this condition is also sufficient to ensure the existence of a continuous displacement field. The need to satisfy the six equations (7) is obviated if the expressions (5) and then (3) are substituted in Eq. (1). This operation leads to displacement equations of equilibrium:

$$\frac{\mu}{1-2\nu}\frac{\partial}{\partial x_i}\left(\frac{\partial u_m}{\partial x_m}\right) + \mu \frac{\partial^2 u_i}{\partial x_m \partial x_m} + b_i = 0 \tag{8}$$

The commonly encountered boundary conditions may also be expressed in terms of the displacement and the displacement gradient. Two possible sets of boundary conditions take the forms

$$u_i = f_i \quad \text{on } \partial\beta_1 \tag{9}$$

and

$$n_j\sigma_{ij} = n_j\left[\lambda \frac{\partial u_k}{\partial x_k}\delta_{ij} + \mu\left(\frac{\partial u_i}{\partial x_j} + \frac{\partial u_j}{\partial x_i}\right)\right] = g_i \quad \text{on } \partial\beta_2 \tag{10}$$

n_j being the outward unit normal vector, f_i and g_i being piecewise smooth prescribed functions, and $\partial\beta_1$, $\partial\beta_2$, being separate portions of the boundary $\partial\beta$ which, together, form the entire boundary. The problems associated with these boundary conditions are known, respectively, as the first and second boundary value problems of elastostatics.

The earliest solutions of these field equations were those in which the elastic state entails at most one varying component of stress or one varying component of displacement. These "one-dimensional" solutions were, and still are, of considerable technological interest. They included the hollow circular cylinder and sphere under uniform pressure, the pure flexure of prismatic bars, and the torsion of the circular cylinder. From a mathematical point of view these problems all reduce to the solution of an ordinary linear differential equation.

Of greater interest are the cases of plane strain:

$$u_1 = u_1(x_1, x_2), \quad u_2 = u_2(x_1, x_2), \quad u_3 = 0 \tag{11}$$

and anti-plane stress†:

$$\sigma_{11} = \sigma_{22} = \sigma_{12} = 0 \tag{12}$$

The first of these equations implies that

$$\begin{gathered} \sigma_{11} = \sigma_{11}(x_1, x_2), \quad \sigma_{22} = \sigma_{22}(x_1, x_2), \quad \sigma_{12} = \sigma_{12}(x_1, x_2) \\ \sigma_{33} = \nu(\sigma_{11} + \sigma_{22}), \quad \sigma_{13} = \sigma_{23} = 0 \end{gathered} \tag{13}$$

and the second implies that

$$\begin{gathered} \sigma_{13} = \sigma_{13}(x_1, x_2), \qquad \sigma_{23} = \sigma_{23}(x_1, x_2) \\ \sigma_{33} = a_1x_1x_3 + a_2x_2x_3 + a_3x_1 + a_4x_2 + a_5x_3 + a_6 \end{gathered} \tag{14}$$

the quantities a_i being constants. Plane strain corresponds to the elastic state in the middle portion of a long cylindrical object (the axis of the cylinder being taken parallel to x_3) when the object is subject to body forces and to surface tractions or deformations that are uniform along the generators. Filon has shown [20] that the same analysis will serve for a thin plate subjected to edge loadings in its plane, provided σ_{11}, σ_{22}, and σ_{12} are interpreted as average values through the thickness. This parallel state has been termed generalized plane stress. Anti-plane stress corresponds to the long cylinder of arbitrary cross section loaded on its end faces—the so-called de St. Venant problem of extension, torsion, and bending. In view of Eqs. (13) and (14) both plane strain and anti-plane stress may be termed two-dimensional stress states. Owing to their technological importance they were the subject of many fruitful and ingenious analyses throughout the latter nineteenth and the early twentieth centuries. Then Kolosov [21] and Muskhelishvili [22] brought both two-dimensional cases within the scope of complex function theory‡. The full resources of analytic function theory could then be brought to bear upon the subject. Broadly speaking, any cylinder cross-section boundary shape that could be mapped on the unit circle was open to systematic analysis for the planar stress state corresponding to imposed edge tractions or imposed edge displacements or even to a combination thereof. The de St. Venant problem of torsion and flexure was solved under similar broad conditions.

The development of systematic procedures for the analysis of the full three-dimensional displacement equations of equilibrium (8) was at first undertaken

† A somewhat more general definition is also found in the literature [19].

‡ This work appears at first to have escaped widespread notice outside the USSR. Some of the results were later independently recovered by Stevenson [23], Green [24], and Milne-Thomson [25].

along two distinct lines. In one, a class of functions satisfying Eqs. (8) was sought and the boundary conditions satisfied by taking a series of these solutions with suitably adjusted coefficients. This approach generally required that the governing equations be expressed in orthogonal coordinates. The belief that it would be a fruitful approach led Lamé to develop the theory of curvilinear coordinates to what is virtually its present state [26]. The method has found application in the analysis of spherical shells, hyperbolic notches, and ellipsoidal cavities among other special shapes. In a second approach displacements and stresses are expressed as integrals that represent a distribution of singularities over the surface or throughout the volume of the body. This line of analysis was originated by Betti [27]. He started from his reciprocal theorem and used the singularities corresponding to a center of compression. The two methods are to a considerable extent fused in the fundamental potential solution of the three-dimensional field Eqs. (8). The complex variable technique of planar elasticity can also then be represented as an outgrowth of this three-dimensional solution. In this process a certain unity in the subject as a whole is exhibited.

Referring to the displacement equations of equilibrium (8), one can observe that all three unknown functions u_i appear in each equation. A systematic approach would replace these displacements by other functions, perhaps less physically transparent in meaning, whose governing equations are separable. Early attempts in this direction were made by Maxwell and by Morera. In the period 1875–1882 J. V. Boussinesq [28] developed a number of functions that could be used to generate stresses and displacements satisfying Eqs. (8) provided the functions themselves satisfied the three-dimensional form of Laplace's equation. The entire resources of potential theory could be brought to bear upon the determination of these functions. But although Boussinesq used his potential functions to analyze the fundamental problem that is associated with his name—the concentrated normal force acting on a plane surface—the underlying potential solutions were neglected. In the period 1932–1935 Papkovitch [29], Grodski [30], and Neuber [31] independently published a solution of the three-dimensional displacement equations of equilibrium in terms of four potential functions. Mindlin established in 1936 [M5]† that these four functions provide a complete solution of Eqs. (8) in the sense that every elastic state can be represented by an appropriate choice of them, at least in the absence of body forces. We therefore refer to this as the fundamental potential solution of elasticity. In 1955 Mindlin [M46] extended his demonstration to include body forces. As the latter reference is not widely available and as the argument exposes certain peculiarities of the fundamental solution it is recovered here.

The basic step in Mindlin's completeness proof is the Helmholtz separation

† Numbers preceded by the letter M in brackets refer to the Publications of R. D. Mindlin listed after the Preface to this volume.

of u_i, expressing it as the sum of a gradient and a curl:

$$u_i = \frac{\partial P}{\partial x_i} + \epsilon_{ikp} \frac{\partial H_p}{\partial x_k} \tag{15}$$

This separation is always possible in view of the assumed smoothness of u_i. Furthermore, as u_i is supposed to satisfy the displacement Eqs. (8), it follows that

$$\frac{\mu}{1 - 2\nu} \frac{\partial}{\partial x_i} \left(\frac{\partial^2 P}{\partial x_m \partial x_m} \right) + \mu \frac{\partial^2}{\partial x_m \partial x_m} \left(\frac{\partial P}{\partial x_i} + \epsilon_{ikp} \frac{\partial H_p}{\partial x_k} \right) + b_i = 0 \tag{16}$$

$$\mu \frac{\partial^2}{\partial x_m \partial x_m} \left(\frac{2 - 2\nu}{1 - 2\nu} \frac{\partial P}{\partial x_i} + \epsilon_{ikp} \frac{\partial H_p}{\partial x_k} \right) + b_i = 0 \tag{17}$$

Now let

$$\psi_i = \frac{2 - 2\nu}{1 - 2\nu} \frac{\partial P}{\partial x_i} + \epsilon_{ikp} \frac{\partial H_p}{\partial x_k} \tag{18}$$

In view of Eq. (17), the three functions ψ_i are functions whose Laplacians are known throughout the solid:

$$\frac{\partial^2 \psi_i}{\partial x_m \partial x_m} = -\mu^{-1} b_i \tag{19}$$

Also, if we differentiate Eq. (18) with respect to x_i,

$$\frac{\partial \psi_i}{\partial x_i} = \frac{2 - 2\nu}{1 - 2\nu} \frac{\partial^2 P}{\partial x_i \partial x_i} \tag{20}$$

The complete solution of Eq. (20) is

$$P = \frac{1 - 2\nu}{4 - 4\nu} (x_n \psi_n + \varphi) \tag{21}$$

where

$$\frac{\partial^2 \varphi}{\partial x_m \partial x_m} = \mu^{-1} x_n b_n \tag{22}$$

The Laplacian of φ is therefore also known throughout the solid. If Eq. (21) is substituted in Eq. (15) it appears that u_i can be represented in the form

$$u_i = \frac{1 - 2\nu}{4 - 4\nu} \left(\frac{\partial}{\partial x_i} x_n \psi_n + \frac{\partial \varphi}{\partial x_i} \right) + \epsilon_{ipk} \frac{\partial H_p}{\partial x_k} \tag{23}$$

But from Eq. (18)

$$\epsilon_{ikp} \frac{\partial H_p}{\partial x_k} = \psi_i - \frac{2 - 2\nu}{1 - 2\nu} \frac{\partial P}{\partial x_i} \tag{24}$$

$$= \psi_i - \frac{1}{2} \frac{\partial}{\partial x_i} (x_n \psi_n + \varphi) \tag{25}$$

so that

$$u_i = \psi_i - \frac{1}{4 - 4\nu} \frac{\partial}{\partial x_i} (x_n \psi_n + \varphi) \tag{26}$$

The solution (26), subject to the governing Eqs. (19) and (22), constitutes the fundamental potential solution of elasticity in the presence of body forces. The strain and stress components are readily expressed in terms of the four functions ψ_i, φ:

$$e_{ij} = \frac{1 - 2\nu}{4 - 4\nu} \left(\frac{\partial \psi_i}{\partial x_j} + \frac{\partial \psi_j}{\partial x_i} \right) - \frac{1}{4 - 4\nu} \left(x_n \frac{\partial^2 \psi_n}{\partial x_i \partial x_j} + \frac{\partial^2 \varphi}{\partial x_i \partial x_j} \right) \tag{27}$$

$$\sigma_{ij} = \frac{4\mu\nu}{4 - 4\nu} \frac{\partial \psi_m}{\partial x_m} \delta_{ij} + \frac{2\mu}{4 - 4\nu} \left[(1 - 2\nu) \left(\frac{\partial \psi_i}{\partial x_j} + \frac{\partial \psi_j}{\partial x_i} \right) - x_n \frac{\partial^2 \psi_n}{\partial x_i \partial x_j} - \frac{\partial^2 \varphi}{\partial x_i \partial x_j} \right] \tag{28}$$

The function φ was employed by Lamé [34]. Boussinesq [28] introduced the function ψ_3 but in place of ψ_1 and ψ_2 he employed other functions better adapted to the particular problems he treated.

Apart from its intrinsic importance in demonstrating the completeness of the fundamental potential solution of the equations of linear elasticity, Mindlin's analysis exposes a number of interesting aspects of the solution.

(a) *The Representation* (26, 19, 22) *is Not Unique*

This situation arises from the fact that the Helmholtz separation (15) is not unique. The Helmholtz separation can be made unique by imposing an additional condition on the function triad H_i. The additional condition usually satisfied in proofs of the Helmholtz theorem (e.g., [32, p. 44]) is $\partial H_i/\partial x_i = 0$. But in the development of the fundamental potential solution presented in the preceding paragraph no such condition is imposed. It is not difficult, in fact, to exhibit harmonic function sets that lead to a null state of stress. One such is

$$\psi_i = x_i(x_m x_m)^{-3/2}, \qquad \varphi = -(5 - 4\nu)(x_m x_m)^{-1/2} \tag{29}$$

In application, it is completeness rather than uniqueness of representation that is critical. Often it is of advantage to be able to reserve the possibility of imposing an additional condition upon the four fundamental functions.

(b) *The Representation* (26, 19, 22) *is Closely Related to Other Forms*

In particular, a general solution of the displacement equations of equilibrium published by Galerkin in 1930–1931 takes the form [33]:

$$u_i = \frac{1-\nu}{\mu}\frac{\partial^2 F_i}{\partial x_m \partial x_m} - \frac{1}{2\mu}\frac{\partial^2 F_m}{\partial x_i \partial x_m} \tag{30}$$

This form satisfies Eqs. (8)—as may be verified by direct substitution—provided the three functions F_i satisfy the biharmonic relation

$$\frac{\partial^2}{\partial x_k \partial x_k}\left(\frac{\partial^2 F_i}{\partial x_m \partial x_m}\right) = -\frac{1}{1-\nu}\, b_i \tag{31}$$

Mindlin has shown [M5] that the biharmonic Galerkin functions F_i are related to the fundamental potential functions ψ_i, φ by the relations

$$\psi_i = \frac{1-\nu}{\mu}\frac{\partial^2 F_i}{\partial x_m \partial x_m} \tag{32}$$

$$\varphi = \frac{1-\nu}{\mu}\left(2\frac{\partial F_m}{\partial x_m} - x_k \frac{\partial^2 F_k}{\partial x_m \partial x_m}\right) \tag{33}$$

It is easy to see that if ψ_i satisfies Eq. (19) and φ satisfies Eq. (22), F_i will automatically satisfy Eq. (31).

Westergaard has shown [46] that the Galerkin functions, like the fundamental potential solution, are complete but not unique. They are a generalization of a stress function introduced earlier by Love [11, p. 275] for the analysis of radially symmetric states of stress. In fact by taking $F_1 = F_2 = 0$ and F_3 a function of x_3 and of $r = (x_1^2 + x_2^2)^{1/2}$, Galerkin's strain functions become identical with Love's.

Similarly, if ψ_3 is set equal to zero throughout the solid and the other three functions are regarded as functions of x_1, x_2, England has shown [18, p. 45] that Mindlin's development given above leads to the Kolosov–Muskhelishvili formulation of the plane strain-generalized plane stress analysis.

(c) *Representation by Three Functions*

The fact that the fundamental potential solution is not unique suggests that the four functions ψ_i, φ are not independent. This question has been extensively investigated. Eubanks and Sternberg [35] have shown that if β is a star-shaped region†, $b_i = 0$ and $\nu \neq \frac{1}{4}$, there is no loss of generality in dropping the func-

† A region, β, is said to be star shaped with respect to an interior point, Q, if the straight line joining Q to any other point of β lies entirely within β.

tion φ. The restriction on Poisson's ratio is not necessary if β is convex rather than merely star shaped. Proceeding along different lines, Naghdi and Hsu [36] have considered multiply connected domains. In practice, as previously remarked, it is desirable to retain four functions so as to facilitate the satisfaction of boundary conditions.

(d) *Curvilinear Coordinate Representation*

The formulation (26) may be written in vectorial form:

$$\mathbf{u} = \boldsymbol{\psi} - \frac{1}{4 - 4\nu} \,\mathbf{grad}\, (\mathbf{x} \cdot \boldsymbol{\psi} + \varphi) \tag{34}$$

and the requirement (22) on φ may be correspondingly generalized:

$$\text{div}\,\mathbf{grad}\,\varphi = \mu^{-1}\mathbf{x} \cdot \mathbf{b} \tag{35}$$

The correct generalization of Eq. (19) is

$$\mathbf{grad}\,\text{div}\,\boldsymbol{\psi} - \mathbf{curl}\,\mathbf{curl}\,\boldsymbol{\psi} = -\mu^{-1}\mathbf{b} \tag{36}$$

This formulation is frequently of advantage when dealing with boundaries of relatively simple shape. In spherical polar coordinates, to take a simple example, the displacements having point symmetry correspond to

$$\mathbf{u} = \psi_R(R)\mathbf{e}_R \tag{37}$$

In the absence of body forces, substitution in Eq. (36) yields the differential equation

$$\psi_R'' + 2R^{-1}\psi_R' - 2R^{-2}\psi_R = 0 \tag{38}$$

whence

$$\psi_R = C'R + DR^{-2} \tag{39}$$

and†

$$\mathbf{u} = (CR + DR^{-2})\mathbf{e}_R \tag{40}$$

Appropriate choice of the constants C and D will satisfy the boundary conditions of normal traction or specified displacement on the faces of a spherical shell. The results of Lamé [11, p. 142] are at once recovered.

The fundamental potential solution of Boussinesq, Papkovitch, Neuber, and Grodski effects a separation of variables in the second-order partial differential field equations of elastostatics. Furthermore, the displacement functions in terms of which the solution is expressed are governed by Poisson's (or Laplace's) equation. It might be thought, therefore, that elastostatics had been brought to the same state of completeness as electrostatics. That this is not quite the case is due to the relative complexity of the elastostatic boundary

† The function $\varphi = \varphi(R)$ adds nothing more and may therefore be neglected.

conditions. Broadly speaking, if the Laplacian of a function is known throughout a region and if the Green's function for Laplace's equation is known for the region, the value of the function throughout the region can be expressed in terms of its values on the boundary. The key theorem to this effect is [32, p. 30]

$$-4\pi V(x_i) = \int_{\partial\beta} V(\xi_i) \frac{\partial G(\xi_i, x_i)}{\partial n} d^2\xi_i + \int_{\beta} G(\xi_i, x_i) \frac{\partial^2 V(\xi_i)}{\partial \xi_i \partial \xi_i} d^3\xi_i \quad (41)$$

Here $V(x_i)$ is the wanted function; $G(\xi_i, x_i)$ is the Green's function for the region β and $\partial G/\partial n$ is its normal derivative at the boundary; x_i are the coordinates of a field point and ξ_i the coordinates of a source point. The Green's function for Laplace's equation depends only on the shape of the region β. Its existence and determination have been the subject of extensive study in potential theory, e.g., [37, p. 248]. The effect of Eq. (41) in conjunction with Eqs. (19) and (22) is to reduce three-dimensional stress analysis to the discovery of combinations of the ψ_i, φ and their derivatives, that are known on the boundary. This process is by no means self-evident.

4 Fundamental Problems

Certain problems occupy key places in three-dimensional elastostatics because they lead to the solution of a wide variety of special cases that are important in science and engineering. In the history of the subject these key problems have often exposed and resolved theoretical difficulties. We consider them here, taking as point of departure the fundamental potential solution of the previous section. This is in contrast, for the most part, to the differing ways in which their analysis was originally conceived. System follows art.

The simplest fundamental three-dimensional problem is that of a concentrated force in the interior of a large (infinite) elastic solid. Following Kelvin [38] the concentrated force P_3 acting at the origin in the direction of the x_3-axis is defined by considering a small domain β' surrounding the origin. In β' the body-force component $b_3(\xi_i)$ has a nonzero value; elsewhere it vanishes. The domain β' shrinks to a smaller domain centered on the origin. Then

$$P_3 = \lim_{\beta' \to 0} \int_{\beta'} b_3 d^3\xi_i \quad (42)$$

Now apply Eq. (41) taking $\psi_3(x_i)$ for the function V. Since the Green's function for the infinite domain is

$$G(x_i, \xi_i) = p^{-1} = [(x_1 - \xi_1)^2 + (x_2 - \xi_2)^2 + (x_3 - \xi_3)^2]^{-1/2} \quad (43)$$

it follows from Eqs. (41) and (19) that

$$-4\pi\psi_3 = \int_{\partial\beta} \psi_3(\xi_i) \frac{\partial p^{-1}}{\partial n} d^2\xi_i + \int_{\beta'} p^{-1}(-\mu^{-1}b_3) d^3\xi_i \quad (44)$$

The first of these integrals vanishes because ψ_3 must approach zero as distance from the origin increases if displacements are to diminish with distance from the point of application of the force and because $\partial p^{-1}/\partial n = 0(\xi^{-2})$ whilst $d^2\xi_i = 0(\xi^2)$. The second integral also vanishes outside the region β'. As the region β' shrinks down to the origin

$$\lim_{\beta' \to 0} p^{-1} = R^{-1} = (x_m x_m)^{-1/2} \tag{45}$$

and so we have at once

$$\psi_3 = \frac{P_3}{4\pi\mu} R^{-1} \tag{46}$$

Similar reasoning leads to the conclusion that ψ_1 and ψ_2 vanish everywhere. In the case of φ we have

$$-4\pi\varphi = \int_{\partial\beta} \varphi(\xi_i) \frac{\partial p^{-1}}{\partial n} d^2\xi_i - \mu^{-1} \int_{\beta'} p^{-1}\xi_3 b_3 d^3\xi_i \tag{47}$$

The first integral vanishes as before and the second integral also vanishes when β' shrinks down to the origin ($\xi_3 \to 0$). Consequently ψ_3 as given by Eq. (46) represents the complete solution of this fundamental problem. It follows from Eq. (26) that for a force having components P_j the displacement is given by the expression

$$u_i = \frac{P_j}{16\pi\mu(1-\nu)} \left[\frac{3-4\nu}{R} \delta_{ij} + \frac{x_i x_j}{R^3} \right] \tag{48}$$

Kelvin's problem is interesting from a theoretical standpoint because it introduces the idea of a concentrated force, properly conceived as the limiting case of a body force that is applied to a region. After the publication of de St. Venant's memoir on torsion, it became common to assume that the correctness of a displacement set such as Eq. (48) could be confirmed by computing the corresponding stress state and verifying (a) that it satisfied the stress equilibrium Eq. (1) and (b) that the resultant traction on the region exterior to any sphere surrounding the origin was equipollent to a force in the x_3-direction. Only recently has it been appreciated that boundary conditions of the type (b), sometimes termed "de St. Venant boundary conditions," are insufficient to assure the correctness of the solution. The Kirchhoff uniqueness theorem in linear elasticity breaks down when the stress state includes a singularity. We owe to E. Sternberg and his colleagues Eubanks, Turteltaub, and Wheeler an appreciation of the importance of regarding the concentrated force as the limiting case of a distributed traction [39–41].

The singularity in the stress and displacement fields that arises in the presence of a concentrated force is understandable. Other such singular elastic states—termed "nuclei of strain"—may be obtained from Kelvin's problem. For example, if to the displacements generated by Eq. (46) are added those corresponding to a force $-P_3$ acting in the direction of the negative x_3-axis and located at $(0, 0, c)$, and c is allowed to become vanishingly small while P_3 increases in such a way that P_3c approaches a constant value, C, the limiting state is

$$u_i = \frac{C}{16\pi\mu(1-\nu)}[x_iR^{-3} - (2-4\nu)x_3R^{-3} - 3x_ix_3^2R^{-5}] \tag{49}$$

This state is known as a double force without moment, of strength C. It may be derived from Eq. (48) by differentiation with respect to x_3. The corresponding fundamental potential solution is

$$\psi_3 = \frac{C}{4\pi\mu} - x_3R^{-3}, \qquad \varphi = \frac{C}{4\pi\mu}R^{-1} \tag{50}$$

By adding three such double forces of equal strength directed along the coordinate axes an elastic state is reached in which

$$u_i = -\frac{C(1-2\nu)}{8\pi\mu(1-\nu)}x_iR^{-3} \tag{51}$$

corresponding to

$$\varphi = -[C(1-2\nu)/2\pi\mu]R^{-1} \tag{52}$$

This state is termed a center of compression if $C > 0$ and a center of dilatation if $C < 0$. Other nuclei of strain for the infinite space are given by Love [11, p. 186].

Kelvin's problem is interesting from a point of view oriented toward applications of the theory to particular problems because it formed the basis for the first systematic effort to express elastic states in terms of quantities known on the boundary. This step is due to Betti† [42]. He employed the reciprocal theorem with which his name is associated, taking for one of the two elastic states the center of compression (51) and for the other the wanted elastic state. Betti's method was extended to nonisotropic solids by Fredholm [43]. Cerruti used it to analyze the problem of the plane‡ for the isotropic solid [44], Michell for the anisotropic case as well [45].

The next in complexity of the fundamental problems of three-dimensional elastostatics are those analyzing the stress field created in a half-space by a concentrated force acting at a plane surface. The case of a force normal to the

† It is remarkable that the work of Betti is not mentioned in standard histories of the subject such as Ref [1].

‡ The so-called "problem of the plane" dealing with the elastic states produced in the solid $x_3 \geq 0$ by surface loading or by body forces has an extensive history, owing to its technological importance. Reference may be made to Ref. [10] or to Ref. [11, Section 167], *inter alia.*

plane surface is usually associated with the name of Boussinesq [28] and the case of a force parallel to the surface with that of Cerruti [44]. Both these problems are special cases of what has come to be known as Mindlin's problem: the concentrated force acting at a point below the plane surface of the half-space. This fundamental boundary-value problem was first solved by Mindlin in 1936† [M4]. He added to Kelvin's solution a set of nuclei of strain for the infinite solid so chosen that (a) their singularities occurred outside the half-space and (b) they annulled the tractions on the plane boundary resulting from Kelvin's solution. It is more interesting, however, to approach the solution as Mindlin did in 1955 [M46] when he reviewed the matter and introduced the systematic method that has been outlined in Section 3 and exemplified in Section 4 by the treatment of Kelvin's problem. The salient points of the analysis are instructive.

Consider first the case in which the force P_3 applied at $(0, 0, c)$ below the plane boundary $x_3 = 0$ is directed along the positive x_3-axis. As the state of stress is rotationally symmetric it may be inferred (as in Kelvin's problem) that $\psi_1 = \psi_2 = 0$. On the boundary $x_3 = 0$, the traction components

$$\sigma_{33} = \frac{\mu}{2 - 2\nu}\left[(2 - 2\nu)\frac{\partial\psi_3}{\partial x_3} - \frac{\partial^2\varphi}{\partial x_3^2}\right] \tag{53}$$

$$\sigma_{31} = \frac{\mu}{2 - 2\nu}\frac{\partial}{\partial x_1}\left[(1 - 2\nu)\psi_3 - \frac{\partial\varphi}{\partial x_3}\right] \tag{54}$$

$$\sigma_{32} = \frac{\mu}{2 - 2\nu}\frac{\partial}{\partial x_2}\left[(1 - 2\nu)\psi_3 - \frac{\partial\varphi}{\partial x_3}\right] \tag{55}$$

obtained from Eq. (28) must vanish. The two expressions in parentheses on the right-hand side of Eqs. (53–55) are successively taken as the functions V in Eq. (41). The surface integral in Eq. (41) then vanishes. The Laplacian of V is known through Eqs. (19) and (22). The Green's function for the half-space $x_3 \geq 0$ is

$$G(\xi_i, x_i) = p_1^{-1} - p_2^{-1} \tag{56}$$

where

$$p_1^2 = (x_1 - \xi_1)^2 + (x_2 - \xi_2)^2 + (x_3 - \xi_3)^2 \tag{57}$$

$$p_2^2 = (x_1 - \xi_1)^2 + (x_2 - \xi_2)^2 + (x_3 + \xi_3)^2 \tag{58}$$

The integrand of the volume integral in Eq. (41) is therefore expressed entirely in terms of known quantities. As a matter of fact it is unnecessary to carry out the integration; as in the Kelvin problem, it is only the limit approached by the integral as the region β' shrinks to the point 0, 0, c that matters. Once the bracketed quantities in Eq. (53–55) are thus determined it is a relatively

† A preliminary statement appears in Ref. [M3].

straightforward matter to find

$$\psi_3 = \frac{P_3}{4\pi\mu}\left[R_1^{-1} + (3 - 4\nu)R_2^{-1} + 2c(x_3 + c)R_2^{-3}\right] \tag{59}$$

$$\varphi = \frac{P_3}{4\pi\mu}\left[(1 - 2\nu)(4 - 4\nu)\log(R_2 + x_3 + c) - cR_1^{-1} - c(3 - 4\nu)R_2^{-1}\right] \tag{60a}$$

where

$$R_1^2 = x_1^2 + x_2^2 + (x_3 - c)^2, \qquad R_2^2 = x_1^2 + x_2^2 + (x_3 + c)^2 \tag{60b}$$

From a formal point of view this completes the solution.

The case in which the force is parallel to the surface is treated by Mindlin in the same systematic way. In that case one retains ψ_1, ψ_3, and φ. The boundary conditions on $x_3 = 0$ are now

$$\sigma_{33} = \frac{\mu}{2 - 2\nu}\left[(2 - 2\nu)\frac{\partial\psi_3}{\partial x_3} + 2\nu\frac{\partial\psi_1}{\partial x_1} - x_1\frac{\partial^2\psi_1}{\partial x_3^2} - \frac{\partial^2\varphi}{\partial x_3^2}\right] = 0 \tag{61}$$

$$\sigma_{31} = \frac{\mu}{2 - 2\nu}\left[(1 - 2\nu)\frac{\partial\psi_3}{\partial x_1} + (1 - 2\nu)\frac{\partial\psi_1}{\partial x_3} - x_1\frac{\partial^2\psi_1}{\partial x_1\partial x_3} - \frac{\partial^2\varphi}{\partial x_1\partial x_3}\right] = 0 \tag{62}$$

$$\sigma_{32} = \frac{\mu}{2 - 2\nu}\left[(1 - 2\nu)\frac{\partial\psi_3}{\partial x_2} - x_1\frac{\partial^2\psi_1}{\partial x_2\partial x_3} - \frac{\partial^2\varphi}{\partial x_2\partial x_3}\right] = 0 \tag{63}$$

If the third of these is differentiated with respect to x_1 and subtracted from the derivative of the second with respect to x_2 it appears that on $x_3 = 0$

$$\frac{\partial^2\psi_1}{\partial x_2\partial x_3} = 0 \tag{64}$$

and this implies that on $x_3 = 0$

$$\frac{\partial\psi_1}{\partial x_3} = 0 \tag{65}$$

The left-hand side of Eq. (65) is therefore a function the Laplacian V which is known everywhere and which vanishes on the boundary of the half-space. When this function plays the role of V in Eq. (41) the surface integral vanishes and the volume integral, in the limit, takes the value $-(P_1/\mu)(\partial/\partial x_3)(R_1^{-1} + R_2^{-1})$. It follows that

$$\psi_1 = \frac{P_1}{4\pi\mu}(R_1^{-1} + R_2^{-1}) \tag{66}$$

This result may now be used in Eq. (63). Both Eqs. (63) and (62) will be satisfied provided

$$(1 - 2\nu)\psi_3 - \frac{\partial \varphi}{\partial x_3} = 0 \tag{67}$$

on $x_3 = 0$. The expression is again one whose Laplacian is known throughout the solid. Proceeding as before, it is found that the left-hand side of Eq. (67) must vanish everywhere. Equation (61) can now be written

$$(2 - 2\nu)\frac{\partial \psi_3}{\partial x_3} - \frac{\partial^2 \varphi}{\partial x_3{}^2} - (1 - 2\nu)\frac{\partial \psi_1}{\partial x_1} - \frac{2P_1 c}{4\pi\mu}\frac{\partial^2 R_2{}^{-1}}{\partial x_1 \partial x_3} = 0 \tag{68}$$

on $x_3 = 0$. This expression also has a known Laplacian (the Laplacian of $R_2{}^{-1}$ vanishes, of course) and one finds, with the aid of Eq. (66) that

$$(2 - 2\nu)\psi_3 - \frac{\partial \varphi}{\partial x_3} = \frac{P_1}{4\pi\mu}[2(1 - 2\nu)x_1 R_2{}^{-1}(R_2 + x_3 + c)^{-1} - 2cx_1 R_2{}^{-3}] \tag{69}$$

Equations (66), (67), and (69) lead to the solution

$$\psi_1 = \frac{P_1}{4\pi\mu}(R_1{}^{-1} + R_2{}^{-1}) \tag{66}$$

$$\psi_3 = \frac{P_1}{4\pi\mu}[2(1 - 2\nu)x_1 R_2{}^{-1}(R_2 + x_3 + c)^{-1} - 2cx_1 R_2{}^{-3}] \tag{70}$$

$$\varphi = \frac{P_1}{4\pi\mu}[-2(1 - 2\nu)^2 x_1 (R_2 + x_3 + c)^{-1} - 2(1 - 2\nu)cx_1 R_2{}^{-1}(R_2 + x_3 + c)^{-1}] \tag{71}$$

Mindlin and Cheng have used this solution and Eqs. (59–61) to generate nuclei of strain for the half-space [M33], [47]. Rongved [48] has used the method to determine the stress field due to a force at a point in the interior of a semi-infinite solid with a fixed plane boundary. That elastic state is given by

$$\psi_1 = \frac{P_1}{4\pi\mu}(R_1{}^{-1} - R_2{}^{-1}) \tag{72}$$

$$\psi_3 = \frac{P_1}{4\pi\mu}[2(3 - 4\nu)^{-1}cx_1 R_2{}^{-3}] + \frac{P_3}{4\pi\mu}[R_1{}^{-1} - R_2{}^{-1} - 2c(3 - 4\nu)^{-1}(x_3 + c)R_2{}^{-3}] \tag{73}$$

$$\varphi = \frac{P_3}{4\pi\mu}(cR_2{}^{-1} - cR_1{}^{-1}) \tag{74}$$

The notation is the same as in the previous treatment of Mindlin's problem (for which the plane surface is free). In particular the force P_i acts at 0, 0, c

and the origin of coordinates is taken at the plane surface. Rongved finds that for $\nu = 0.3$ the peak normal stress component on the fixed boundary due to P_3 is

$$\sigma_{33}(0, 0, 0) = 0.336 \frac{P_3}{c^2} \tag{75}$$

while that due to P_1 is

$$\sigma_{33}\left(\frac{\pm c}{2, 0, 0}\right) = \pm 0.076 \frac{P_1}{c^2} \tag{76}$$

So-called fundamental problems are valuable as the genesis of other analyses that have applications in science and technology. The stress field of Mindlin's problem has been applied directly by Mindlin to the analysis of the pressure distribution on retaining walls [M6] and the distribution of stress around a pile [M7]. Cheng and Mindlin have used it to analyze the stress field created by thermal expansion of a heated spheroidal inclusion below the surface of a half-space [M34]. Using the Fourier transform mathematical technique Dean, Parsons, and Sneddon [49] have analyzed the stress field corresponding to pressure normal to the surface, uniformly distributed over a circular area and a circular line that lie in a plane parallel to the surface of the half-space.

5 Applications of Complex Potential Methods

As Love has remarked, the methods developed for the analysis of the governing field equations of elasticity tend to fall into two classes. In both classes typical fundamental solutions are developed and then generalized so as to be adaptable to the solution of particular boundary-value problems. One class of methods generalizes by expansion in an infinite series the coefficients of which are later adjusted to satisfy boundary conditions; the other class generalizes by integration. The former approach has been successfully applied in the analysis of the stress field created in a sphere by surface displacements or tractions that can be expanded in a series of surface or solid harmonics†.

† The so-called problem of the sphere has an extensive literature, a complete review of which would be inappropriate here. Galerkin [50] has listed the particular solutions known in 1942. The earliest of these worth noting is that of Lamé [26]. Except for the simple case of a hollow spherical shell loaded internally and externally by uniform pressure his series expansions are too clumsy to be effective. Kelvin, whose analysis [51] is reproduced in many texts, expresses the displacements in terms of three harmonic polynomials. Each of these polynomials is then determined by the boundary conditions, which have to be written as a series of spherical surface harmonics. The analysis is satisfactory if the boundary conditions consist of prescribed displacements; it is awkward for prescribed surface tractions. Lur'e [4, Chapter 8] has presented an improved version of Kelvin's analysis, starting from the fundamental potential solution of elastostatics. Readers interested in using the theory of elasticity for stress analysis of bodies having spherical boundaries should also consult the papers by Sternberg, Eubanks, and Sadowsky [52] and by Fichera [52]. Their results, though limited to the axisymmetric case, are presented in a readily usable form.

The latter approach, which has been of wide application in elasticity theory, includes integral transform methods and the use of three-dimensional complex potential functions.

Integral transform methods were introduced in elastostatic analysis by Lamb [54]. Lamb's idea was further developed by Terazawa [55] to cover the case of the smooth, symmetrically loaded semi-infinite solid. Muki has recently extended the method to asymmetric loadings [56]. The scope of Hankel transform applications has been greatly extended by Sneddon. Some of the problems discussed in this section are treated by that method in Sneddon's excellent didactic book [57] and in a briefer monograph by Tranter [58]. It seems unnecessary to duplicate these presentations here, especially as the approach now to be developed follows naturally from the fundamental potential solution of elastostatics that was developed in Section 3 and from the basic problems of Section 4.

What may be termed the method of complex potential functions can be traced to early work of Smirnov and Sobolev [59–61]† and Weber [63]. This work appears not to have received wide attention. Rostovtzev [64] employed a similar technique in 1953. As early as 1949, however, A. E. Green [65] had independently developed the complex potential function approach and used it to analyze a number of important indentation and crack stress fields. The seminal treatise of Green and Zerna [3] contains an exposition of these problems. For the purposes of the present discussion it is convenient to begin by noting that if c is set equal to zero in Eq. (60), $R_1 = R_2 = R$, say. Mindlin's problem reduces to Boussinesq's. The solution given by Eqs. (59) and (60) for the concentrated force acting at the origin of coordinates at the surface of the half-space $x_3 \geq 0$ in the x_3 direction is

$$\psi_1 = \psi_2 = 0, \qquad \psi_3 = \frac{P_3(1-\nu)}{\pi\mu} R^{-1} \tag{77}$$

$$\varphi = \frac{P_3(1-\nu)}{\pi\mu} (1-2\nu) \log (R + x_3) \tag{78}$$

That is,

$$\psi_3 = \frac{1}{1-2\nu} \frac{\partial\varphi}{\partial x_3} \tag{79}$$

Equation (79), through Eqs. (23) and (28), implies that

$$u_3 = \frac{1}{(1-2\nu)} \frac{1}{2} \frac{\partial\varphi}{\partial x_3} - x_3 \frac{1}{(1-2\nu)(4-4\nu)} \frac{\partial^2\varphi}{\partial x_3{}^2} \tag{80}$$

† These references deal with two-dimensional dynamical problems. English-speaking readers may prefer to consult Ref. [62] where the material of Refs. [59, 60] is recapitulated.

$$\sigma_{13} = \frac{-\mu}{(2-2\nu)(1-2\nu)} x_3 \frac{\partial^3 \varphi}{\partial x_1 \partial x_3{}^2} \tag{81}$$

$$\sigma_{23} = \frac{-\mu}{(1-2\nu)(1-2\nu)} x_3 \frac{\partial^3 \varphi}{\partial x_2 \partial x_3{}^2} \tag{82}$$

$$\sigma_{33} = \frac{\mu}{(2-2\nu)(1-2\nu)} \left(\frac{\partial^2 \varphi}{\partial x_3{}^2} - x_3 \frac{\partial^3 \varphi}{\partial x_3{}^3}\right) \tag{83}$$

and also that

$$u_1 = \frac{-1}{4-4\nu} \frac{\partial}{\partial x_1} \left[\varphi + \frac{1}{1-2\nu} x_3 \frac{\partial \varphi}{\partial x_3}\right] \tag{84}$$

$$u_2 = \frac{-1}{4-4\nu} \frac{\partial}{\partial x_2} \left[\varphi + \frac{1}{1-2\nu} x_3 \frac{\partial \varphi}{\partial x_3}\right] \tag{85}$$

$$\sigma_{12} = \frac{\mu}{2-2\nu} \left[\frac{-x_3}{1-2\nu} \frac{\partial^3 \varphi}{\partial x_1 \partial x_2 \partial x_3} - \frac{\partial^2 \varphi}{\partial x_1 \partial x_2}\right] \tag{86}$$

$$\sigma_{11} = \frac{\mu}{2-2\nu} \left[\frac{2\nu}{1-2\nu} \frac{\partial^2 \varphi}{\partial x_3{}^2} - \frac{x_3}{1-2\nu} \frac{\partial^3 \varphi}{\partial x_1{}^2 \partial x_3} - \frac{\partial^2 \varphi}{\partial x_1{}^2}\right] \tag{87}$$

$$\sigma_{22} = \frac{\mu}{2-2\nu} \left[\frac{2\nu}{1-2\nu} \frac{\partial^2 \varphi}{\partial x_3{}^2} - \frac{x_3}{1-2\nu} \frac{\partial^3 \varphi}{\partial x_2{}^2 \partial x_3} - \frac{\partial^2 \varphi}{\partial x_2{}^2}\right] \tag{88}$$

We shall be interested at first in three types of problem. In each of them (a) the body is the half-space $x_3 \geq 0$, (b) all stresses and displacements decay to zero "at infinity," (c) no shear stresses act on the bounding plane $x_3 = 0$, i.e., $\sigma_{13} = \sigma_{23} = 0$ on $x_3 = 0$. The third boundary condition on $x_3 = 0$ is mixed:

A. Pressure problems $\quad \sigma_{33} = s_3(r), \quad r = (x_1{}^2 + x_2{}^2)^{1/2} < a \quad$ (89a)

$\sigma_{33} = 0, \quad r > a \quad$ (89b)

B. Indentation problems $\quad u_3 = U_3(r), \quad r < a \quad$ (90a)

$\sigma_{33} = 0, \quad r > a \quad$ (90b)

C. Crack problems $\quad \sigma_{33} = s_3(r), \quad r < a \quad$ (91a)

$u_3 = 0, \quad r > a \quad$ (91b)

Considering cases A and B it appears from Eq. (83) that φ must be a harmonic function whose second derivative vanishes for $x_3 = 0$, $r > a$. Such a function is

$$\varphi = \operatorname{Im}\,[x_3' \log (R' + x_3') - R'] \tag{92a}$$

where

$$x_3' = x_3 + it, \qquad R' = [x_1^2 + x_2^2 + (x_3 + it)^2]^{1/2} \tag{92b}$$

and "Im" means "imaginary part of." We generalize this solution by taking

$$\varphi = \operatorname{Im} \int_0^a n(t)[x_3' \log (R' + x_3') - R']dt \tag{93}$$

$n(t)$ being an, as yet, arbitrary function that will later be adjusted so as to satisfy the first of boundary conditions (89) or (90). When Eq. (93) is substituted in Eq. (83) and x_3 is set equal to zero, one finds

$$\sigma_{33}(x_3 = 0) = \frac{-\mu}{(2 - 2\nu)(1 - 2\nu)} \int_r^a \frac{n(t)\,dt}{\sqrt{t^2 - r^2}}, \qquad r < a \tag{94a}$$

$$= 0, \qquad r > a \tag{94b}$$

Conditions (89b) and (90b) are satisfied automatically. To satisfy Eq. (89a) it is necessary that

$$\frac{-\mu}{(2 - 2\nu)(1 - 2\nu)} \int_r^a \frac{n(t)\,dt}{\sqrt{t^2 - r^2}} = s_3(r) \tag{95}$$

It is easy to see that the solution of this integral equation is

$$n(t) = \frac{(4 - 4\nu)(1 - 2\nu)}{\pi\mu} \frac{d}{dt} \int_t^a \frac{r s_3(r)\,dr}{\sqrt{r^2 - t^2}} \tag{96}$$

This completes the solution of case A. The example of greatest technological importance is $s_3(r) = -p$, a uniform pressure over the circle of radius a. In this example Eq. (96) gives

$$n(t) = \frac{p}{\pi\mu}(4 - 4\nu)(1 - 2\nu)t(a^2 - t^2)^{-1/2} \tag{97}$$

All the components of displacement and stress can now be written down. Of greatest interest is the pressure distribution along the axis of symmetry, $\sigma_{33}(r = 0)$. From Eqs. (83), (92a), and (97).

$$\sigma_{33}(r = 0) = -p\left[1 - \left(1 - \frac{a^2}{x_3^2}\right)^{-3/2}\right] \tag{98}$$

This result is the basis for the widely used "Boussinesq chart" by means of which foundation engineers compute the pressure at any location below the surface of an elastic half-space due to an arbitrary surface loading. The idea for the construction of these charts (which can be extended to displacements or any other radially symmetric quantity) is due to N. M. Newmark whose papers [66, 67] should be consulted for details.

The solution for case A reveals the physical meaning of the stress function $n(t)$. If, as before, we let P_3 denote the resultant external force acting in the direction of the x_3-axis

$$P_3 = 2\pi \int_0^a [-\sigma_{33}(x_3 = 0)] r dr$$

$$P_3 = 2\pi \int_0^a r dr \int_r^a \frac{\mu}{(2 - 2\nu)(1 - 2\nu)} \frac{n(t)\, dt}{\sqrt{t^2 - r^2}}$$

$$P_3 = \frac{2\pi\mu}{(2 - 2\nu)(1 - 2\nu)} \int_0^a t n(t)\, dt \tag{99}$$

In case B the same function given in Eq. (93) will serve. Satisfaction of boundary condition (90b) is automatic; (90a) requires

$$U_3(r) = \frac{1}{2(1 - 2\nu)} \left[\frac{\pi}{2} \int_0^a n(t)\, dt - \int_0^r n(t) \cos^{-1}\frac{t}{r}\, dt \right] \tag{100}$$

The complete solution of this integral equation is

$$n(t) = \frac{-4(1 - 2\nu)}{\pi} \left[C\delta(a - t) + \frac{1}{t}\frac{d}{dt} \int_0^t \frac{dU_3}{dr} \frac{r^2 dr}{\sqrt{t^2 - r^2}} \right] \tag{101}$$

Here $\delta(a - t)$ is the Dirac delta function of argument $a - t$ and C is an arbitrary constant. The physical significance of C becomes apparent upon considering a few elementary (though technologically important) examples†. For the flat-ended smooth, rigid die shown in Fig. 1, $U_3(r)$ = constant, say w_0. Then from Eq. (101)

$$n(t) = -\frac{4}{\pi}(1 - 2\nu) C\delta(a - t) \tag{102}$$

and from Eq. (99)

$$P_3 = \frac{-4\mu}{1 - \nu} Ca \tag{103}$$

Also, from Eq. (100), for $r < a$,

$$w_0 = -C \tag{104}$$

† References to the original treatments of some of these examples may be found in Refs. [57, and 65].

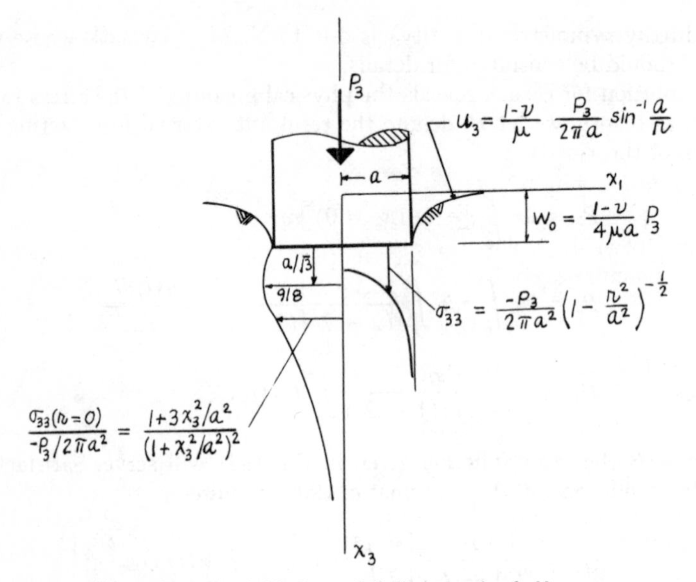

Fig. 1. Smooth, rigid, flat-ended die indenting half-space.

The magnitude of either the indentation or the indenting force must be specified. Either determines the constant C. The relationship between P_3 and w_0 in this case is linear,

$$w_0 = P_3 \frac{1-\nu}{4\mu a} \tag{105}$$

because the contact area does not change as the load P_3 is applied. From Eq. (94a) one finds at once that

$$\sigma_{33}(x_3 = 0, r < a) = \frac{-2\mu w_0}{\pi(1-\nu)a}\left(1 - \frac{r^2}{a^2}\right)^{-1/2} \tag{106}$$

The singularity in stress at the edge of the contact region is due to the abrupt change in slope which occurs there:

$$u_3(x_3 = 0, r > a) = \frac{1-\nu}{\mu} \frac{P_3}{2\pi a} \sin^{-1}\frac{a}{r} \tag{107}$$

This stress singularity is signalized analytically by the nonvanishing of the constant C. It is associated with what may be termed the "complete" contact problem in which the contact region does not change with load. The surface lateral displacement is

$$u_r(x_3 = 0, r < a) = -\frac{P_3}{2\pi a}\frac{1-2\nu}{2\mu}\frac{a}{r}\left[1 - \left(1 - \frac{r^2}{a^2}\right)^{1/2}\right] \tag{108a}$$

$$u_r(x_3 = 0, r > a) = -\frac{P_3}{2\pi r}\frac{1-2\nu}{2\mu} \tag{108b}$$

Equation (108a) shows why the surface must be presumed smooth. The case in which the boundary condition is appropriate to a perfectly rough die,

$$u_1(r < a, x_3 = 0) = u_2(r < a, x_3 = 0) = 0 \tag{109}$$

is considerably more involved analytically. It has been successfully treated by Mossakovsky [68] and (later) by Keer [69] and by Ufliand [70]. The solution [68] was obtained by transforming the half-space problem into a plane potential problem. This was solved by the method of linear relationship and the result returned to the half-space by inverse operators. In Ref. [70] the problem is treated in toroidal coordinates and the solution obtained through the Mehler–Fok integral transform. Reference [69] follows and extends the complex potential function solution used here. It makes possible the direct calculation of the quantities that are of greatest interest. Of these the most important, as given in Refs. [68, 69], is the pressure beneath the die:

$$\sigma_{33}(x_3 = 0, r < a) = -\alpha\frac{1}{r}\frac{d}{dr}\int_0^r \frac{t\,dt}{\sqrt{r^2 - t^2}} \sin\left[\beta \log\frac{a-t}{a+t}\right] \tag{110}$$

$$\alpha = \frac{8\mu(1-\nu)w_0}{\pi(1-2\nu)(3-4\nu)^{1/2}}, \qquad \beta = (2\pi)^{-1}\log(3-4\nu)$$

This is to be contrasted with Eq. (106). The oscillating divergence at the tip of the contact implies that the physical significance of the solution requires careful interpretation in that region†.

The "incomplete" contact in which the contact area grows progressively with load may be illustrated by the example of the conical indenter shown in Fig. 2. In this case we take $C = 0$ because $\sigma_{33}(x_3 = 0, r = a)$ must vanish; this is the condition that defines the (as yet unknown) contact radius a.

With the notation indicated in Fig. 2,

$$U_3 = w_0 - r\cot\theta \tag{111}$$

Equation (101) then makes

$$n(t) = 2(1-2\nu)\cot\theta \tag{112}$$

† A similar effect is found in two-dimensional problems; see Refs. [71–73] and [22, p. 476].

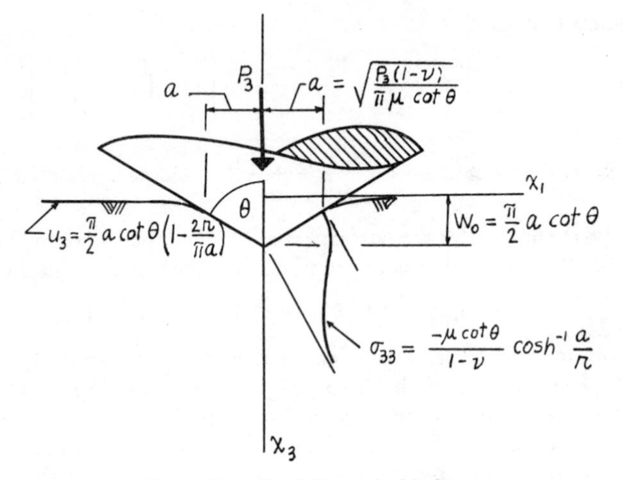

Fig. 2. Smooth, rigid, conical indenter.

and Eq. (99) makes

$$P_3 = \pi\mu(1 - \nu)^{-1}a^2 \cot\theta \tag{113}$$

This determines a. Now from Eq. (100)

$$u_3(r < a, x_3 = 0) = \frac{\pi}{2} a \cot\theta - r \cot\theta \tag{114}$$

Comparing this with Eq. (111) it is evident that

$$w_0 = \frac{\pi}{2} a \cot\theta \tag{115}$$

This implies that

$$w_0^2 = P_3 \frac{\pi(1 - \nu) \cot\theta}{4\mu} \tag{116}$$

so that load is not proportional to displacement. Finally we note that the surface displacement and slope exhibit no discontinuity in the incomplete contact case. Here

$$u_3(r > a, x_3 = 0) = \left[a \sin^{-1}\frac{a}{r} + (r^2 - a^2)^{1/2} - r\right] \cot\theta \tag{117}$$

and

$$\sigma_{33}(r < a, x_3 = 0) = -\mu(1 - \nu)^{-1} \cot\theta \cosh^{-1}\frac{a}{r} \tag{118}$$

other interesting examples of indentation problems could be cited. A quadratic variation in U_3 leads to the Hertz analysis of contact stresses for spheres; U_3 proportional to r^{2n} leads to Shtaerman's [74] generalization of the Hertz problem†. The entire subject of contact stresses has blossomed in the past twenty years. Separate monographs in this area, such as those of Galin [75] and Ling [108] and the dissertation of Kalker [110] are well justified.

Boundary-value problems of class C are solved by taking for φ the real rather than the imaginary part of the function used in Eq. (92a) for classes A and B. That is, after generalization,

$$\varphi = \text{Re} \int_0^a n(t)[x_3' \log (R' + x_3') - R']dt \tag{119}$$

From Eqs. (80) and (83) it then follows that on the boundary $x_3 = 0$

$$u_3(r < a) = \frac{1}{2 - 4\nu}\left[\log r \int_0^a n(t)\,dt + \int_r^a n(t) \log \frac{\sqrt{t^2 - r^2} + t}{r}\,dt\right] \tag{120a}$$

$$u_3(r > a) = \frac{1}{2 - 4\nu} \log r \int_0^a n(t)\,dt \tag{120b}$$

$$\sigma_{33}(r < a) = \frac{\mu}{(2 - 2\nu)(1 - 2\nu)} \int_0^r n(t)\,(r^2 - t^2)^{-1/2}dt \tag{121a}$$

$$\sigma_{33}(r > a) = \frac{\mu}{(2 - 2\nu)(1 - 2\nu)} \int_0^a n(t)\,(r^2 - t^2)^{-1/2}dt \tag{121b}$$

In order to satisfy Eq. (91b) and make u_3 vanish outside the circle $r = a$ it is necessary to have

$$\int_0^a n(t)\,dt = 0 \tag{122}$$

† The physical validity of "improvements" on the Hertz theory is open to question. Recently Goodman and Keer [76] have solved the problem of spherical contact using a method appropriate to the sphere. They show that only in the case of a sphere pressed into a slightly larger spherical cavity (like a ball-and-socket joint) are the corrections to the Hertz theory significant. In other cases the corrections which result when higher-order curvature terms such as Shtaerman's are introduced into the boundary conditions while a method of analysis appropriate to the problem of the plane is retained are of the same order of magnitude as the corrections which result from treating the bodies as *bona fide* spheres. Torvik has shown [77] that the neglect of products of displacement gradients in the classical theory of elasticity can be as serious as the approximation to the boundary conditions embodied in the conventional Hertz analysis. On this question reference may also be made to a paper by Mow, Chow, and Ling [80] and to the paper by Kalker and van Randen [109].

The remaining boundary requirement (91a) is that

$$\frac{\mu}{(2-2\nu)(1-2\nu)}\int_0^r n(t)\,(r^2-t^2)^{-1/2}dt = s_3(r) \tag{123}$$

This is satisfied by taking

$$n(t) = \frac{(4-4\nu)(1-2\nu)}{\pi\mu}\left[\frac{d}{dt}\int_0^t \frac{s_3(r)r}{\sqrt{t^2-r^2}}\,dr + C\delta(a-t)\right] \tag{124}$$

where now the arbitrary constant C is to be chosen so that Eq. (122) is satisfied. The case C is now completely solved. As an illustration the example of the penny-shaped crack, first analyzed by Sneddon [78] and by Sneddon and Elliot [79] using the Hankel transform method, later by Green [65] using essentially the present technique, will serve. Consider an infinite solid on the plane $x_3 = 0$ of which there exists a crack of radius a. The crack is held open by an internal fluid pressure p_0. The essential boundary conditions are those of Eq. (91) with $s_3(r) = -p_0$. We have from Eqs. (122) and (123)

$$n(t) = (4-4\nu)(1-2\nu)\frac{p_0}{\pi\mu}[a\delta(a-t)-1] \tag{125}$$

The significant results are obtained from Eqs. (120a) and (121b):

$$u_3(r < a, x_3 = 0) = \frac{p_0}{\pi\mu}(2-2\nu)(a^2-r^2)^{1/2} \tag{126}$$

$$\sigma_{33}(r > a, x_3 = 0) = \frac{2p_0}{\pi}\left[a(r^2-a^2)^{-1/2} - \sin^{-1}\frac{a}{r}\right] \tag{127}$$

The physical implications of this result and the corresponding one in which the crack is opened by an external tension field† have provided many fruitful insights to the fracture of materials. Considering the hydraulic fracture of mineral-bearing rock, Khristianovich and Zheltov [81] were led to speculate that the stress singularity at the tip of the crack predicted by Eq. (127) would disappear if the fluid pressure failed to penetrate to the tip of the crack. This is indeed the case. Subsequent work by the authors cited, and by Barenblatt, has given rise to the theory of equilibrium cracks [82]. This theory assumes that forces of cohesion act on the crack surfaces so as to annul the singularity in stress which would otherwise occur at the tip. The width of the edge region in which the cohesive forces are appreciable is assumed to be small compared with the radius of the crack and the distribution of the cohesive forces is assumed to be a function of temperature and material properties but

† The solution is the same as given by Eqs. (126) and (127) except that the far-field tension T replaces p_0 and the inverse sine becomes an inverse cosine.

independent of load. In these circumstances the shape of the cross section of the crack is also independent of load. It has the form shown in Fig. 3a rather than that shown in Fig. 3b which is predicted by Eq. (126). The only characteristic of the putative forces of cohesion that enters into the question of crack propagation is the so-called modulus of cohesion, K. It may be calculated from the condition that, considering points in the plane of the crack distant s from the tip of the crack, the normal stress component perpendicular to the plane of the crack takes the form

$$\lim_{s\to 0} \sigma_{33}(r > a,\ x_3 = 0) = K(\pi^2 s)^{-1/2} + 0(1) \tag{128}$$

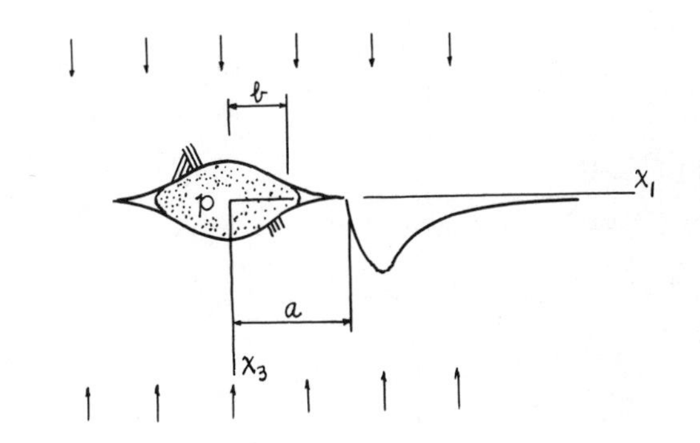

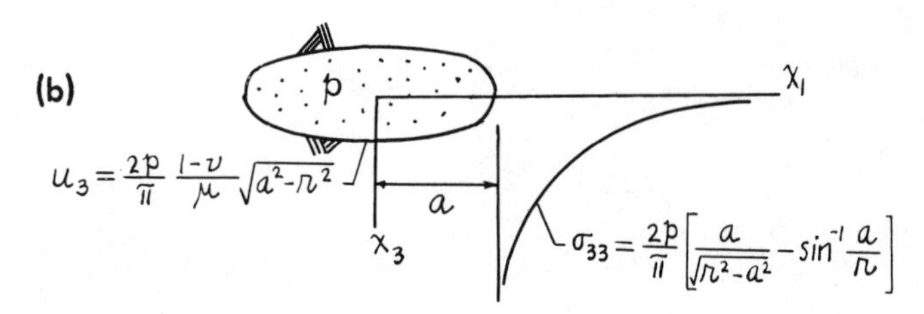

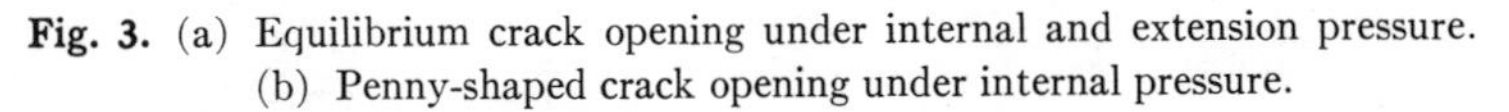

Fig. 3. (a) Equilibrium crack opening under internal and extension pressure. (b) Penny-shaped crack opening under internal pressure.

In the case of the penny-shaped crack $s = r - a$ and it follows from Eq. (127) that (with a far-field tension T replacing p_0)

$$\lim_{r\to a} \sigma_{33}(r > a,\ x_3 = 0) = (2a)^{1/2} T(\pi^2 s)^{-1/2} \tag{129}$$

whence

$$K = T(2a)^{1/2} \tag{130}$$

Keer [85] has investigated the stress distribution at the edge of an equilibrium crack and computed in terms of K the distance over which the cohesive forces act. The older Griffith theory has also been used by Sack in conjunction with the analysis of the penny-shaped crack to arrive at a crack propagation criterion [83, 84]. The presence of the stress singularity does not disturb the Griffith theory because this happens to be an integrable singularity and Griffith's is an energy criterion.

The Sobolev–Green approach is by no means limited to radially symmetric stress states. In analogy with Eqs. (77)–(79) we may write the fundamental potential solution for the half-space subjected to shearing tractions in terms of harmonic functions F_1, F_2:

$$\psi_1 = -\frac{1}{\mu}\frac{\partial^2 F_2}{\partial x_3^2}, \qquad \psi_3 = -\frac{4-4\nu}{2\mu}\frac{\partial F_1}{\partial x_3} + \frac{1}{\mu}\frac{\partial^2 F_2}{\partial x_1 \partial x_3} \tag{131}$$

$$\varphi = \frac{1}{\mu}\left[x_1\frac{\partial^2 F_2}{\partial x_3^2} - x_3\frac{\partial^2 F_2}{\partial x_1 \partial x_3} + (4-4\nu)\frac{\partial F_2}{\partial x_1} - (2-2\nu)^2 F_1\right] \tag{132}$$

Then ψ_1, ψ_3, and φ are harmonic functions and

$$\mu u_1 = (1-\nu)\frac{\partial F_1}{\partial x_1} + \frac{\partial^2 F_2}{\partial x_2^2} + \tfrac{1}{2}x_3\frac{\partial^2 F_1}{\partial x_1 \partial x_3} \tag{133}$$

$$\mu u_2 = (1-\nu)\frac{\partial F_1}{\partial x_2} - \frac{\partial^2 F_2}{\partial x_1 \partial x_2} + \tfrac{1}{2}x_3\frac{\partial^2 F_1}{\partial x_2 \partial x_3} \tag{134}$$

$$\mu u_3 = -\tfrac{1}{2}(1-2\nu)\frac{\partial F_1}{\partial x_3} + \tfrac{1}{2}x_3\frac{\partial^2 F_1}{\partial x_3^2} \tag{135}$$

$$\sigma_{33} = x_3\frac{\partial^3 F_1}{\partial x_3^3} \tag{136}$$

$$\sigma_{13} = x_3\frac{\partial^3 F_1}{\partial x_3^2 \partial x_1} + \frac{\partial^2 F_1}{\partial x_1 \partial x_3} + \frac{\partial^3 F_2}{\partial x_2^2 \partial x_3} \tag{137}$$

$$\sigma_{23} = x_3\frac{\partial^3 F_1}{\partial x_3^2 \partial x_2} + \frac{\partial^2 F_1}{\partial x_2 \partial x_3} - \frac{\partial^3 F_2}{\partial x_1 \partial x_2 \partial x_3} \tag{138}$$

If the surface $x_3 = 0$ is free of traction in (say) the x_2-direction one takes $F_1 = \partial F_2/\partial x_1$. The nonvanishing superficial stress and displacement components are then

$$\mu u_1 = (1 - \nu)\frac{\partial^2 F_2}{\partial x_1^2} + \frac{\partial^2 F_2}{\partial x_2^2}, \qquad \mu u_2 = -\nu \frac{\partial^2 F_2}{\partial x_1 \partial x_2} \tag{139a,b}$$

$$\mu u_3 = -\tfrac{1}{2}(1 - 2\nu)\frac{\partial^2 F_2}{\partial x_1 \partial x_3}, \qquad \sigma_{13} = -\frac{\partial^3 F_2}{\partial x_3^3} \tag{139c,d}$$

Corresponding to Eq. (93), the plane $x_3 = 0$ is now cleared of traction outside the circle $r = a$ by taking

$$F_2 = \operatorname{Im} \int_0^a n_1(t)\{[\tfrac{1}{2}(x_3')^2 - \tfrac{1}{4}r^2]\log(R' + x_3') - \tfrac{3}{4}R'x_3' + \tfrac{1}{4}r^2\}dt \tag{140}$$

Then

$$\sigma_{13}(r < a, 0) = \int_r^a \frac{n_1(t)\,dt}{\sqrt{t^2 - r^2}} \tag{141}$$

$$\mu u_1(r < a, 0) = -\frac{2-\nu}{2}\frac{\pi}{2}\int_0^a n_1(t)\,dt + \int_r^a n_1(t)\left[\frac{2-\nu}{2}\cos^{-1}\frac{t}{r} - \tfrac{1}{2}\nu r^{-2}\cos 2\theta\, t(r^2 - t^2)^{1/2}\right]dt \tag{142}$$

$$\mu u_2(r < a, 0) = -\tfrac{1}{2}\nu r^{-2}\int_0^r n_1(t)\,t(r^2 - t^2)^{1/2}\sin 2\theta\, dt \tag{143}$$

$$\mu u_3(r < a, 0) = -\tfrac{1}{2}(1 - 2\nu)r^{-1}\cos\theta\left[\int_0^a n_1(t)\,t\,dt - \int_r^a n_1(t)(t^2 - r^2)^{1/2}dt\right] \tag{144}$$

The corresponding quantities for $r > a$ are obtained on replacing the symbol r, where it appears in the limit of an integral by the symbol a. Equations (141)–(144) may be inverted to provide expressions for $n_1(t)$ in terms of the specified stress or displacement distribution. For example, if $\sigma_{13}(r < a, x_3 = 0) = s_1(r)$

$$n_1(t) = -\frac{2}{\pi}\frac{d}{dt}\int_t^a r s_1(r)(r^2 - t^2)^{-1/2}dr \tag{145}$$

These ideas may be illustrated by a variety of technologically important instances. It is clear from Eqs. (142) and (143) that if n_1 is proportional to the Dirac delta function of argument $a - t$ the horizontal motion within the circle

$r = a$ will be a rigid-body displacement in the direction of the x_1-axis. That is

$$n_1(t) = -\mu u_0 \frac{4}{\pi} (2 - \nu)^{-1} \delta(a - t) \tag{146}$$

makes the rigid-body displacement of magnitude u_0. Then the required surface shear stress is Eq. (141)

$$\sigma_{13}(r < a, 0) = -\mu u_0 \frac{4}{\pi} (2 - \nu)^{-1} (a^2 - r^2)^{-1/2} \tag{147a}$$

and

$$u_2(r > a, 0) = u_0 \frac{2\nu}{\pi} (2 - \nu)^{-1} \frac{a}{r^2} (r^2 - a^2)^{1/2} \sin 2\theta \tag{147b}$$

All the surface values may be written down at once by reference to Eqs. (142)–(144). This result was first obtained by Westmann [87] using the Hankel transform procedure. It is interesting to compare the techniques. In this connection the previously cited monograph of Muki [56] should also be consulted. Muki treats, *inter alia*, the case in which a uniform shear stress $\sigma_{13} = Q/\pi a^2$ is applied over the circle $r < a$. He finds that the peak shear stress is very approximately $1.2Q/\pi a^2$ in magnitude and that it occurs at the leading and trailing edges of the loaded circle. Considerable shear stresses occur only in a thin layer near the loaded surface. Hamilton and Goodman have investigated in detail the stress field created by a circular sliding contact [88] using the complex potential method. Thompson and Robinson [86] have used it to find exact solutions of dynamic problems of indentation and transient loading.

The problem of rigid-body, shear-induced, lateral displacement of a circular region would, in the absence of slip, serve also to determine the tangential compliance of elastic spheres pressed together. Since the total tangential force is given by the expression

$$P_1 = -2\pi \int_0^a n_1(t)\, dt \tag{148}$$

it follows from Eq. (146) that

$$P_1 = 8\mu u_0 (2 - \nu)^{-1} \tag{149}$$

The compliance of one body is therefore

$$C' = \frac{du_0}{dP_1} = \frac{2 - \nu}{8\mu a} \tag{150}$$

The other member of the two-body contact experiences a similar displacement so that for a pair of bodies having elastic constants denoted by subscripts 1 and 2, respectively

$$C' = \frac{1}{8a}\left(\frac{2-\nu_1}{\mu_1} + \frac{2-\nu_2}{\mu_2}\right) \tag{151}$$

Mindlin, after deriving this result in Ref. [M29] has pointed out that in the case of perfect roughness there is a stress singularity at the edge of the contact region. This cannot occur in reality. Instead, as P_1 increases from zero, local microslip develops in an annulus $a' < r < a$. When P_1 reaches a limiting value fP_3, a' vanishes and sliding occurs. At that point

$$\sigma_{13}(r < a, 0) = -\frac{3fP_3}{2\pi a^2}\left(1 - \frac{r^2}{a^2}\right)^{1/2} \tag{152}$$

This last follows from the Hertz theory of contact stresses together with the assumption that the Amontons–Coulomb friction law holds on a microscopic scale. The corresponding stress function (145) is

$$n_1(t) = -\frac{3fP_3}{2\pi a^3}t \tag{153}$$

The displacement in the x_1-direction given by Eq. (153) is, from Eq. (142),

$$2\mu u_1(r < a, 0) = \frac{3fP_3}{32a^3}[4(2-\nu)a^2 - 2(2-\nu)r^2 + \nu r^2 \cos 2\theta] \tag{154}$$

At partial slip, $P_1 < fP_3$, a shear stress distribution inside the circle $r = a'$ must be added so as to annul the r-dependent part of Eq. (154) and leave only a constant displacement in the no-slip region. This stress distribution is given by a stress function n_1' that is the negative of n_1 as given by Eq. (153). Consequently, in the unslipped interior of the contact region $r < a'$ the total displacement is

$$2\mu u_1(r < a, 0) = \frac{3fP_3}{8a^3}(2-\nu)[a^2 - (a')^2] \tag{155}$$

and

$$\sigma_{13}(r < a', 0) = -\frac{3fP_3}{2\pi a^3}\{[a^2 - r^2]^{1/2} - [(a')^2 - r^2]^{1/2}\} \tag{156}$$

To find the magnitude of a' one may write, from Eq. (148)

$$P_1 = 2\pi\frac{3fP_3}{2a^3}\left(\int_0^a t^2 dt - \int_0^{a'} t^2 dt\right) \tag{157}$$

or

$$\frac{a'}{a} = \left(1 - \frac{P_1}{fP_3}\right)^{1/3} \tag{158}$$

Substitution of Eq. (158) in Eq. (155) yields

$$u_1(r < a', 0) = \frac{3fP_3}{16\mu a}(2 - \nu)\left[1 - \left(1 - \frac{P_1}{fP_3}\right)^{2/3}\right] \tag{159}$$

The tangential compliance of one body, C, is the derivative of this expression with respect to P_1:

$$C = \frac{2 - \nu}{8\mu a}\left(1 - \frac{P_1}{fP_3}\right)^{-1/3} \tag{160}$$

For a pair of bodies having like elastic constants the compliance of the two bodies is $2C$. The value of C given by Eq. (160) is to be compared with that of a perfectly rough contact, C', given by Eq. (150). These results are due to Mindlin [M29]. They are confirmed by the classic experiments of K. L. Johnson as described in another section of this volume to which the reader must be referred for details.

6 Role of the Classical Theory in Continuum Mechanics

The development of the classical theory of elasticity in recent years has been marked by two related movements. One of these is the inclusion of strain-inducing phenomena other than loading by external forces. The second is a reexamination and reformation of fundamental principles along formally rigorous lines.

The first of these movements has really been present since the inception of the classical theory of elasticity. Temperature effects were introduced by Duhamel and by F. E. Neumann [1, vol. 2, pt. 2, p. 6]. All that is required to include them is to modify stress–strain law (5) by adding a term $-(3\lambda + 2\mu)\alpha T\delta_{ij}$ proportional to the linear coefficient of thermal expansion, α, and to the field temperature T. When this is done a term analogous to a body force

$$-\alpha\mu\frac{2 - 2\nu}{1 - 2\nu}\frac{\partial T}{\partial x_i} \tag{161}$$

appears on the left-hand side of the displacement equations of equilibrium (8). A similar "effective body force" term appears in the Eqs. (19) and (22) governing the fundamental potential solution of the equations of equilibrium. Of course the temperature too is governed by a field theory. In fact†

$$\kappa\frac{\partial^2 T}{\partial x_m \partial x_m} = \rho c_v\frac{\partial T}{\partial t} + (3\lambda + 2\mu)\alpha T\frac{\partial e_{mm}}{\partial t} \tag{162}$$

† Reference [92, p. 82]. The inertia terms $\rho\partial^2 u_i/\partial t^2$ should also be included in Eqs. (8). These are often omitted on the grounds that temperature changes are so slow as to make inertia forces negligible. The last term in Eq. (162) too is often omitted, the argument being that heat generated by small displacements will itself be small.

Here t is the time variable, κ denotes the thermal conductivity of the material, ρ is the density and c_v is the specific heat at constant volume. The appearance of the last, nonlinear, term is awkward because it couples dynamic strain and temperature effects. In the static case however this complication does not arise. Static thermal stresses can be treated in the same way as a body force with a potential $-\alpha\mu T(2 - 2\nu)(1 - 2\nu)^{-1}$. For detailed treatment of many thermal stress problems reference may be made to the previously cited book of Lur'e [4] or to the more specialized books of Boley and Wiener [89], Maizel [90], or Parkus [91].

Another important extension of the classical theory has been to so-called poroelastic or fluid-filled bodies. This theory was developed by Biot [93] to generalize the consolidation theory of Terzaghi. In this formulation Darcy's flow law replaces the heat flow Eq. (162) and a term proportional to the divergence of the average fluid displacement field appears in the displacement equations of equilibrium. The assumption here is that flow occurs at very low Reynolds numbers. More complex models have been developed by Green, Naghdi, Crochet, and Steel [94, 95]. The Biot model has found extensive application in soil mechanics and in secondary recovery of petroleum.

The solution of an elastostatic problem also yields the solution to a viscoelastic problem. The two problems are related through a correspondence principle. The "corresponding" viscoelastic solution is obtained on replacing the actual loads by their Laplace transforms and the elastic constants by quotients of operator polynomials. The modern theory of viscoelastic leans heavily upon classical elastostatics.

Many other extensions of the classical theory of elasticity are described in other articles of this volume. They are concerned with piezoelectricity, electroelasticity, photoelasticity, and vibrations and waves in plates and bars. The interrelationships between these field theories provide interesting and useful analogies, mostly two-dimensional in character. These are described in the review article of Mindlin and Salvadori [M32] which includes an extensive annotated bibliography. The vitality of the classical theory of elasticity arises from its ability to form connections with other field theories.

Two extensions of the classical theory along somewhat different lines may be mentioned. One of them, due to Biot, is the treatment of the effect of small elastic strains superposed on finite strains by means of a method of incremental deformations. This analysis is of great importance in geophysical questions. Many of Biot's valuable papers are recovered in his monograph [13]. Another such extension is the treatment of the elastic states that are created in a body in a gravitational field when it reaches its final size by accretion of material. The conventional treatment of body forces in continuum mechanics implies that these forces are applied to a structure having the dimensions of the final body. This treatment, which is quite satisfactory when the body forces arise from the presence of an electromagnetic or acceleration field, fails when the body grows to its final shape by gradual accretion of material. Then the weight of each new layer loads and deforms the earlier-deposited material before hardening. Successive layers are applied to a body that is in a state of initial stress. The final gravitational stresses in such a body depend on the order and manner in which the final shape is attained. In this final state there is, in general, a non-vanishing dislocation tensor. The appropriate method of computing this final state when the final stresses and deformations are small enough to justify using the constitutive and geometrical equations of the classical theory of elasticity has been worked out by Brown and Goodman [96].

The second movement of recent years requiring notice has been the reformulation of the foundations of continuum mechanics along rigorous axiometric lines. This has been largely the achievement of Truesdell, Toupin, Noll, Gurtin, and colleagues. Their work is well documented in the previously cited Refs. [9, 10, 16]. It deals with continuous media, fluid and solid, the behavior of which conforms to the conservation laws of mechanics and thermodynamics. The linear elastic material is then shown to occupy a legitimate place within this larger framework.

In the spirit of this development there has been a renewed appreciation of the importance of uniqueness theorems and of the need for a precise statement of de St. Venant's principle. The uniqueness question centers on establishing the conditions under which there can exist no more than one displacement set of class C^2 corresponding to an elastic state with given boundary conditions. Kirchhoff early established limits on the elastic constants that assure uniqueness for a bounded region β subject to static traction, displacement, or "mixed" boundary conditions, assuming that no nuclei of strain are present. These limits were broadened significantly by E. and F. Cosserat who showed that there can be at most one classical solution to the displacement boundary-value problem of elastostatics in a bounded, homogenous, isotropic region when

$$\mu \neq 0, \qquad -\infty < \nu < \tfrac{1}{2} \quad \text{or} \quad 1 < \nu < \infty \tag{163}$$

These limits† have recently been thoroughly examined for anisotropic materials, for bodies containing complete or incomplete regions of infinity, for bodies of special shape, for two-dimensional elastostatic states, for elastodynamic states, and for special boundary conditions. The monograph [97] by Knops and Payne describes these extensions in detail and includes valuable references to recent work of Bramble, Fichera, Ericksen, Diaz, Coleman and others. As mentioned earlier, in Section 4, the limitation of the Kirchhoff–Cosserat uniqueness theorem to nonsingular elastic states excludes idealizations such as concentrated forces or other nuclei of strain. These extremely convenient abstractions may, however, be brought within the scope of a rigorous mathematical theory. To accomplish this the stress distribution due to a concentrated force (for example)

† They have been extended to domains containing a complete region at infinity by Duffin and Noll [98] and to the dynamic case by Gurtin and Sternberg [99].

must be defined as the limiting solution when a distributed traction equipollent to the concentrated force is applied over a region of ever-dimishing extent. The distributed traction (or body force) produces a regular stress field. Once the existence of the limiting solution has been established the orders of the associated stress singularities can be displayed. They can then be used as the basis for a direct reformulation of the concentrated force problem that will satisfy a uniqueness criterion. This program has been carried out by Sternberg in collaboration with Rosenthal [100], Eubanks [101, 39], Turteltaub [40], and Wheeler [41].

The boundary conditions that arise in applications of the mathematical theory of elasticity represent a species of simplification and idealization. While the resultant force and torque applied to a surface is often known with a high degree of accuracy, the detailed distribution of surface tractions can, in practice, rarely be closely controlled. As a consequence, the alteration of the stress field that occurs in a solid when the tractions on a limited surface element are replaced by another statically equivalent set is a matter of great importance. The suggestion that this disturbance is slight was first advanced by de St. Venant [1, vol. 2, pt. 1, p. 12] in connection with his investigation of torsion. A statement of what has come to be known as *de St. Venant's Principle* was enunciated by Boussinesq† in 1885. This long-accepted formulation was questioned by von Mises in 1945. Generalizing from specific examples, he proposed an amended form the correctness of which was subsequently established by Sternberg [102].

The viewpoint taken by von Mises and by Sternberg is that de St. Venant's principle concerns the order of magnitude‡ of stress components $\sigma_{ij}(Q)$ at a point Q in the interior of an elastic body at rest. The amended principle asserts that if distributed tractions T_i are applied to a surface region $s(\epsilon)$ of radius ϵ and if

$$\int_s T_i ds = 0, \qquad \int_s T_i x_j ds = 0 \tag{164}$$

then

$$\sigma_{ij}(Q) = 0(\epsilon^4) \tag{165}$$

For concentrated forces the order of magnitude decreases by 2. Force systems that satisfy both of Eqs. (164) are said to be in astatic equilibrium. Berg has

† Reference [28, p. 298]. "Des forces extérieures, qui se font en équilibre, sur un solide élastique et dont les points d'application se trouvent tous a l'intérieur d'une sphère donnée, ne produisent pas de déformations sensibles à des distances de cette sphère qui sont d'une certaine grandeur par rapport à son rayon."

‡ We say that $f(x)$ is of order of magnitude x^n if $\lim_{x\to 0} |f(x)/x^n| < N$ where N is independent of x. This is written $f(x) = 0(x^n)$.

shown [103] that satisfaction of Eqs. (164) implies that the system of loads on s produces zero mean strain in the neighborhood of s. Stronger results are possible for solids having special dimensions. This approach—a return to the original point of view of de St. Venant—has been developed by Toupin [104], Roseman and Knowles [105]. Toupin and Roseman show that in a long cylinder loaded at one end, stresses decay exponentially with distance from the loaded end. Knowles achieves a similar result for the case of plane strain.

7 Closure

Regrettable omissions are almost inevitable in any brief exposition of a large subject. Some of these omissions have been mentioned in Section 1. They have been made in order to emphasize the theme of this presentation: the dominating role played by the fundamental three-dimensional potential solutions. Other omissions will occur to every reader who is conversant with the achievements of the classical theory of elasticity. It can only be hoped that the reference to lengthier treatises may supply these deficiencies, or at least meet the needs of those who will wish to make application of the theory. No claim of completeness can be made for the list of references. Indeed, a complete bibliography of the past decade would include thousands of items and would be useless unless carefully organized.

In the past decade, also, the use of the finite element technique for the analysis of plate and shell structures has become widespread. This approach suffers a severe handicap in three-dimensional stress analysis. As Zienkiewicz has pointed out [106], an adequate analysis of a two-dimensional square region requires solution of some 800 simultaneous equations; the equivalent analysis of a three-dimensional cube would require solving some 24,000 simultaneous equations. Bandwidth increases in nearly the same ratio. Nevertheless, with the development of increasingly sophisticated elements, the finite element method can be used for three-dimensional stress analysis where it is possible to take advantage of symmetry and where the stress field does not vary too rapidly. Dealing always with particular cases, the finite element method can hope to utilize the understanding of stress-field behavior in the vicinity of singularities and stress concentrations that is provided by the classical theory.

John von Neumann has remarked [107]:

> As a mathematical field travels far from its empirical source, or still more, if it is a second and third generation only indirectly inspired by ideas coming directly from reality, it is beset with very grave dangers. It becomes more and more pure aestheticizing, more and more purely *l'art pour l'art*. This need not be bad if the field is surrounded by correlated subjects which still have close empirical connections or if the discipline is under the influence of men with an exceptionally well-developed taste. But there is a grave danger that the subject will develop along the line of least resistance, that the stream, so far from its source, will separate into a multitude of insignificant branches, and that the

discipline will become a disorganized mass of details and complexities. In other words, at a great distance from its empirical source, or after much "abstract" inbreeding, a mathematical subject is in danger of degeneration. At the inception the style is usually classical; when it shows sign of becoming baroque, then the danger signal is up.

It is to be hoped that the classical theory of elasticity will avoid these pitfalls as well in the future as it has in the past.

References

1. I. Todhunter and K. Pearson, *A History of the Theory of Elasticity*, Cambridge University Press, 1886–1893.
2. C. Truesdell, The mechanical foundations of elasticity and fluid dynamics, *J. Rat. Mech. Anal.* **1**, 125 (1952).
3. A. E. Green and W. Zerna, *Theoretical Elasticity*, 2nd edition, Oxford University Press, 1968.
4. A. I. Lur'e *Three-Dimensional Problems of the Theory of Elasticity*, (in Russian), Gos. Izdat. Tech. Lit., Moscow, 1955. German translation, Akad-Verlag, Berlin, 1963. English translation by D. B. McVean, Interscience, New York, 1964.
5. I. S. Sokolnikoff, *The Mathematical Theory of Elasticity*, 2nd edition, McGraw-Hill, New York, 1956.
6. V. V. Novozhilov, *Theory of Elasticity*, (in Russian), Gos. Izdat., Leningrad, 1958. English translation by J. J. Shor-Kon, Office of Technical Services, Washington, D.C., 1961.
7. E. Trefftz, Math. elastzitätstheorie, *Handbuch der Physik*, Springer-Verlag, Berlin, 1928, Vol. 6, Chap. 2.
8. I. N. Sneddon and D. S. Berry, The classical theory of elasticity, *Handbuch der Physik*, Springer-Verlag, Berlin, 1958, Vol. 6, Chap. 1.
9. C. Truesdell and R. Toupin, The classical field theories, *Handbuch der Physik*, Springer-Verlag, Berlin, 1960, Vol. 3, Chap. 1.
10. M. E. Gurtin, Theory of Elasticity, *Handbuch der Physik*, 2nd edition, Springer-Verlag, Berlin—Heidelberg—New York, 1971, Vol. 6.
11. A. E. H. Love, *The Mathematical Theory of Elasticity*, 4th edition, Cambridge University Press, 1927.
12. S. Timoshenko and J. N. Goodier, *Theory of Elasticity*, 3rd edition, McGraw-Hill, New York, 1970.
13. M. A. Biot, *Mechanics of Incremental Deformation*, Wiley, New York, 1965.
14. L. Solomon, *Élasticité Linéaire*, Masson, Paris, 1968.
15. A. E. Green and J. E. Adkins, *Large Elastic Deformations*, Oxford University Press, 1960.
16. C. Truesdell and W. Noll, Non-linear field theories of mechanics, *Handbuch der Physik*, Springer-Verlag, Berlin, 1965, Vol. 3, Chap. 3.
17. H. Ziegler, *Principles of Structural Stability*, Blaisdell, Waltham, Mass., 1968.
18. A. H. England, *Complex Variable Methods in Elasticity*, Wiley-Interscience, London, 1971.
19. L. M. Milne-Thomson, *Antiplane Elastic Systems*, Springer, Berlin, 1962.

20. E. G. Coker and L. N. G. Filon, *A Treatise on Photo-Elasticity*, 2nd edition revised by H. T. Jessop, Cambridge University Press, 1957, pp. 125–128.
21. G. V. Kolosov, Sur les problèmes d'élasticité à deux dimensions, *C.R. Acad. Sci.* **146** (1908) and **148** (1909). See also Sur le problème plan de la théorie de l'élasticité, *Atti IV Congr. Int'l Mat.* **3,** Rome (1909); *On the application of complex function theory to a plane problem of the mathematical theory of elasticity* (in Russian), Thesis Dorpat (Yuriev) Univ. (1909); Über einige Eigenschaften des ebenen Problems der Elastizitätstheorie *Zeitschrift f. Math. u. Physik* **62** (1914); Sur une application des formules de Schwarz, de Villat et de Dini au problème plan de l'élasticité, *C.R. Acad. Sci.* **193** (1931); Sur une transformation des équations de l'élasticité, *C.R. Acad. Sci.* **184** (1927); *Application of Complex Functions in the Theory of Elasticity* (in Russian), Gostekizdat, Moscow, 1935.
22. N. I. Muskhilishvili, *Some Basic Problems of the Mathematical Theory of Elasticity* (in Russian), 4th edition, Akademikniga, Moscow, 1966; English translation of 3rd edition by J. R. M. Radok, P. Noordhoff, Groningen, Netherlands, 1953. The list of references at the end of this book may be consulted for other references to the work of N. I. Muskhelishvili and his school.
23. A. C. Stevenson, *Proc. Roy. Soc.* **A184,** 129, 218 (1945).
24. A. E. Green, *Proc. Roy. Soc.* **A180,** 173 (1942).
25. L. M. Milne-Thomson, *J. Lond. Math. Soc.* **17,** 115 (1942).
26. G. Lamé, *Leçons sur les Coordonnées Curvilignes. . .*, Mallet-Bachelier, Paris, 1859.
27. E. Betti, Teoria dell Elastica, *Il Nuovo Cimento*, Ser. 2, **7, 8,** 5–21, 69–97, 158–180 (1872); **9,** 34–43; **10,** 58–84 (1873).
28. J. V. Boussinesq, *Applications des potentials. . .*, Gauthier-Villars, Paris, 1885, especially pp. 63, 72.
29. P. F. Papkovitch, Expressions générales des composantes des tensions. . ., *C.R. Acad. Sci.*, Paris, **195,** 754 (1932). See also Solution générale des équations différentielles fundamentals d'élasticité exprimée par trois fonctions harmoniques, *C.R. Acad. Sci.*, Paris, **195,** 513 (1932); The representation of the general integral of the fundamental equations of the theory of elasticity in terms of harmonic functions (in Russian), *Izv. Akad. Nauk. SSSR*, Fiz.-Mat. Ser. **10,** 1425 (1932).
30. G. D. Grodski, The integration of the general equation of equilibrium of an isotropic elastic body with the aid of the Newtonian potentials and harmonic functions (in Russian), *Izv. Akad. Nauk. SSSR*, 587 (1935).
31. H. Neuber, Ein neuer Anstaz zur Lösung räumlicher Probleme der Elastizitätstheorie Der Hohlkegel unter Einzellast als Beispiel, *Z. angew. Math. Mech.* **14,** 203 (1934). See also *Kerbspannungslehre*, Springer, Berlin, 1958.
32. C. E. Weatherburn, *Advanced Vector Analysis*, Bell, London, 1944.
33. B. G. Galerkin, On an investigation of stresses and deformations in elastic isotropic solids (in Russian), *Doklad. Akad. Nauk. SSSR* **14,** 353 (1930). See also, On the general solution of a problem in the theory of elasticity in three dimensions by means of stress and displacement functions (in Russian), *Doklad. Akad. Nauk. SSSR* **10,** 281 (1931); *Prik. Mat. Mekh. Akad. Nauk. SSSR* **6,** 487 (1942).
34. G. Lamé, *Leçons sur la Theorie Mathématique de l'Élasticité des Corps Solides*, Bachelier, Paris, 1852, p. 72.
35. R. A. Eubanks and E. Sternberg, On the completeness of the Boussinesq-Papkovitch stress functions, *J. Rat. Mech. Anal.* **5** (1956).

36. P. M. Naghdi and H. S. Hsu, On a representation of displacements in linear elasticity in terms of three stress functions, *J. Math. Mech.* **10** (1961).
37. W. D. MacMillan, *The Theory of the Potential*, Dover Publications, New York, 1958.
38. Kelvin (W. Thomson), Note on the integration of the equations of equilibrium of an elastic solid, *Cambridge and Dublin Math. J.* **3** (1848). See also *Mathematical and Physical Papers*, Cambridge University Press, Vol. 1, p. 97.
39. E. Sternberg and R. A. Eubanks, On the concept of concentrated loads and an extension of the uniqueness theorem in the linear theory of elasticity, *J. Rat. Mech. Anal.* **4,** 135 (1955).
40. M. J. Turteltaub and E. Sternberg, On concentrated loads and Green's functions in elastostatics, *Arch. Rat. Mech. Anal.* **29,** 193 (1968).
41. L. T. Wheeler and E. Sternberg, Some theorems in classical elastodynamics, *Arch. Rat. Mech. Anal.* **31,** 51 (1968).
42. E. Betti, Teoria dell' Elastica, *Il Nuovo Cimento*, Ser. 2, **7, 8,** 5, 69, 158 (1872); **9, 10,** 34, 58 (1873) or see *Opere* **2,** 291.
43. I. Fredholm, Sur les équations de l' équilibre d'un corps solide élastique, *Acta Math.* **23,** 1.
44. V. Cerruti, Ricerche intorno all' equilibrio de' corpi elastici isotropi, *Reale Acc. dei Lincei*, Ser. 3a, **13,** 81 (1882).
45. J. H. Michell, The stress in an aelotropic elastic solid with an infinite plane boundary; *Proc. London Math. Soc.*, 1st ser, **32,** 247 (1900).
46. H. M. Westergaard, *Theory of Elasticity and Plasticity*, Harvard University Press, Cambridge, Mass., 1952, Sec. 69.
47. D. H. Cheng, Displacements and stresses due to nuclei of strain in an elastic half-space, Columbia University, Department of Civil Engineering and Mechanics, Tech. Rep. No. 47 (1961).
48. L. Rongved, Force at point in the interior of a semi-infinite solid with fixed boundary *J. appl. Mech.* **22,** 545–546 (1955).
49. W. R. Dean, H. W. Parsons, and I. N. Sneddon, *Proc. Cambridge Phil. Soc.* **40** (1955).
50. B. G. Galerkin, Equilibrium of an elastic spherical shell (in Russian), *Prikl. Mat. & Mekh.* **6,** 487 (1942).
51. Kelvin (W. Thomson), *Mathematical and Physical Papers*, Cambridge University Press, 1892, Vol. III, p. 351. (The analysis was first published in 1863.)
52. E. Sternberg, R. A. Eubanks, M. Sadowsky, On the axisymmetric problem of elasticity for a region bounded by two concentric spheres, *Proceedings of the First National Congress on Applied Mechanics*, U.S.A., 1953.
53. G. Fichera, Sul calcolo della deformazioni, dotate di simmetria assiale, di una strato sferico elastica, *Atti. accad. Naz. Lincei, Classe sci. fis.*, Ser. 8, **6,** 583 (1949).
54. H. Lamb, On Boussinesq's problem, *Proc. London Math. Soc.* **34,** 276–284 (1902).
55. K. Terezawa, *Phil. Trans. Roy. Soc.*, Ser A, 35 (1916). See also *J. Coll. Sci. Imp. Uni. Tokyo* **37** (1916).
56. R. Muki, Asymmetric problems of the theory of elasticity for a semi-infinite solid and a thick plate, *Progress in Solid Mech.*, North Holland, Netherlands, 1960, Vol. 1.
57. I. N. Sneddon, *Fourier Transforms*, McGraw-Hill, New York, 1951.
58. C. J. Tranter, *Integral Transforms in Mathematical Physics*, Methuen, London, 1951.
59. V. I. Smirnov and S. L. Sobolev, Sur une méthode nouvelle dans dans le problème plan des vibrations élastiques, *Trud. Inst. Seism. Akad. Nauk. SSSR* **20** (1932).

60. V. I. Smirnov and S. L. Sobolev, On the application of a new method of investigation of elastic vibrations in space with axial symmetry (in Russian), *Trud. Inst. Seism. Akad. Nauk, SSSR* **29** (1933).
61. S. L. Sobolev, Russian language translation of Frank and von Mises, *Differential Equations of Mathematical Physics*, O.N.T.I., Moscow–Leningrad, 1937, Chap. 12.
62. V. I. Smirnov, *A Course of Higher Mathematics*. English translation by D. E. Brown, Addison-Wesley, Reading, Mass., 1964, Vol. 3, Part 12, Secs. 52, 53, 55.
63. C. Weber Zur Umwandlung von Rotationssymmetrischen Problemen in Zweidimensionale und Umgekehrt, *Z. angew. Math. Mech.* **20,** 117–118 (1940).
64. N. A. Rostovtzev, Complex stress functions in the axisymmetric contact problem of the theory of elasticity, *Prikl. Mat. i. Mekh.* **17,** 611–614 (1953).
65. A. E. Green, On Boussinesq's problem and penny-shaped cracks, *Proc. Cambridge Phil. Soc.* **45,** 251 (1949).
66. N. M. Newmark, Influence charts for computation of stresses in elastic foundations, *Univ. of Illinois Eng'g Exp't Sta. Bull.*, Ser. 338, 1–28 (1942).
67. N. M. Newmark, Influence charts for computation of vertical displacements in elastic foundations, *Univ. of Illinois Eng'g Exp't. Sta. Bull.*, Ser. 367, 1–14.
68. V. I. Mossakovskii, *Prikl. Mat. Mekh.* **18,** 187–196 (1954).
69. L. M. Keer, Mixed boundary-value problems for an elastic half-space, *Proc. Cambridge Phil. Soc.* **63,** 1379.
70. Ia. S. Ufliand, *Prikl. Mat. Mekh.* **20,** 187–196 (1956).
71. A. H. England, A crack between dissimilar media, *J. appl. Mech.* **32,** 400–402 (1965).
72. F. Erdogan, stress distribution in bonded dissimilar materials with cracks, *J. appl. Mech.* **32,** 403–410 (1965).
73. M. L. Williams, The stresses around a fault or crack in dissimilar media, *Bull. Seis. Soc. Amer.* **49,** 199–204 (1959).
74. I. Ya. Shtaerman, On a certain generalization of the Hertz problem (in Russian), *Prikl. Mat. i Mekh.* **5** (1941).
75. L. A. Galin, *Contact Problems in the Theory of Elasticity*. English translation by H. Moss, North Carolina State College, Raleigh, N.C., 1961.
76. L. E. Goodman and L. M. Keer, The contact stress problem for an elastic sphere indenting an elastic cavity, *Int. J. Solids Struct.* **1,** 407–415 (1965).
77. P. J. Torvik, Note on the boundary conditions for contact problems, *J. appl. Mech.* **34,** 761–762 (1967).
78. I. N. Sneddon, The distribution of stress in the neighborhood of a crack in an elastic solid, *Proc. Roy. Soc.* **A187,** 187–229 (1946).
79. I. N. Sneddon and H. A. Elliot, The opening of a Griffith crack under internal pressure, *Quart. Appl. Math.* **4,** 262 (1946).
80. V. C. Mow, P. L. Chow, 2nd, F. F. Ling, Microslips between contacting paraboloids, *J. appl. Mech.* **34,** 321–328 (1967).
81. S. A. Khristianovitch and Y. P. Zheltov, Formation of vertical fractures by means of highly viscous liquid, *Proc. 4th World Petroleum Congress,* Sec. II/T.O.P., paper 3, 579 (1954).
82. G. I. Barenblatt, The mathematical theory of equilibrium cracks in brittle fracture, *Advances in Applied Mechanics*, Academic Press, New York, 1962, Vol. 7, pp. 55–129.
83. R. A. Sack, Extension of Griffith's theory of rupture to three dimensions. *Proc. Phys. Soc.* **58,** 729 (1946). See also Ref. [84].
84. I. N. Sneddon, *Crack Problems in the Mathematical Theory of Elasticity*, North Carolina State College, Raleigh, N.C., 1961, p. 130 *et seq.*

Generalized Elastic Continua

H. F. TIERSTEN

Mechanics Division, Rensselaer Polytechnic Institute, Troy, New York 12181

and

J. L. BLEUSTEIN

AMF Incorporated, Morehead Patterson Center, Stamford, Connecticut 06907

1 Introduction

Essential to the development of the theory of elasticity, or any of the usual three-dimensional theories of inelastic continua for that matter, is the definition of the mechanical traction vector which acts across any surface of separation in the continuum and is the mechanism of interaction between adjacent elements of the continuum. An elastic continuum responds to the traction (or consequent stress) with a (possibly finite) deformation, which consists of a local translation, a local rotation, and a strain [1]. In the proper form of the theory the strain is related to the symmetric part of the stress [2], but the local rotation, which varies from point to point, takes place freely [3]. These facts may be regarded as a consequence of either the conservation of angular momentum or the invariance of the stored energy function in a rigid rotation [4], i.e., of the principle of material objectivity [5]. One consequence of the foregoing considerations is that in the linear theory of elasticity the six independent components of stress are proportional to the six independent components of infinitesimal strain, which consists of the symmetric part of the displacement gradient tensor; and the antisymmetric part of the displacement gradient tensor, which is the small local rotation, is unrestrained [6]. It follows then that in the linear theory for an arbitrarily anisotropic (triclinic) elastic solid twenty-one independent elastic constants exist.

In the early and middle 1950s two independent groups of scientists [7, 8] proposed, for somewhat different reasons, a theory of anisotropic linear elasticity in which all nine independent components of the stress tensor were related to the nine independent components of the displacement gradient tensor. The resulting theory contains forty-five independent elastic constants in place of the aforementioned twenty-one independent elastic constants in the case of the triclinic crystal. During the late 1950s the new theory received considerable notoriety, to the extent that it was considered for inclusion in a revision of the IEEE Standards on vibrating anisotropic crystals, which was then in progress. One member of the committee asked Professor Mindlin's advice on this matter. After a brief investigation Mindlin recommended that the committee ignore the forty-five constant theory. His recommendation was based on the observation that both groups had neglected to satisfy the law of the conservation of angular momentum. However, each group had a somewhat different reason for ignoring the law. One group [7] simply presumed that the conservation of angular momentum need not be satisfied since the conservation of linear momentum had been satisfied, and in point (particle) mechanics the conservation of angular momentum is a consequence of the conservation of linear momentum. This presumption is in error because in a continuum a differential element of volume contains a nondenumerable infinity of points as opposed to the single point occupied by a particle in point mechanics, and, as a consequence, in continuum mechanics the law of the conservation of angular momentum is independent of the law of the conservation of linear momentum. The other group [8] postulated the existence of local mystical couples which could not be identified quantitatively as variables and simply served to permit the angular momentum of a differential element of volume to be unbalanced. Although these considerations are essentially without precise mathematical meaning, they motivated Mindlin to think of generalizing the theory of elasticity by including a couple-traction [9] in addition to the usual force-traction mentioned earlier. This investigation resulted in the paper "Effects of Couple-Stresses in Linear Elasticity" [M76]†. During this independent investigation, Mindlin became aware of the recently published work of Truesdell and Toupin [10], in which they consider the nonlinear couple-stress theory during a portion of their general treatment of classical fields, and recalled earlier work by the Cosserat brothers [11] on generalized elastic continua. It turned out that couple-stress elasticity could be interpreted as a Cosserat continuum with constrained rotations. However, in his usual thorough manner Mindlin had pursued the linear theory sufficiently far to have found both static and dynamic potential representations and solutions in the isotropic case and uncovered many properties of the linear equations.

While engaged in the aforementioned work, Mindlin acquainted Toupin with an error in Ref. [10], which motivated Toupin to correct the error and do additional work on nonlinear couple-stress theory [12]. In this work Toupin employs a variational principle, which leads to the generalization of couple-stress theory known as the full strain-gradient theory. In Toupin's nonlinear couple-stress theory the stored energy function depends on the eight independent components of the material curl of the finite strain, while in his strain-gradient theory the stored energy function depends on all eighteen independent

† Numbers preceded by the letter M in brackets refer to the Publications of R. D. Mindlin listed after the Preface to this volume.

components of the material gradient of the finite strain instead of the above-mentioned eight independent combinations appearing in the curl of the strain. Since the material coefficients of the additional terms in the stored energy function, resulting from the inclusion of the additional ten independent components in the strain-gradient theory, are of the same order of magnitude as the coefficients of the terms in the stored energy function of the couple-stress theory, the latter theory must be deemed incomplete. Toupin's variational considerations produced a significant improvement on the defined field vector determined couple-stress theory, which included only the unbalanced double-force components acting on a surface, by automatically including all the self-equilibrating double-force components acting on surfaces, which do work.

In 1964 Green and Rivlin [13] published a theory of monopolar stress fields, which we prefer to call multi-stress fields, in which they employ a defined vector field rate of working procedure rather than a variational principle. An important characteristic of Green and Rivlin's treatment is that they include multi-stress components that do no work and, consequently, are not contained in an ordinary variational principle. In the dipolar or, preferably, double-stress case Green and Rivlin's theory is a generalization of Toupin's strain-gradient theory, in that all twenty-seven components of the double-stress tensor are included in the description instead of only the eighteen which do work. The generalization is important even though the additional nine components do not enter any of the constitutive relations or differential equations because they do enter the boundary conditions and are required for the proper and complete specification of many types of boundary-value problems. In addition Green and Rivlin's treatment makes clear the physical meaning of Toupin's generalized forces. These facts underline the importance of defining the fundamental field vectors acting on surfaces in constructing complete field theories for mechanical continua, rather than simply postulating a suitable energy dependence in a variational principle. The point is that there can exist force components which do no work, and the physical definition of force field vectors is required for the proper specification of nontrivial boundary conditions. Using results of Green and Rivlin, it is possible to extend Toupin's variational principle for the strain-gradient theory to include those double-force components which do no work and enter no differential equations but do enter the boundary conditions. This extension is detailed in Section 2 of this article. The two theories are then identical. An interesting discussion of the relation between the dipolar theory of Green and Rivlin and the strain-gradient theory of Toupin has been given in a paper by Green and Naghdi [14], in which they solve a problem within the framework of the two theories.

Even though the couple-stress theory is incomplete in view of the double-stress theory, or even the strain-gradient theory for that matter, the couple-stress theory is simpler; sufficiently simpler that there are interesting and very revealing boundary-value problems that are tractable within the framework of the couple-stress theory for isotropic materials but not within the framework of the strain-gradient theory. Some very interesting and enlightening boundary-value problems of this nature were treated by Sternberg and those who worked with him. Sternberg started with the works of Mindlin [M76, M77] in which he developed the linear equations and derived the three-dimensional potential functions [M76] and two-dimensional stress functions [M77, M78] for the isotropic continuum. In brief Sternberg found that although the existence of couple-stresses removed or mitigated singularities in many problems [15, 16] in isotropic linear elasticity, in certain other problems [17] their existence enhanced singularities which previously were present. In particular a most interesting problem treated by Bogy and Sternberg [16] was that of a quarter plane subject to a uniform shear stress on one surface and free on the other surface. The singularity at the edge, which exists in ordinary elasticity, disappeared when couple-stresses were included. This is as expected since the surface couples, which do not exist in ordinary elasticity, permit the asymmetric traction at the edge. Many others [18] throughout the world published solutions of couple-stress boundary-value problems because problems in the area were interesting and not abnormally difficult. Some problems in linear isotropic couple-stress theory are discussed in Section 3.

In 1965 Mindlin extended the linear strain-gradient theory to include the second gradient of strain [M85]. This theory is the first continuum description to include surface energy of deformation in a natural manner, i.e., without making the usual assumption of the existence of a vanishingly thin surface membrane. This theory is discussed briefly in Section 4. In the interim between the initial work on the couple-stress theory and his work on the second gradient of strain, Mindlin conceived of a different and very far reaching idea, that of a three-dimensional continuum with microstructure. The basic idea consists of endowing each point of a continuum with an *internal* displacement field, which is expanded as a power series in *internal* coordinate variables. The lowest-order theory is obtained by retaining only the first, or linear terms. The result is a theory of elasticity including among other things, the long wavelength limit of the optical lattice modes [M81]. This theory and some of its consequences are discussed in some detail in Section 4. Motivated by Mindlin's work on the linear theory of elastic materials with microstructure, Toupin [19] presented a non-linear theory of elastic materials with microstructure, in which he utilized the concept of deformable directors due to Ericksen and Truesdell [20]. When the directors are rigid, the theory reduces to that of a Cosserat continuum [11]. In the linear case Toupin relates the deformable director concept to the micro-structure idea of Mindlin. Around the same time as Mindlin, Eringen and Suhubi [21] published a nonlinear theory of a microelastic solid, which utilizes essentially the same basic idea as Mindlin. Also, in 1964 Green and Rivlin [22] presented a nonlinear mechanical theory containing internal velocity and force variables, which they call multipolar continuum mechanics. Although the resulting equations are similar in certain respects to those of Mindlin [M81], Toupin [19], and Eringen and Suhubi [21], Green and Rivlin's physical point of departure is somewhat different than the others in that they use internal moments of position coordinates as their basic internal variables. The general treatment of Green and Rivlin is carried further than that of the others [M81, 19, 21], who treat only the lowest microstructure expansion in detail. However, Mindlin [M81, M93] and Eringen [21] carry the expansion they treat in detail further than Green and Rivlin [22] carry their work. On account of the aforementioned differences between the general treatment of Green and Rivlin and the work of others, Green, Naghdi, and Rivlin [23] and Green [24] have discussed the relation of multipolar continuum mechanics to the director description of Toupin [19] and the expansion of Eringen and Suhubi [21], respectively. Very few boundary-value problems have been treated within the framework of the linear theory of elastic materials with microstructure [25]. The theory of elastic materials with microstructure has influenced developments in the mechanics of laminated and fiber reinforced materials and framed structures [26]. In particular, the microstructure idea was applied to the mechanics of material composites by Herrmann and others [27] who followed his lead.

After originating and working in the area of microstructure in linear elasticity for a number of years, Mindlin investigated ionic lattice theory [M93] in order to compare the long wavelength limit with linear microstructure theory. Mindlin found that the system of equations differed in certain respects and concluded that the long wavelength limit of the ionic lattice equations were equivalent to the linear continuum equations for two interpenetrating solid continua interacting by mutual forces and couples. After this discovery, Mindlin halted his effort on microstructure theory and devoted his time to other research, which turned out to be highly productive and fruitful. Thus it seems clear that just as specific physical considerations led Mindlin to investigate both couple-stress theory and microstructure theory, they caused him to abandon his effort on microstructure theory because of the particular physical problems in which Mindlin was interested. This thread of intense concern with the description of physical reality in mathematical terms seems to pervade all of Mindlin's research.

In Section 2 of this article the developments of nonlinear couple-stress theory, strain-gradient theory and double-stress theory are reviewed in light of each other and interrelated. Particular emphasis is given to the proper setting of boundary conditions in terms of prescribed quantities within the framework of the theories because the existence of some confusion about the boundary conditions has been brought to the attention of the authors. In Section 3 the linearization of the couple-stress equations is reviewed along with the uniqueness theorem for the isotropic case. Both the static and dynamic potential formulations of linear, isotropic couple-stress elasticity are reviewed, the plane-strain formulation is presented and the solutions of some problems are discussed. This amount of detail on linear, isotropic couple-stress elasticity is contained because, as noted earlier, solutions are more readily obtained within the framework of couple-stress theory than the more complicated double-stress theory. In Section 4 a few further generalized continua treated by Mindlin are discussed from a linear standpoint. These include the theory of the second gradient of strain and its relation to surface-tension and microstructure theory.

2 Couple-Stress, Strain-Gradient, and Double-Stress

If, on an arbitrary surface separating a portion of a material continuum from the remainder, a couple per unit area $\mathbf{m}$ is taken to act in addition to the usual force per unit area $\mathbf{t}$, the equations of the conservation of mass, linear momentum, angular momentum and energy take the integral forms, respectively,

$$\frac{d}{dt}\int_V \rho dV = 0 \tag{2.1}$$

$$\int_S \mathbf{t} dS + \int_V \rho \mathbf{g} dV = \frac{d}{dt}\int_V \rho \mathbf{v} dV \tag{2.2}$$

$$\int_S (\mathbf{y} \times \mathbf{t} + \mathbf{m}) dS + \int_V (\mathbf{y} \times \mathbf{g} + \mathbf{l}) \rho dV = \frac{d}{dt}\int_V \mathbf{y} \times \rho \mathbf{v} dV \tag{2.3}$$

$$\frac{d}{dt}\int_V \rho(\tfrac{1}{2}\mathbf{v}\cdot\mathbf{v} + \mathcal{E}) dV = \int_S (\mathbf{t}\cdot\mathbf{v} + \tfrac{1}{2}\mathbf{m}\cdot\nabla \times \mathbf{v}) dS + \int_V \rho(\mathbf{f}\cdot\mathbf{v} + \tfrac{1}{2}\mathbf{l}\cdot\nabla \times \mathbf{v}) dV \tag{2.4}$$

where d/dt is the material time-derivative, ρ is the mass density, $\mathbf{y}$ is the spatial position vector, $\mathbf{v}$ is the material velocity, $d\mathbf{y}/dt$, $\mathbf{g}$ and $\mathbf{l}$ are the body force and couple per unit mass, respectively, $\mathcal{E}$ is the internal energy per unit mass, ∇ is the spatial gradient, S denotes a surface enclosing a volume V, and $\frac{1}{2}\nabla \times \mathbf{v}$ is the local angular velocity of the material continuum. It is assumed that a material point at $\mathbf{y}$ at time t was initially at $\mathbf{X}$, the material coordinate or reference position, where [28]

$$y_i = y_i(X_L, t), \qquad \mathbf{y} = \mathbf{y}(\mathbf{X}, t) \tag{2.5}$$

which is one-to-one and differentiable as often as required. We consistently use the convention that capital indices denote the Cartesian components of $\mathbf{X}$ and lowercase indices, the Cartesian components of $\mathbf{y}$. Both dyadic and Cartesian tensor notation are used interchangeably. A comma followed by an index denotes partial differentiation with respect to a coordinate, i.e.,

$$y_{i,L} = \frac{\partial y_i}{\partial X_L}, \qquad X_{K,j} = \frac{\partial X_K}{\partial y_j} \tag{2.6}$$

and the summation convention for repeated tensor indices is employed. It should be noted that the spatial gradient $\boldsymbol{\nabla}$ can be written

$$\boldsymbol{\nabla} = \mathbf{e}_i \frac{\partial}{\partial y_i} \tag{2.7}$$

where $\mathbf{e}_i$ denotes a unit base vector in the ith Cartesian direction.

Application of Eq. (2.2) to an elementary tetrahedron in the usual manner yields the definition of the classical force-stress tensor $\boldsymbol{\tau}$:

$$\mathbf{t} = \mathbf{n} \cdot \boldsymbol{\tau} \tag{2.8}$$

Similarly, application of Eq. (2.3) to an elementary tetrahedron yields the definition of the couple-stress tensor $\boldsymbol{\mu}$:

$$\mathbf{m} = \mathbf{n} \cdot \boldsymbol{\mu} \tag{2.9}$$

From Eqs. (2.8), (2.2), and (2.1), with the aid of the divergence theorem and the arbitrariness of V, we obtain

$$\boldsymbol{\nabla} \cdot \boldsymbol{\tau} + \rho \mathbf{g} = \rho \frac{d\mathbf{v}}{dt} \tag{2.10}$$

which is the usual force-stress equation of motion. From Eqs. (2.8), (2.9), (2.3), and (2.1), with the aid of the divergence theorem, Eq. (2.10) and the arbitrariness of V, we obtain the couple-stress equation of motion

$$\mu_{ji,j} + \rho l_i + e_{ijk}\tau_{jk} = 0 \tag{2.11}$$

where, for simplicity, we have introduced Cartesian tensor notation and e_{ijk} is the spatial alternating tensor [29]. From Eq. (2.11) we obtain $\boldsymbol{\tau}^A$, the antisymmetric part of $\boldsymbol{\tau}$, which may be written in the component form

$$\tau_{kl}{}^A = \tfrac{1}{2} e_{jlk}\mu_{ij,i} + \tfrac{1}{2} e_{jlk}\rho l_j \tag{2.12}$$

Since

$$\boldsymbol{\tau} = \boldsymbol{\tau}^S + \boldsymbol{\tau}^A \tag{2.13}$$

the substitution of Eqs. (2.12) and (2.13) into Eq. (2.10) yields

$$\boldsymbol{\nabla} \cdot \boldsymbol{\tau}^S + \tfrac{1}{2} \boldsymbol{\nabla} \times \boldsymbol{\nabla} \cdot \boldsymbol{\mu} + \rho \mathbf{g} + \tfrac{1}{2} \boldsymbol{\nabla} \times \rho \mathbf{l} = \rho \frac{d\mathbf{v}}{dt} \tag{2.14}$$

which is a more useful form of the stress equations of motion when couples and/or couple-stresses are present, and we have returned to invariant dyadic notation. From Eqs. (2.8), (2.9), (2.4), and (2.1), with the aid of the divergence theorem, Eqs. (2.10), (2.12), (2.13), and the arbitrariness of V, we obtain

$$\rho \frac{d\mathcal{E}}{dt} = \tau_{ij}{}^S d_{ij} + \tfrac{1}{2}\mu_{ij} e_{jkl} v_{l,ki} \tag{2.15}$$

where the rate of deformation tensor d_{ij} is defined by

$$d_{ij} = \tfrac{1}{2}(v_{i,j} + v_{j,i}) \tag{2.16}$$

If $\boldsymbol{\mu}^D$, the deviator of $\boldsymbol{\mu}$, is defined by

$$\mu_{ij}{}^D = \mu_{ij} - \tfrac{1}{3}\mu_{kk}\delta_{ij} \tag{2.17}$$

where δ_{ij} is the spatial Kronecker delta, Eq. (2.15) may be written as

$$\rho \frac{d\mathcal{E}}{dt} = \tau_{ij}{}^S d_{ij} + \tfrac{1}{2}\mu_{ij}{}^D e_{jkl} v_{l,ki} \tag{2.18}$$

since $e_{jkl}v_{l,kj} \equiv 0$. Equation (2.18) may be called the first law of thermodynamics (energetics) for the continuum with couple-stresses. Moreover, since $e_{lmj}\mu_{kk,jm} \equiv 0$, Eq. (2.14) can be written in the form

$$\nabla \cdot \boldsymbol{\tau}^S + \tfrac{1}{2}\nabla \times \nabla \cdot \boldsymbol{\mu}^D + \rho\mathbf{g} + \tfrac{1}{2}\nabla \times \rho\mathbf{l} = \rho \frac{d\mathbf{v}}{dt} \tag{2.19}$$

Thus we see that the scalar of $\boldsymbol{\mu}$ does not appear in the equations of motion and, on account of Eq. (2.18), will not appear in the constitutive equations either. Consequently, it appears that the scalar $\mu \equiv \mu_{kk}/3$ of the couple-stress $\boldsymbol{\mu}$ is indeterminate and, by virtue of Eq. (2.12), the antisymmetric part of the force-stress is left indeterminate. However, consideration of the boundary conditions at material surfaces of discontinuity reveals that when force- and couple-tractions are prescribed on a boundary surface, the scalar μ is determined at the boundary surface from the prescribed value of $\mathbf{m}$.

Application of Eqs. (2.2) and (2.3) across a material surface of discontinuity in the usual way yields the jump conditions

$$n_i[\tau_{ij}] + \bar{t}_j = 0, \qquad n_i[\mu_{ij}] + \bar{m}_j = 0 \tag{2.20}$$

where $\bar{t}_j$ and $\bar{m}_j$ are the components of the prescribed force- and couple-tractions, and we have introduced the conventional notation $[a_i]$ for $a_i{}^+ - a_i{}^-$ and n_i denotes the components of a unit normal directed from the $-$ to the $+$ side of the surface of discontinuity. If there is no solid matter on the $+$ side, from Eq. (2.20) we obtain the boundary equations

$$n_i\tau_{ij}{}^- = \bar{t}_j, \qquad n_i\mu_{ij}{}^- = \bar{m}_j \tag{2.21}$$

and from here on we drop the minus superscript. For traction boundary conditions $\bar{t}_j$ and $\bar{m}_j$ are prescribed. Substituting from Eqs. (2.12), (2.13), and (2.17) into Eq. (2.21), we obtain

$$n_i\tau_{ij}{}^S + \tfrac{1}{2}n_i e_{kji}\mu_{lk,l}{}^D + \tfrac{1}{2}n_i e_{kji}\mu_{,k} + \tfrac{1}{2}n_i e_{kji}\rho\bar{l}_k = \bar{t}_j \tag{2.22}$$

$$n_i\mu_{ij}{}^D + n_j\mu = \bar{m}_j \tag{2.23}$$

which constitute six boundary equations. However, the scalar of the couple-stress, μ, can be eliminated from the boundary Eqs. (2.22) and (2.23) while reducing them from six to five. To this end we form the scalar equation

$$n_i\mu_{ij}{}^D n_j + \mu = n_j\bar{m}_j \tag{2.24}$$

and substitute into Eq. (2.22) to obtain

$$n_i\tau_{ij}{}^S + \tfrac{1}{2}n_i e_{kji}\mu_{lk,l}{}^D - \tfrac{1}{2}n_i e_{kji}(n_l\mu_{lm}{}^D n_m)_{,k} = \bar{t}_j - \tfrac{1}{2}n_i e_{kji}\rho\bar{l}_k - \tfrac{1}{2}n_i e_{kji}(n_l\bar{m}_l)_{,k} \tag{2.25}$$

Since the normal component of Eq. (2.23) has already been incorporated in Eq. (2.25), from Eq. (2.23) we form

$$e_{lkj}n_k n_i\mu_{ij}{}^D = e_{lkj}n_k\bar{m}_j \tag{2.26}$$

which are two independent equations. Equations (2.25) and (2.26) constitute five boundary equations independent of μ, which, with the equations of motion (2.19) and the constitutive equations for $\tau_{ij}{}^S$ and $\mu_{ij}{}^D$, may *in principle* be solved for the spatial position $\mathbf{y} = \mathbf{y}(\mathbf{X}, t)$ without determining μ. However, after a solution has been obtained, μ at any boundary on which $\bar{\mathbf{m}}$ is prescribed may be determined from Eq. (2.24). Nevertheless, at points not on the boundary *any* smooth scalar function μ which meets the boundary values is allowed. Thus μ is really meaningful physically only at the boundaries where the $\bar{m}_j$ are prescribed. In the foregoing discussion it was implicitly assumed that the bounding surface was smooth, i.e., that $\mathbf{n}$ was a continuous function of position on the surface, in taking the spatial gradient of the scalar Eq. (2.24). However, if an edge is formed by the intersection of two smooth surfaces, there is a discontinuity in $\mathbf{n}$ across the edge, and the gradient of Eq. (2.24) cannot be taken. Nevertheless, Eq. (2.24) holds on either side of the edge, and the difference yields

$$[\mathbf{n}\cdot\boldsymbol{\mu}^D\cdot\mathbf{n}] = [\mathbf{n}]\cdot\bar{\mathbf{m}} \tag{2.27}$$

which condition must be satisfied across every edge along with Eqs. (2.25) and (2.26) when tractions are prescribed. In a position (and normal derivative of position) type of boundary-value problem μ remains completely indeterminate. From Eqs. (2.12) and (2.17) it is clear that any indeterminacy in μ results in an indeterminacy in $\tau_{kl}{}^A$. Moreover, in a position type of boundary-value problem, Eq. (2.27), which is not used in obtaining a solution, determines *a posteriori* a force per unit length that must be exerted in the direction of the edge.

Toupin [12] has shown that in couple-stress theory the constitutive equations may be obtained by assuming that

$$\mathcal{E} = \mathcal{E}(E_{LM}, K_{LM}) \tag{2.28}$$

where

$$E_{LM} = \tfrac{1}{2}(y_{i,L}y_{i,M} - \delta_{LM}), \qquad K_{ML} = e_{NKL}E_{MK,N} \tag{2.29}$$

where δ_{LM} and e_{NKL} are the material Kronecker delta and material alternating tensor, respectively, and where it is understood that $\mathcal{E}$ is constructed in such a way that $\partial\mathcal{E}/\partial E_{KL} = \partial\mathcal{E}/\partial E_{LK}$ and $\partial\mathcal{E}/\partial K_{RR} = 0$. Since

$$d_{ij} = X_{K,i}X_{L,j}\frac{dE_{KL}}{dt}$$

$$\tfrac{1}{2}e_{jkl}v_{l,ki} = J^{-1}y_{j,P}X_{M,i}\frac{dK_{MP}}{dt} - J^{-1}[e_{NLP}X_{K,r}y_{j,P}X_{M,i}y_{r,NM}]^S\frac{dE_{KL}}{dt} \tag{2.30}$$

where $J = \det y_{l,R}$ and the superscript S stands for symmetric, from Eq. (2.18) we may write

$$\rho\frac{d\mathcal{E}}{dt} = \left[\tau_{ij}{}^S X_{K,i}X_{L,j} - \mu_{ij}{}^D\frac{1}{J}(e_{NLP}X_{K,r}X_{M,i}y_{j,P}y_{r,NM})^S\right]\frac{dE_{KL}}{dt}$$
$$+ \mu_{ij}{}^D\frac{1}{J}y_{j,L}X_{M,i}\frac{dK_{ML}}{dt} \tag{2.31}$$

Since $dK_{RR}/dt \equiv 0$, we introduce the Lagrangian multiplier λ and multiply it by dK_{RR}/dt and add the result to the right-hand side of Eq. (2.31) while differentiating $\mathcal{E}$ to obtain

$$\left[\rho\frac{\partial\mathcal{E}}{\partial E_{KL}} - \tau_{ij}{}^S X_{K,i}X_{L,j} + \mu_{ij}{}^D\frac{1}{J}(e_{NLP}X_{K,r}X_{M,i}y_{j,P}y_{r,NM})^S\right]\frac{dE_{KL}}{dt}$$
$$+ \left[\rho\frac{\partial\mathcal{E}}{\partial K_{ML}} - \mu_{ij}{}^D\frac{1}{J}y_{j,L}X_{M,i} - \lambda\delta_{ML}\right]\frac{dK_{ML}}{dt} = 0 \tag{2.32}$$

Since we have introduced the Lagrangian undetermined multiplier λ, we may treat all dK_{ML}/dt as if they are independent, and we assume that $\boldsymbol{\tau}^S$ and $\boldsymbol{\mu}^D$ are independent of dE_{KL}/dt and dK_{ML}/dt, we obtain

$$\tau_{ij}{}^S = \rho y_{i,K}y_{j,L}\frac{\partial\mathcal{E}}{\partial E_{KL}} + \rho\left[y_{i,L}\frac{\partial\mathcal{E}}{\partial K_{MP}}e_{NLP}y_{j,NM}\right]^S$$

$$\mu_{ij}{}^D = \rho J y_{i,M}X_{L,j}\frac{\partial\mathcal{E}}{\partial K_{ML}} \tag{2.33}$$

where we have solved for $\tau_{ij}{}^S$ and $\mu_{ij}{}^D$ and found that $\lambda = 0$ because $\mu_{kk}{}^D = 0$

and $\mathcal{E}$ has been constructed so that $\partial\mathcal{E}/\partial K_{RR} = 0$. Equations (2.33) are one form of Toupin's constitutive equations. This completes the formulation of the nonlinear couple-stress theory. The equations may be reduced to the three Eqs. (2.19) in the three dependent variables y_i, and the boundary conditions (2.25) and (2.26) and jump condition (2.27) may be expressed in terms of the same three dependent variables.

When Toupin obtained the couple-stress constitutive Eqs. (2.33), he realized that the resulting theory was unduly restrictive (incomplete) in that it arbitrarily excluded ten of the eighteen independent components of the strain-gradient, which clearly produce terms in the stored energy function of the same order of magnitude as those included. Toupin obtains his more general equations of static strain-gradient theory from a variational principle of the form

$$\delta \int_V \rho\mathcal{E}dV = \int_V F_i^1\delta y_i dV + \int_S F_i^2\delta y_i dS + \int_S F_i^3 D\delta y_i dS + \oint_C F_i^4\delta y_i ds \quad (2.34)$$

where S is a closed surface enclosing a volume V, C represents the edges of intersections of smooth portions of S, the F_i^N represent *generalized forces*, D represents the normal derivative $n_k\partial_k$, and

$$\mathcal{E} = \mathcal{E}(E_{KL}, B_{LMN}) \quad (2.35)$$

where E_{KL} is defined in Eq. $(2.29)_1$, and

$$B_{LMN} = y_{i,L}y_{i,MN} \quad (2.36)$$

Taking the variations [12] in Eq. (2.34) and letting the coefficients of the independent variations δy_i in V on S and on C and $D\delta y_i$ on S vanish, we obtain

$$-T_{ij,i} + \eta_{(ki)j,ki} = F_j^1 \quad (2.37)$$

$$n_iT_{ij} - n_in_kD\eta_{(ki)j} - n_iD_k\eta_{(ki)j} - D_i(n_k\eta_{(ki)j}) = F_j^2 \quad (2.38)$$

$$n_in_k\eta_{(ki)j} = F_j^3 \quad (2.39)$$

$$[m_in_k\eta_{(ki)j}] = F_j^4 \quad (2.40)$$

where $\eta_{(ki)j} = \eta_{(ik)j}$ and

$$T_{ij} = \rho y_{i,M}y_{j,L}\frac{\partial\mathcal{E}}{\partial E_{LM}} + \rho(y_{i,M}y_{j,RN} + y_{j,M}y_{i,RN})\frac{\partial\mathcal{E}}{\partial B_{MRN}} \quad (2.41)$$

$$\eta_{(ki)j} = \rho y_{i,M}y_{k,L}y_{j,K}\frac{\partial\mathcal{E}}{\partial B_{KML}} \quad (2.42)$$

and D_k is the surface gradient operator defined by

$$D_k = \partial_k - n_kD \quad (2.43)$$

$$m_i = e_{ijk}s_jn_k \quad (2.44)$$

and **s** denotes a unit vector tangent to the curve C and we have introduced the conventional notation $[\mathbf{a}] = \mathbf{a}^+ - \mathbf{a}^-$. In order to use this theory one must relate the generalized forces F_j^N to physically defined quantities. Although Toupin [12] relates the generalized forces to physically defined quantities in the case of couple-stress theory, the relation of the generalized forces to appropriate physical quantities in the general strain-gradient theory was not established by means of the variational approach.

However, Green and Rivlin [13] have presented a very powerful defined vector field rate of working procedure in which field vectors are defined physically—in the same way as the traction vector of ordinary continuum theories—and which enables the difficulty with the physical interpretation of the generalized forces inherent in the variational approach to be removed. Although Green and Rivlin [13, 30] have shown that all the conservation equations can be obtained from the single equation of the conservation of energy and appropriate invariance conditions, we prefer to consider the conservation of mass (2.1) and linear momentum (2.2) as fundamental laws in the conventional manner. Although the law of the conservation of angular momentum (2.3) remains valid, it is not included with the other two equations because it does not yield any information which is useful in the field equations when the continuum is energetically sensitive to double-forces without moment. When double-force systems are present, the law of the conservation of angular momentum *must* be replaced by the more powerful principle of the invariance of the stored energy function in a rigid rotation, which assures that the law of the conservation of angular momentum is satisfied. At this point we note that although Green and Rivlin [13] treated high-order self-equilibrating, rate of working force systems, we discuss only what they call the dipolar and we call the double-force case. The stress equations of motion remain the same and are given in Eq. (2.10). The equation of the conservation of energy takes the integral form

$$\frac{d}{dt}\int_V \rho(\tfrac{1}{2}v_jv_j + \mathcal{E})\,dV = \int_S (t_jv_j + \Delta_{ij}v_{j,i})\,dS + \int_V \rho(g_jv_j + \psi_{ij}v_{j,i})\,dV \qquad (2.45)$$

in place of the form given in Eq. (2.4) for the couple-stress case, where $\Delta_{ij}v_{j,i}$ is the rate at which work is done per unit surface area by the surface double-forces per unit area Δ_{ij} and $\rho\psi_{ij}v_{j,i}$ is the rate at which work is done per unit volume by the double-forces per unit mass ψ_{ij}. It should be noted that this reduces to the couple-stress case when

$$\Delta_{ij} = \tfrac{1}{2}e_{ijk}m_k, \qquad \psi_{ij} = \tfrac{1}{2}e_{ijk}l_k \qquad (2.46)$$

Following Green and Rivlin [13], we take the material time-derivative of Eq. (2.45), employ Eq. (2.1), and apply the resulting equation to an elementary tetrahedron with volume approaching zero, to obtain

$$(t_j - n_i\tau_{ij})v_j + (\Delta_{ij} - n_k\eta_{kij})v_{j,i} = 0 \tag{2.47}$$

which, with Eq. (2.8) which still holds, enables us to write

$$(\Delta_{ij} - n_k\eta_{kij})v_{j,i} = 0 \tag{2.48}$$

Now, if we assume—as did Green and Rivlin—that Δ_{ij} and η_{kij} are independent of $v_{j,i}$, we obtain

$$\Delta_{ij} = n_k\eta_{kij} \tag{2.49}$$

where η_{kij} is a third rank tensor with twenty-seven independent components. Taking the material time-derivative of Eq. (2.45), substituting from Eqs. (2.8) and (2.49) into Eq. (2.45), applying the divergence theorem and employing Eq. (2.10), some kinematic relations, and the arbitrariness of V, we obtain

$$\rho \frac{d\mathcal{E}}{dt} = [X_{M,i}(\tau_{ij} + \eta_{kij,k} + \rho\psi_{ij}) + X_{M,ik}\eta_{kij}] \frac{d}{dt}(y_{j,M}) + X_{M,i}X_{L,k}\eta_{kij} \frac{d}{dt}(y_{j,ML}) \tag{2.50}$$

from which it is clear that

$$\mathcal{E} = \mathcal{E}(y_{j,M}; y_{j,ML}) \tag{2.51}$$

and since $\mathcal{E}$ must be invariant in a rigid rotation, we find, from Cauchy's theorem [31] on invariant functions of vectors, that $\mathcal{E}$ may be reduced to the form shown in Eq. (2.35) in place of the form in Eq. (2.51). From Eqs. $(2.29)_1$, (2.35), (2.36), and (2.50) we obtain the constitutive equations

$$\tau_{ij} = T_{ij} - \eta_{kij,k} - \rho\psi_{ij} \tag{2.52}$$

and (2.42), where T_{ij} is given in Eq. (2.41), and we note that $\eta_{[ki]j} = -\eta_{[ik]j}$ does not occur in the differential equations. However, as Green and Rivlin [13] noted, the $\eta_{[ki]j}$ play an important role in the satisfaction of the boundary conditions.

To clarify the situation we now suppose that at points in the interior of a body we have

$$g_j = \bar{g}_j, \qquad \psi_{ij} = \bar{\psi}_{ij} \tag{2.53}$$

and on the bounding surface we have

$$t_j = \bar{t}_j, \qquad \Delta_{ij} = \bar{\Delta}_{ij} \tag{2.54}$$

where the bars in Eqs. (2.53) and (2.54) indicate that the quantities are prescribed. In this traction boundary-value problem we now follow Green and

Rivlin and observe that in order for Eqs. (2.8) and (2.49) to be satisfied at the bounding surface, we must have

$$\bar{t}_j = n_i \tau_{ij} \tag{2.55}$$

$$\bar{\Delta}_{ij} = n_k \eta_{kij} = n_k(\eta_{(ki)j} + \eta_{[ki]j}) \tag{2.56}$$

where $\eta_{(ki)j}$ may be expressed in terms of the dependent field variables y_i through Eqs. (2.42) and (2.36), and from Eqs. (2.52) and (2.53) we can write

$$\tau_{ij} = \sigma_{ij} - \rho\bar{\psi}_{ij} - \eta_{[ki]j,k} \tag{2.57}$$

where

$$\sigma_{ij} = T_{ij} - \eta_{(ki)j,k} \tag{2.58}$$

and σ_{ij} may be expressed in terms of the dependent field variables y_i through Eqs. (2.58), (2.41), and (2.42), and $\bar{\psi}_{ij}$ is prescribed. Substituting from Eq. (2.57) into Eq. (2.55), we obtain

$$\bar{t}_j = n_i\sigma_{ij} - n_i\rho\bar{\psi}_{ij} - n_i\eta_{[ki]j,k} \tag{2.59}$$

We now repeat the observations of Green and Rivlin to the effect that Eqs. (2.56) and (2.59) constitute twelve boundary equations for the determination of the y_i and the six unknown functions $n_k\eta_{[ki]j}$ at the bounding surface. Now, the boundary quantities $n_k\eta_{[ki]j}$, which do not appear in the differential Eqs. (2.10), can be eliminated from the boundary conditions (2.56) and (2.59) while reducing them from twelve to six. To this end from Eq. (2.56) we form

$$D_i\bar{\Delta}_{ij} = D_i(n_k\eta_{(ki)j}) + D_i(n_k\eta_{[ki]j}) \tag{2.60}$$

$$n_i\bar{\Delta}_{ij} = n_i n_k \eta_{(ki)j} \tag{2.61}$$

and note that

$$D_i(n_k\eta_{[ki]j}) = -n_i\eta_{[ki]j,k} \tag{2.62}$$

since $Dn_k = 0$, $D_i n_k = D_k n_i$ and $\eta_{[ki]j} = -\eta_{[ik]j}$. Now, substituting from Eq. (2.62) into Eq. (2.60) and subtracting Eq. (2.60) from Eq. (2.59), we obtain

$$\bar{t}_j - D_i\bar{\Delta}_{ij} + n_i\rho\bar{\psi}_{ij} = n_i\sigma_{ij} - D_i(n_k\eta_{(ki)j}) \tag{2.63}$$

which with Eq. (2.61) constitute six boundary equations independent of $\eta_{[ki]j}$, which with the differential Eqs. (2.10) enables the solution for the field variables y_i to be obtained in principle. After the field variables have been obtained, the six quantities $n_k\eta_{[ki]j}$ may readily be determined from Eq. (2.56). Nevertheless, as with the scalar of the couple stress, at points not on the boundary *any* smooth tensor field $\eta_{[ki]j}$ which meets the boundary values $n_k\eta_{[ki]j}$ is allowed. Thus the $\eta_{[ki]j}$ are physically meaningful only at the boundaries where the $\bar{\Delta}_{ij}$ are prescribed. Equations (2.61) and (2.63) correspond with Eqs. (9.22) and (9.21), respectively, of Green and Rivlin [13]. In the foregoing discussion it was implicitly assumed that the bounding surface was smooth in taking the surface gradients in Eqs. (2.60), (2.62), and (2.63). However, if an edge is formed by the intersection of two smooth surfaces, the gradient cannot be taken because of the discontinuity in $\mathbf{n}$ across the edge. Nevertheless, since Eq. (2.56) holds on either side of an edge, Eq. (2.61) is satisfied at each point of the surface by a solution, the normal $\mathbf{n}$ to the surfaces is continuous along the direction $\mathbf{s}$ of the intersection and the $\eta_{[ki]j}$ can be represented in the form $e_{kil}A_{lj}$, where A_{lj} is independent of $\mathbf{n}$, the difference of the components of Eq. (2.56) in the surfaces on either side of the intersection normal to the intersection yields

$$[m_i n_k \eta_{(ki)j}] = [m_i]\bar{\Delta}_{ij} \tag{2.64}$$

which condition must be satisfied across every edge along with Eqs. (2.61) and (2.63) when tractions are prescribed. The importance of Eq. (2.64) in the solution of a problem with an edge is discussed by Green and Naghdi [14]. In addition to providing a complete formulation of the traction boundary-value problem of the theory of elastic materials with double-stresses, Eqs. (2.57), (2.58), (2.61), (2.63), and (2.64) with Eqs. (2.10), (2.41), and (2.42), enable us to give expressions for Toupin's generalized forces F_j^1, F_j^2, F_j^3, and F_j^4 of the strain-gradient theory in terms of physically defined prescribed quantities. A comparison of Eqs. (2.10), (2.57), (2.58), (2.63), (2.61), and (2.64) with Eqs. (2.37), (2.38), (2.39), and (2.40) reveals that

$$F_j^1 = -\rho\bar{g}_j + \rho\bar{\psi}_{ij,i} \tag{2.65}$$

$$F_j^2 = \bar{t}_j - D_i\bar{\Delta}_{ij} + n_i\rho\bar{\psi}_{ij} \tag{2.66}$$

$$F_j^3 = n_i\bar{\Delta}_{ij} \tag{2.67}$$

$$F_j^4 = [m_i]\bar{\Delta}_{ij} \tag{2.68}$$

As noted by Green and Rivlin, considerations similar to those discussed above apply when the surface position $\bar{y}_i$ and surface double-forces $\bar{\Delta}_{ij}$ are given. The required six boundary conditions for the determination of the field variables y_i are the $\bar{y}_i$, Eqs. (2.61) and (2.64). When the boundary conditions are on the position y_i and its normal derivative Dy_i, Eq. (2.64), which is not required for a solution, determines *a posteriori* a force per unit length that must be exerted on the edge.

It is both interesting and enlightening to obtain the foregoing formulation of boundary-value problems from Green and Rivlin's field vector point of view in a way that is closely related to Toupin's variational approach. To this end we return to the equation of the conservation of energy (2.45). Taking the material time-derivative on the left-hand side of Eq. (2.45) and employing Eqs. (2.1) and (2.10), certain kinematic relations, Eqs. $(2.29)_1$, (2.36), (2.41), (2.42), (2.57), and (2.58) and rearranging terms, we are able to bring Eq. (2.45) to the form

$$\int_S [t_j - n_i(\sigma_{ij} - \rho\psi_{ij})]v_j dS + \int_S [\Delta_{ij} - n_k\eta_{(ki)j}]v_{j,i} dS = 0 \quad (2.69)$$

Introducing the surface gradient operator D_j defined in Eq. (2.43) and employing the surface divergence theorem [32], we obtain

$$\int_S [\Delta_{ij} - n_k\eta_{(ki)j}]v_{j,i} dS$$

$$= \int_S [\Delta_{ij} - n_k\eta_{(ki)j}]n_i D v_j dS - \int_S [D_i(\Delta_{ij} - n_k\eta_{(ki)j})]v_j dS$$

$$+ \int_{S_1+S_2=S} (D_l n_l) n_i(\Delta_{ij} - n_k\eta_{(ki)j}) v_j dS + \oint_C [m_i(\Delta_{ij} - n_k\eta_{(ki)j}) v_j] ds \quad (2.70)$$

where C denotes an edge formed by the intersection of two portions S_1 and S_2, each with continuous normal, of the closed surface S. Substituting from Eq. (2.70) into Eq. (2.69), we obtain

$$\int_S [t_j - n_i(\sigma_{ij} - \rho\psi_{ij}) - D_i(\Delta_{ij} - n_k\eta_{(ki)j}) + (D_l n_l) n_i(\Delta_{ij} - n_k\eta_{(ki)j})]v_j dS$$

$$+ \int_S n_i(\Delta_{ij} - n_k\eta_{(ki)j}) D v_j dS + \oint_C [m_i(\Delta_{ij} - n_k\eta_{(ki)j}) v_j] ds = 0 \quad (2.71)$$

If the velocity field is arbitrary, Eq. (2.71) yields the conditions

$$\bar{t}_j - n_i(\sigma_{ij} - \rho\bar{\psi}_{ij}) - D_i(\bar{\Delta}_{ij} - n_k\eta_{(ki)j}) + (D_l n_l) n_i(\bar{\Delta}_{ij} - n_k\eta_{(ki)j}) = 0 \quad (2.72)$$

$$n_i(\bar{\Delta}_{ij} - n_k\eta_{(ki)j}) = 0 \quad (2.73)$$

$$[m_i(\bar{\Delta}_{ij} - n_k\eta_{(ki)j})] = 0 \quad (2.74)$$

and from Eq. (2.73), Eq. (2.72) can be replaced by

$$\bar{t}_j - n_i(\sigma_{ij} - \rho\bar{\psi}_{ij}) - D_i(\bar{\Delta}_{ij} - n_k\eta_{(ki)j}) = 0 \quad (2.75)$$

and Eqs. (2.73), (2.75), and (2.74) are identical with Eqs. (2.61), (2.63), and (2.64), respectively, in which the $\eta_{[ki]j}$ do not appear. Although the $\eta_{[ki]j}$ do not appear in the foregoing energetic formalism, they can be introduced and serve to modify the formalism slightly. To see this note that

$$\int_V [(\eta_{[ki]j} v_{j,i})_{,k} - (\eta_{[ki]j,k} v_j)_{,i}] dV = 0 \quad (2.76)$$

and add Eq. (2.76) to the left-hand side of Eq. (2.45). Under these circumstances, in place of Eq. (2.69) we obtain

$$\int_S [t_j - n_i(\sigma_{ij} - \rho\psi_{ij} - \eta_{[ki]j,k})]v_j dS + \int_S [\Delta_{ij} - n_k\eta_{kij}]v_{j,i} dS = 0 \quad (2.77)$$

Then, while performing the operations associated with the equivalent of Eq. (2.70), we note that

$$\begin{aligned} D_i(n_k\eta_{[ki]j}) &= (D_i n_k)\eta_{[ki]j} + n_k D_i \eta_{[ki]j} \\ &= n_k\eta_{[ki]j,i} - n_i n_k D\eta_{[ki]j} \\ &= -n_i\eta_{[ki]j,k} \end{aligned}$$

Hence, in place of Eq. (2.71) we obtain

$$\int_S [\bar{t}_j - n_i(\sigma_{ij} - \rho\bar{\psi}_{ij}) - D_i(\bar{\Delta}_{ij} - n_k\eta_{(ki)j}) + (D_l n_l)n_i(\bar{\Delta}_{ij} - n_k\eta_{(ki)j})]v_j dS$$

$$+ \int_S n_i[\bar{\Delta}_{ij} - n_k\eta_{(ki)j}]Dv_j dS + \oint_C [m_i(\bar{\Delta}_{ij} - n_k\eta_{kij})v_j]ds = 0 \quad (2.78)$$

Now, if the velocity field is arbitrary, we obtain Eqs. (2.72), (2.73), and

$$[m_i(\bar{\Delta}_{ij} - n_k\eta_{kij})] = 0 \quad (2.79)$$

in place of Eq. (2.74); and, as before, from Eq. (2.73), Eq. (2.72) can be replaced by Eq. (2.75). However, as noted earlier, the $\eta_{[ki]j}$ can be eliminated from Eq. (2.79) because $[m_i n_k \eta_{[ki]j}] = 0$, thereby reducing Eq. (2.79) to Eq. (2.74), and the six $n_k\eta_{[ki]j}$ can be determined *a posteriori* from Eq. (2.56) when the $\bar{\Delta}_{ij}$ are prescribed. In a position type of boundary-value problem, i.e., when y_j and Dy_j are prescribed on S, the equations indicate that after a solution is obtained, we may determine t_j and $n_i\Delta_{ij}$ at the surface S from either Eqs. (2.61) and (2.63) or Eqs. (2.73) and (2.75), and a force per unit length along C from Eq. (2.74). Clearly, then, in a complete position type of boundary-value problem the $n_k\eta_{[ki]j}$ at the boundary remain undetermined, while in a traction type of boundary-value problem the $n_k\eta_{[ki]j}$ may be determined from *known* boundary values.

The previous considerations indicate how Toupin's variational principle for the strain-gradient theory may be extended in order to account for the entire double-stress theory. From the foregoing it seems clear that Toupin's variational principle may be extended by writing

$$\delta \int_V \rho \mathcal{E} dV + \int_S [n_k\eta_{[ki]j}(\delta y_j)_{,i} - n_i\eta_{[ki]j,k}\delta y_j]dS$$

$$= \int_V \rho[\bar{f}_j\delta y_j + \bar{\psi}_{ij}(\delta y_j)_{,i}]dV + \int_S [\bar{t}_j\delta y_j + \bar{\Delta}_{ij}(\delta y_j)_{,i}]dS, \quad (2.80)$$

where $\mathcal{E}$ is given in Eq. (2.35), the surface integral on the left-hand side is identically zero and the barred variables represent the physically defined quantities described earlier, and generalized forces have not been employed. When the variations are carried out for arbitrary δy_j in V on C and on S and $D\delta y_j$ on S, the theory will consist of the differential Eqs. (2.10), the constitutive Eqs. (2.52), (2.41), and (2.42), the relations $(2.29)_1$ and (2.36), the boundary conditions (2.73) and (2.75), and the condition (2.79) at an edge. If the variation in the stored energy integral is carried out and the volume integral in Eq. (2.80) is written as a coefficient of δy_j, which must vanish, and, following Green and Rivlin [13], we require invariance under *homogeneous* translations, we obtain Eq. (2.55). Then, if we assume that $\bar{\Delta}_{ij}$ and η_{kij} are independent of $(\delta y_j)_{,i}$, we obtain Eq. (2.56).

3 Linear Couple-Stress Theory

The equations of couple-stress elasticity are linearized in essentially the same way as those of ordinary elasticity. Specifically, the *small* displacement field $\mathbf{u}$ is introduced by means of the equation

$$y_i = \delta_{iK}(X_K + u_K) = x_i + u_i \tag{3.1}$$

in which on the far right we have eliminated the requirement that capital letters and indices refer to material coordinates and use lowercase letters and indices exclusively. We then substitute from Eq. (3.1) into the pertinent equations in Section 2 and neglect products of first and second derivatives of $\mathbf{u}$ wherever they occur in the differential equations and boundary conditions. When we do this we find

$$\begin{aligned} \frac{\partial}{\partial y_k} &\approx \frac{\partial}{\partial x_k}, \quad \rho \approx \rho_0, \quad v_i = \frac{\partial u_i}{\partial t} \\ E_{KL} &\approx \tfrac{1}{2}(u_{k,l} + u_{l,k}) = \epsilon_{kl} \\ K_{ML} &\approx e_{nkl}\epsilon_{mk,n} = \tfrac{1}{2}e_{nkl}u_{k,nm} = \kappa_{ml} \end{aligned} \tag{3.2}$$

where ρ_0 is the reference mass density, which is constant in a homogeneous material. We now expand $\mathcal{E}$ in Eq. (2.28) in a power series in E_{LM} and K_{LM}, thus

$$\rho_0\mathcal{E} = \tfrac{1}{2}c_{KLMN}E_{KL}E_{MN} + b_{KLMN}E_{KL}K_{MN} + \tfrac{1}{2}a_{KLMN}K_{KL}K_{MN} + \text{h.o.t.} \tag{3.3}$$

where we have omitted terms linear in $\mathbf{E}$ and $\mathbf{K}$ because we assume that the undeformed state is stress-free. When we introduce our linearizing assumptions on $\mathbf{u}$ and substitute from Eq. (3.2), we obtain

$$\rho_0\mathcal{E} \equiv U = \tfrac{1}{2}c_{klmn}\epsilon_{kl}\epsilon_{mn} + b_{klmn}\epsilon_{kl}\kappa_{mn} + \tfrac{1}{2}a_{klmn}\kappa_{kl}\kappa_{mn} \tag{3.4}$$

which is consistent with our linearizing assumptions because the constitutive

Eqs. (2.33), when linearized, take the form

$$\tau_{ij}{}^{S} = \frac{\partial U}{\partial \epsilon_{ij}}, \qquad \mu_{ij}{}^{D} = \frac{\partial U}{\partial \kappa_{ij}} \tag{3.5}$$

which, with Eq. (3.4), yields

$$\begin{aligned} \tau_{ij}{}^{S} &= c_{ijkl}\epsilon_{kl} + b_{ijkl}\kappa_{kl} \\ \mu_{ij}{}^{D} &= b_{klij}\epsilon_{kl} + a_{ijkl}\kappa_{kl} \end{aligned} \tag{3.6}$$

which are the linearized form of Toupin's constitutive equations and are identical with those of Aero and Kuvshinskii [33]. For a more complete discussion of the nature of the linearization see Section 3 of Ref. [M76].

We now note that ϵ_{ij} is the usual small strain tensor and, from Eq. $(3.2)_5$, that κ_{ij} is the gradient of the small rotation vector w_i since

$$w_i = \tfrac{1}{2}e_{ijk}u_{k,j} \tag{3.7}$$

which is the axial vector of the small rotation tensor w_{ij}, where

$$w_{ij} = \tfrac{1}{2}(u_{j,i} - u_{i,j}) \tag{3.8}$$

An examination [M76] of Eq. (3.4), recalling all the relations among the ϵ_{ij} and κ_{ij} and the conventions for the construction of U, reveals that for arbitrary anisotropy there are thirty-six independent components of **a**, forty-eight independent components of **b**, and the usual twenty-one independent components of **c**. Since $\boldsymbol{\kappa}$ is an axial tensor, **b** must be an axial tensor; and, hence, $\mathbf{b} = 0$ for centrosymmetric, isotropic materials. Moreover, since $\boldsymbol{\varepsilon}$ is symmetric and $\kappa_{kk} = 0$, the internal energy density U for a centrosymmetric, isotropic material may be written in the form

$$U = \tfrac{1}{2}\lambda\epsilon_{kk}\epsilon_{ll} + \mu\epsilon_{kl}\epsilon_{kl} + 2\eta\kappa_{kl}\kappa_{kl} + 2\eta'\kappa_{kl}\kappa_{lk} \tag{3.9}$$

From Eqs. (3.5) and (3.9) we obtain the constitutive equations of linear, isotropic, couple-stress theory in the form

$$\begin{aligned} \tau_{ij}{}^{S} &= \lambda\epsilon_{kk}\delta_{ij} + 2\mu\epsilon_{ij} \\ \mu_{ij}{}^{D} &= 4\eta\kappa_{ij} + 4\eta'\kappa_{ji} \end{aligned} \tag{3.10}$$

Substituting for ϵ_{kl} and κ_{kl} from Eq. (3.2) in Eq. (3.10) and writing the result in invariant dyadic notation, we obtain

$$\begin{aligned} \boldsymbol{\tau}^{S} &= \lambda\boldsymbol{\nabla}\cdot\mathbf{u}\mathbf{I} + \mu(\boldsymbol{\nabla}\mathbf{u} + \mathbf{u}\boldsymbol{\nabla}) \\ \boldsymbol{\mu}^{D} &= 2\eta\boldsymbol{\nabla}\boldsymbol{\nabla}\times\mathbf{u} + 2\eta'\boldsymbol{\nabla}\times\mathbf{u}\boldsymbol{\nabla} \end{aligned} \tag{3.11}$$

where **I** is the idemfactor, which is the dyadic representation of the Kronecker delta.

When the linearizing assumptions on $\mathbf{u}$ are introduced in Eq. (2.19) and we substitute from Eq. (3.2), we obtain the linear stress equations of motion:

$$\nabla \cdot \boldsymbol{\tau}^S + \tfrac{1}{2}\nabla \times \nabla \cdot \boldsymbol{\mu}^D + \mathbf{f} + \tfrac{1}{2}\nabla \times \mathbf{C} = \rho \partial^2 \mathbf{u}/\partial t^2 \tag{3.12}$$

where

$$\mathbf{f} = \rho \mathbf{g}, \qquad \mathbf{C} = \rho \mathbf{l} \tag{3.13}$$

and in the linear theory we have

$$\nabla \approx \mathbf{e}_i \partial/\partial x_i \tag{3.14}$$

Substituting from Eq. (3.11) into Eq. (3.12), we obtain the displacement equations of motion, including the influence of couple-stresses,

$$\mu \nabla^2 \mathbf{u} + (\lambda + \mu)\nabla\nabla \cdot \mathbf{u} + \eta \nabla^2 \nabla \times \nabla \times \mathbf{u} + \mathbf{f} + \tfrac{1}{2}\nabla \times \mathbf{C} = \rho \ddot{\mathbf{u}} \tag{3.15}$$

where we have introduced the conventional dot notation for partial differentiation with respect to time. Note that the isotropic couple-stress constant η' does not appear in the displacement equations of motion (3.15). However, η' does appear in the boundary conditions. The condition of positive definite energy density U in Eq. (3.9) results in the conditions

$$\mu > 0, \quad 3\lambda + 2\mu > 0, \quad \eta > 0, \quad -1 < \frac{\eta'}{\eta} < 1 \tag{3.16}$$

which determine the permissible range of values of the isotropic elastic constants when couple-stresses are present. Since $\mu > 0$ and $\eta > 0$, a material constant may be defined by

$$l = \frac{\sqrt{\eta}}{\sqrt{\mu}} \tag{3.17}$$

where l has the dimension of length, and if $l = 0$ the couple-stresses vanish. At this point it should be noted, as in Section 3 of Ref. [M76], that since the ordinary theory of elasticity has been verified experimentally in great detail, l must be small in comparison with dimensions and wavelengths usually encountered. In the linear theory Eq. (2.12) for $\tau_{kl}{}^A$ takes the form

$$\tau_{kl}{}^A = \tfrac{1}{2} e_{jlk} \mu_{ij,i}{}^D + \tfrac{1}{2} e_{jlk} \mu_{,j} + \tfrac{1}{2} e_{jlk} C_j \tag{3.18}$$

and since the indeterminacy in μ, discussed in Section 2 for the nonlinear theory, exists in the linear theory also, the associated indeterminacy in $\tau_{kl}{}^A$ exists in the linear theory by virtue of Eq. (3.18).

Conditions sufficient for a unique solution of the equations of linear, isotropic, couple-stress elasticity are established in Section 5 of Ref. [M76] in the usual [34] Neumann manner, based on the positive definiteness of U and

$\dot{\mathbf{u}}\cdot\dot{\mathbf{u}}$. The conditions are

1. In V: at each point, $\mathbf{f}$, $\nabla \times \mathbf{C}$ and initial values of $\mathbf{u}$ and $\dot{\mathbf{u}}$.
2. On S: at each point, one member of each of the five products

$$p_\alpha u_\alpha, \quad p_\beta u_\beta, \quad \tau_{\gamma\gamma} u_\gamma, \quad \mu_{\gamma\alpha} w_\alpha, \quad \mu_{\gamma\beta} w_\beta \tag{3.19}$$

where repeated Greek indices are not to be summed and α, β, γ denote orthogonal curvilinear coordinates with γ increasing outward, normal to S, and

$$\mathbf{p} = \mathbf{n}\cdot\boldsymbol{\tau}^S + \tfrac{1}{2}\mathbf{n} \times [\nabla\cdot\boldsymbol{\mu}^D - \nabla(\mathbf{n}\cdot\boldsymbol{\mu}^D\cdot\mathbf{n})] \tag{3.20}$$

so that

$$p_\alpha = \tau_{\gamma\alpha}{}^S - \tfrac{1}{2}\mathbf{e}_\beta\cdot(\nabla\cdot\boldsymbol{\mu}^D - \nabla\mu_{\gamma\gamma}{}^D),$$
$$p_\beta = \tau_{\gamma\beta}{}^S + \tfrac{1}{2}\mathbf{e}_\alpha\cdot(\nabla\cdot\boldsymbol{\mu}^D - \nabla\mu_{\gamma\gamma}{}^D) \tag{3.21}$$

where $\mathbf{e}_\alpha$, $\mathbf{e}_\beta$, $\mathbf{e}_\gamma$ denote orthogonal unit vectors at a point in the direction of the orthogonal curvilinear coordinates. From Eqs. (3.20) and (2.25), we have

$$\mathbf{p} = \bar{\mathbf{t}} - \tfrac{1}{2}\mathbf{n} \times \bar{\mathbf{C}} - \tfrac{1}{2}\mathbf{n} \times \nabla(\mathbf{n}\cdot\bar{\mathbf{m}}) \tag{3.22}$$

where, for traction boundary conditions, $\bar{\mathbf{t}}$, $\bar{\mathbf{C}}$, and $\bar{\mathbf{m}}$ are prescribed.

3. On an edge formed by the intersection of two smooth surfaces: at each point of the edge either

$$[\mu_{\gamma\gamma}{}^D] \qquad \text{or } \mathbf{s}\cdot\mathbf{u} \tag{3.23}$$

where $\mathbf{s}$ is a unit vector in the direction of the edge and the jump in $\mu_{\gamma\gamma}{}^D$ across the edge represents a force per unit length along the edge. From Eqs. (2.24) and (3.23), we have

$$[\mu_{\gamma\gamma}{}^D] = [\mathbf{n}]\cdot\bar{\mathbf{m}} \tag{3.24}$$

which is identical with Eq. (2.27). For traction boundary conditions, Eq. (3.24) must be satisfied. However, for displacement boundary conditions, $[\mu_{\gamma\gamma}{}^D]$ determines *a posteriori* a force per unit length that must be exerted in the direction of the edge.

The above conditions, which are sufficient for uniqueness of solution of the linear couple-stress equations, are in complete conformity with the boundary conditions of the nonlinear couple-stress theory discussed in Section 2. At this point it is perhaps worth noting that since the differential Eqs. (3.15) and traction boundary conditions (3.20) of couple-stress theory reduce to those of classical, infinitesimal, isotropic elasticity as $l \to 0$ and, as already observed, l must be very small for most materials, the effects of couple-stresses must be very small for applied tractions. However, since in the couple-stress theory the conditions on w_α and w_β do not reduce to those of classical elasticity as $l \to 0$, the most severe effects probably exist in the immediate vicinity of surfaces that are essentially rigidly held, i.e., where the boundary conditions are geometric. This should occur in the softer material at interfaces where a soft material abuts a hard material.

The linear displacement equations of motion for isotropic media with couple-stresses, Eq. (3.15), indicate that wave motion differs in some respects from that of the classical theory. To see some of the similarities and differences note that the divergence and curl of Eq. (3.15) yield, respectively,

$$C_1^2\nabla^2\Delta = \ddot{\Delta} \tag{3.25}$$

$$C_2^2(1 - l^2\nabla^2)\nabla^2\mathbf{w} = \ddot{\mathbf{w}} \tag{3.26}$$

where $\mathbf{w}$ is defined in Eq. (3.7), l in Eq. (3.17) and

$$C_1^2 = \frac{(\lambda + 2\mu)}{\rho}, \quad C_2^2 = \frac{\mu}{\rho}, \quad \Delta = \nabla\cdot\mathbf{u} \tag{3.27}$$

Equation (3.25) indicates that the dilatation Δ is propagated nondispersively, with velocity C_1 as in ordinary linear, isotropic elasticity. However, Eq. (3.26) indicates that the rotation $\mathbf{w}$ is propagated quite differently than in ordinary elasticity. Equation (3.26) indicates that the propagation of a plane wave satisfies

$$C^2 = C_2^2(1 + l^2\xi^2), \qquad \omega^2 = \xi^2C_2^2(1 + l^2\xi^2) \tag{3.28}$$

where C is the phase velocity, ξ the propagation wave number, and ω the circular frequency of the plane wave. Since l is positive, Eq. $(3.28)_2$ indicates that for a given ω, there are two ξ, one real and one pure imaginary, which does not propagate. From Eq. (3.28) the propagating wave is dispersive, the phase velocity C, and group velocity $d\omega/d\xi$ both increasing with increasing ξ. The imaginary, or nonpropagating, ξ has a wave number cut-off at zero frequency and $\xi = \pm i/l$. This branch is a major source of difference between both dynamic and static elastic fields with and without couple-stresses. It causes local effects near boundaries and singularities.

It has been shown that the complete solution of the dynamic, isotropic, displacement equations of motion (3.15) can be represented in the form [M76]

$$\mathbf{u} = \nabla\varphi + \nabla \times \mathbf{H}, \qquad \nabla\cdot\mathbf{H} = 0 \tag{3.29}$$

where φ and $\mathbf{H}$ satisfy

$$C_1^2\nabla^2\varphi = \ddot{\varphi}, \qquad C_2^2(1 - l^2\nabla^2)\nabla^2\mathbf{H} = \ddot{\mathbf{H}} \tag{3.30}$$

When these equations along with the appropriate boundary conditions are applied to the thickness-shear vibration of a plate of thickness $2h$, the transcendental frequency equation [M76]

$$\tan\gamma = f^3\tanh f\gamma \tag{3.31}$$

is obtained where

$$\gamma = \xi_1 h, \quad f = \frac{(1+g^2)^{1/2}+1}{g}, \quad g = \frac{\pi\Omega l}{h} \tag{3.32}$$

and

$$\xi_1 = 2^{-1/2} l^{-1} \left[\left(1 + \frac{4l^2\omega^2}{C_2{}^2} \right)^{1/2} - 1 \right]^{1/2}$$

$$\Omega = \frac{2h\omega}{\pi C_2} = 2\gamma\pi^{-1}\left(1 + \frac{\gamma^2 l^2}{h^2}\right)^{1/2} \tag{3.33}$$

In the absence of couple-stress ($l = 0$), Eq. (3.31) reduces to

$$\cot \gamma = 0 \tag{3.34}$$

Thus it is clear that the existence of couple-stresses serves to change the transcendental equation for the resonant frequencies of thickness-shear vibration of an isotropic plate rather severely. However, it turns out [M76] that the resonant frequency of the lowest mode changes only slightly for l large; but the resonant frequencies of all the higher modes change drastically. A similar analysis of the torsional vibrations of an isotropic, circular cylinder [M76] reveals that the resonant frequencies exhibit the same type of behavior, but in this instance the resonant frequencies of all elastic modes change drastically for l large.

Since, in addition to the operator ∇^2, the operators $(1 - l^2\nabla^2)$ and $(1 - l^2\nabla^2)\nabla^2$ arise in static problems in linear, isotropic couple-stress elasticity, it is valuable to have Green's formulas for the operators

$$(1 - l^2\nabla^2)\psi, \qquad (1 - l^2\nabla^2)\nabla^2\psi \tag{3.35}$$

where ψ is a scalar function. The result for the operator in Eq. $(3.35)_1$ takes the form

$$4\pi\psi = \int_S \mathbf{n}\cdot(\psi\nabla\varphi_1 - \varphi_1\nabla\psi)\,dS - l^{-2}\int_V \varphi_1(1 - l^2\nabla^2)\psi\,dV \tag{3.36}$$

where

$$\varphi_1 = -r_1{}^{-1}e^{-r_1/l} \tag{3.37}$$

and

$$r_1 = \sqrt{(x-\xi)^2 + (y-\eta)^2 + (z-\zeta)^2} \tag{3.38}$$

is the distance from the field point $P(x, y, z)$ to the source point $Q(\xi, \eta, \zeta)$ and the integrals in Eq. (3.36) are over Q. The equivalent result for the operator in

Eq. $(3.35)_2$ takes the form

$$4\pi\psi = \int_S \mathbf{n}\cdot[r_1^{-1}\nabla\psi - \psi\nabla r_1^{-1} + l^2\nabla^2\psi\nabla\varphi_2 - l^2\varphi_2\nabla\nabla^2\psi]dS - \int_V \varphi_2(1 - l^2\nabla^2)\nabla^2\psi dV \quad (3.39)$$

where

$$\varphi_2 = r_1^{-1}(1 - e^{-r_1/l}) \quad (3.40)$$

It has been shown [M76] that the complete solution of the displacement equations of equilibrium, i.e., Eq. (3.15) with $\ddot{\mathbf{u}} = 0$, may be represented in the form

$$\mathbf{u} = \mathbf{B} - l^2\nabla\nabla\cdot\mathbf{B} - \alpha\nabla[\mathbf{r}\cdot(1 - l^2\nabla^2)\mathbf{B} + \beta] \quad (3.41)$$

where $\mathbf{B}$ and β satisfy

$$\mu(1 - l^2\nabla^2)\nabla^2\mathbf{B} = -\bar{\mathbf{f}} - \tfrac{1}{2}\nabla \times \bar{\mathbf{C}} \quad (3.42)$$

$$\mu\nabla^2\beta = \mathbf{r}\cdot(\bar{\mathbf{f}} + \tfrac{1}{2}\nabla \times \bar{\mathbf{C}}) \quad (3.43)$$

and α is given by

$$\alpha = \frac{\lambda + \mu}{2(\lambda + 2\mu)} \quad (3.44)$$

With the aid of this formulation, the solutions for the concentrated force and couple in the infinite medium were constructed [M76]. For the concentrated force, the result is

$$\mathbf{B} = \frac{\mathbf{P}}{4\pi\mu r}(1 - e^{-r/l}), \qquad \beta = 0 \quad (3.45)$$

where $\mathbf{P}$ is the concentrated force located at the origin and $r = \sqrt{x^2 + y^2 + z^2}$; and for the concentrated couple, the result is

$$\mathbf{B} = -\frac{1}{8\pi\mu}\mathbf{C} \times \nabla\left(\frac{1}{r} - \frac{e^{-r/l}}{r}\right), \qquad \beta = 0 \quad (3.46)$$

where $\mathbf{C}$ is the concentrated couple located at the origin.

In Ref. [M76] the formulation (3.41)–(3.44) was used to solve problems of stress concentration for a spherical and a cylindrical cavity in a field of simple tension in the presence of couple-stresses.

Mindlin [M77] has formulated the equations of linear, isotropic, static couple-stress elasticity for the case of plane strain in the absence of body forces and couples in terms of two scalar functions. If x_3 is normal to the plane, the force-stresses and couple-stresses can be represented in the form

$$\tau_{11} = \eta_{,22} - \sigma_{,12}, \qquad \tau_{22} = \eta_{,11} + \sigma_{,12}$$

$$\tau_{12} = -\eta_{,12} + \sigma_{,22}, \qquad \tau_{21} = -\eta_{,12} + \sigma_{,11}$$

$$\mu_{13} \equiv \mu_1 = \sigma_{,1}, \qquad \mu_{23} \equiv \mu_2 = \sigma_{,2} \tag{3.47}$$

where η and σ, respectively, satisfy

$$\nabla_1^4\eta = 0, \qquad \nabla_1^2(1 - l^2\nabla_1^2)\sigma = 0 \tag{3.48}$$

and ∇_1^2 is the planar Laplacian. For this plane-strain description, it is convenient to write the constitutive equations in the form

$$\epsilon_{11} = \frac{1+\nu}{E}[\tau_{11} - \nu(\tau_{11} + \tau_{22})]$$

$$\epsilon_{22} = \frac{1+\nu}{E}[\tau_{22} - \nu(\tau_{22} + \tau_{11})]$$

$$\epsilon_{12} = \frac{\tau_{12} + \tau_{21}}{\mu}$$

$$\kappa_{13} \equiv \kappa_1 = \frac{\mu_1}{4\eta}, \qquad \kappa_{23} \equiv \kappa_2 = \frac{\mu_2}{4\eta} \tag{3.49}$$

where Young's modulus E and Poisson's ratio ν are given by

$$E = \frac{\mu(3\lambda + 2\mu)}{\lambda + 2\mu}, \qquad \nu = \frac{\lambda}{\lambda + 2\mu} \tag{3.50}$$

and

$$\kappa_1 = w_{3,1}, \qquad \kappa_2 = w_{3,2} \tag{3.51}$$

$$\tau_{21} - \tau_{12} = \mu_{1,1} + \mu_{2,2} \tag{3.52}$$

and w_3 and ϵ_{11}, ϵ_{22}, ϵ_{12} are defined in Eqs. (3.7) and $(3.2)_4$, respectively; and, of course, $u_3 = 0$, $u_1 = u_1(x_1, x_2)$, $u_2 = u_2(x_1, x_2)$. In plane-strain traction boundary conditions take the form

$$n_1\tau_{11} + n_2\tau_{21} = \bar{t}_1, \qquad n_1\tau_{12} + n_2\tau_{22} = \bar{t}_2$$

$$n_1\mu_1 + n_2\mu_2 = \bar{m} \tag{3.53}$$

where $\bar{t}_1$, $\bar{t}_2$, and $\bar{m}$ are the prescribed force-tractions and couple-traction, respectively, on the boundary S and (n_1, n_2) denote the components of the outward unit normal to S; and displacement boundary conditions are on u_1, u_2, and w_3.

This plane-strain description has been written in the Muskhelishvili complex variable form by Mindlin [M78] and Huilgol [35]. The formulation starts as in ordinary isotropic plane elasticity [36], and we have

$$\eta = \mathrm{Re}\,[\bar{z}\varphi(z) + \chi(z)] \tag{3.54}$$

$$\tau_{11} + \tau_{22} = 4\,\mathrm{Re}\,\Phi(z) \tag{3.55}$$

where $z = x_1 + ix_2$; and when couple-stresses are included, we have the extension [M78], [35]

$$\tau_{21} - \tau_{12} = 4\,\frac{\partial^2\sigma}{\partial z\partial\bar{z}}$$

$$\tau_{22} - \tau_{11} + i(\tau_{21} + \tau_{12}) = 2\left[\bar{z}\Phi'(z) + \Psi(z) + 2i\,\frac{\partial^2\sigma}{\partial z^2}\right]$$

$$\mu_1 - i\mu_2 = 2\,\frac{\partial\sigma}{\partial z} \tag{3.56}$$

where

$$\Phi(z) = \varphi'(z), \qquad \Psi(z) = \psi'(z) = \chi''(z) \tag{3.57}$$

and $\varphi(z)$ and $\chi(z)$ are the usual holomorphic functions of ordinary plane elasticity [36]. In addition it has been found that

$$\tau_{11} - i\tau_{12} = 2\,\mathrm{Re}\,\Phi(z) - [\bar{z}\Phi'(z) + \Psi(z)] + 2i\left[\frac{\partial^2\sigma}{\partial z\partial\bar{z}} - \frac{\partial^2\sigma}{\partial z^2}\right]$$

$$\sigma - l^2\nabla^2\sigma = 8(1-\nu)l^2\,\mathrm{Im}\,\Phi(z) \tag{3.58}$$

In plane strain the displacement can be written in the form

$$2\mu(u_1 + iu_2) = (3-4\nu)\varphi(z) - z\overline{\Phi(z)} - \overline{\psi(z)} - 2\,\frac{\partial\sigma}{\partial\bar{z}} \tag{3.59}$$

Huilgol [35] obtained the expression for the resultant force and moment on an arc AB of a plane curve due to the presence of force-stress and couple-stress in the form

$$T_1 + iT_2 = -i[\varphi(z) + z\overline{\varphi'(z)} + \overline{\psi(z)}]_A^B - 2\left[\frac{\partial\sigma}{\partial\bar{z}}\right]_A^B$$

$$M = \mathrm{Re}\,[\chi(z) - z\psi(z) - z\bar{z}\varphi'(z)]_A^B + i\left[z\,\frac{\partial\sigma}{\partial z} - \bar{z}\,\frac{\partial\sigma}{\partial\bar{z}}\right]_A^B \tag{3.60}$$

where T_1 and T_2 are the x_1 and x_2 components, respectively, of the resultant force and M is the resultant plane moment. Huilgol [35] uses these results to obtain the solution for a concentrated force applied at a point in the infinite plane. The complex potentials take the form

$$\Phi(z) = -\frac{T_1 + iT_2}{2(1+\kappa)}\frac{1}{z}$$

$$\Psi(z) = \frac{\kappa(T_1 - iT_2)}{2\pi(1+\kappa)}\frac{1}{z} + 4(1-\nu)l^2\frac{T_1 + iT_2}{\pi(1+\kappa)}\frac{1}{z^3} \tag{3.61}$$

and σ takes the form

$$\sigma = \frac{b_1}{z} + \frac{\bar{b}_1}{\bar{z}} + K_1(c_1 \sin\theta + d_1 \cos\theta) \tag{3.62}$$

where

$$b_1 = -4i(1-\nu)l^2 a_1, \quad c_1 - id_1 = 8(1-\nu)la_1, \quad \theta = \tan^{-1}\frac{x_2}{x_1}$$

$$a_1 = -\frac{T_1 + iT_2}{2\pi(1+\kappa)}, \qquad \kappa = 3 - 4\nu \tag{3.63}$$

and K_1 is the modified Bessel function of the second kind and first order.

As noted in the Introduction, many problems in linear, isotropic couple-stress theory were solved by a large number of workers [18]. However, Sternberg [15–17], with coworkers Muki and Bogy, solved a number of problems concerning the influence of couple-stresses on stress concentrations and stress singularities under unusually interesting circumstances. Since the geometry in the problems and the associated mathematical analysis are somewhat involved, any meaningful analytical discussion of the problems is not possible in this short survey. Nevertheless, in our opinion some of their results are so interesting and important for an understanding of the influence of couple-stresses in linear elastic solids that a discussion of their work is mandatory in such a survey as this. A most interesting result obtained by Bogy and Sternberg is that the presence of couple-stresses removes the corner singularity, which exists [37] in the absence of couple-stresses, for the quarter plane subject to a uniform shear load on one edge while the other edge remains free. This occurs because the stress, which must be symmetric in the classical theory, can be asymmetric when couple-stresses are present, and the particular distribution of traction applied results in an asymmetric stress condition at the corner element. Since the couple-stress length parameter l is small, the corner asymmetry in stress disappears a short distance from the corner. Another most interesting result obtained by Muki and Sternberg [17] is that although the existence of couple-stresses tends to reduce stress concentration factors and singularities in certain instances, it tends to increase them in others. This means that a finite l, no matter how small, can influence certain quantities adversely. Since a small l undoubtedly always exists, this can be very important physically. This effect probably arises primarily because the couple-stress equations are higher order than the classical equations and, as a consequence, require more boundary conditions, and not all the boundary conditions reduce to those of the classical theory as l approaches zero. Specifically, the conditions on $\mathbf{w}$ do not reduce to the classical conditions as $l \rightarrow 0$ and can give rise to a boundary layer effect unknown in the classical theory.

4 Further Generalized Continua: Linear Theory

Mindlin [M85] has formulated a linear theory of deformation of an elastic solid in equilibrium, in which the stored energy density is a function of the linear strain and the first and second gradients of the linear strain. Mindlin obtains the description from a variational principle of the form

$$\delta \int_V U dV = \int_V f_k \delta u_k dV + \int_S (\overset{1}{t}_k \delta u_k + \overset{2}{t}_k D \delta u_k + \overset{3}{t}_k D^2 \delta u_k) dS \qquad (4.1)$$

where u_k is the small mechanical displacement, D is the normal derivative defined in Section 2, f_k is the body force, the t_k are generalized surface tractions analogous to Toupin's F_j^2 and F_j^3 and U is assumed to be of the form

$$U = U(\overset{1}{\epsilon}_{ij}, \overset{2}{\epsilon}_{ijk}, \overset{3}{\epsilon}_{ijkl}) \qquad (4.2)$$

where

$$\overset{1}{\epsilon}_{ij} = \tfrac{1}{2}(u_{i,j} + u_{j,i}), \qquad \overset{2}{\epsilon}_{ijk} = u_{k,ij}, \qquad \overset{3}{\epsilon}_{ijkl} = u_{l,ijk} \qquad (4.3)$$

and $\overset{2}{\epsilon}_{kij}$ can be replaced by the strain-gradient $\overset{1}{\epsilon}_{ij,k}$ and $\overset{3}{\epsilon}_{klij}$ can be replaced by the second gradient of strain $\overset{1}{\epsilon}_{ij,kl}$ and S is a *smooth* surface enclosing a volume V. Although Mindlin [M85] initially treats smooth surfaces and we consider smooth surfaces only, Mindlin [M85] considers edges and corners in an appendix. The result of taking the variations in Eq. (4.1), operating in the usual manner, introducing Eq. (2.43) and applying the divergence theorem and the surface divergence theorem is

$$
\begin{aligned}
&\int_V [-(\overset{1}{\tau}_{ij} - \overset{2}{\tau}_{kij,k} + \overset{3}{\tau}_{klij,kl})_{,i} - f_j]\delta u_j dV \\
&\quad + \int_S [n_i(\overset{1}{\tau}_{ij} - \overset{2}{\tau}_{kij,k} + \overset{3}{\tau}_{klij,kl}) - \overset{1}{t}_j + L_i n_k(\overset{2}{\tau}_{kij} - \overset{3}{\tau}_{lkij,l}) \\
&\quad + L_i L_k n_l \overset{3}{\tau}_{lkij} - L_i(D_i n_m) n_l n_k \overset{3}{\tau}_{lkmj}]\delta u_j dS \\
&\quad + \int_S [n_k n_i(\overset{2}{\tau}_{kij} - \overset{3}{\tau}_{lkij,l}) + n_i L_k n_l \overset{3}{\tau}_{lkij} + L_i n_l n_k \overset{3}{\tau}_{lkij} - \overset{2}{t}_j] D\delta u_j dS \\
&\quad + \int_S [n_k n_l n_i \overset{3}{\tau}_{klij} - \overset{3}{t}_j] D^2 \delta u_j dS = 0
\end{aligned}
\tag{4.4}
$$

where

$$
\overset{1}{\tau}_{ij} = \partial U/\partial \overset{1}{\epsilon}_{ij}, \quad \overset{2}{\tau}_{kij} = \partial U/\partial \overset{2}{\epsilon}_{kij}, \quad \overset{3}{\tau}_{klij} = \partial U/\partial \overset{3}{\epsilon}_{klij}
\tag{4.5}
$$

$$
L_j = n_j(D_k n_k) - D_j
\tag{4.6}
$$

Since the variations δu_j in V and δu_j, $D\delta u_j$ and $D^2\delta u_j$ on S are arbitrary, there follow the stress equations of equilibrium

$$
(\overset{1}{\tau}_{ij} - \overset{2}{\tau}_{kij,k} + \overset{3}{\tau}_{klij,kl})_{,i} + f_j = 0
\tag{4.7}
$$

and the traction boundary conditions

$$
\begin{aligned}
n_i(\overset{1}{\tau}_{ij} - \overset{2}{\tau}_{kij,k} + \overset{3}{\tau}_{klij,kl}) + L_i[n_k(\overset{2}{\tau}_{kij} - \overset{3}{\tau}_{lkij,l}) + L_k n_l \overset{3}{\tau}_{lkij} \\
- (D_i n_m) n_l n_k \overset{3}{\tau}_{lkmj}] = \overset{1}{t}_j
\end{aligned}
\tag{4.8}
$$

$$
n_k n_i(\overset{2}{\tau}_{kij} - \overset{3}{\tau}_{lkij,l}) + n_i L_k n_l \overset{3}{\tau}_{lkij} + L_i n_l n_k \overset{3}{\tau}_{lkij} = \overset{2}{t}_j
\tag{4.9}
$$

$$
n_k n_l n_i \overset{3}{\tau}_{klij} = \overset{3}{t}_j
\tag{4.10}
$$

Thus, as observed by Mindlin [M85], in this theory there are three scalar equations of equilibrium and nine boundary conditions. Also, as noted by Mindlin, other admissible sets of nine boundary conditions, in terms of the geometric quantities $\mathbf{u}$, $D\mathbf{u}$, and $D^2\mathbf{u}$ or products of appropriate components of $\overset{1}{\mathbf{t}}$ and $\mathbf{u}$ and/or $\overset{2}{\mathbf{t}}$ and $D\mathbf{u}$ and/or $\overset{3}{\mathbf{t}}$ and $D^2\mathbf{u}$, are apparent from the form of Eq. (4.4). However, it should be noted that the foregoing discussion of traction boundary conditions holds for a *smooth* surface only. If the surface has edges and corners, additional traction type jump conditions across edges and corners are required in accordance with Mindlin's treatment in the appendix of Ref. [M85]. At this point it should be mentioned that the specification of prescribed traction conditions in terms of physically defined quantities is unclear to us from the foregoing variational considerations; and consideration of the full triple-stress (tripolar) theory along the lines outlined by Green and Rivlin [13], analogous to the discussion of the double-stress theory presented in Section 2, are required.

Mindlin [M85] specializes the treatment to homogeneous, centrosymmetric, isotropic materials, and, hence, takes U in the form

$$\begin{aligned} U = {} & \tfrac{1}{2}\lambda \overset{1}{\epsilon}_{ii}\overset{1}{\epsilon}_{jj} + \mu \overset{1}{\epsilon}_{ij}\overset{1}{\epsilon}_{ji} + a_1 \overset{2}{\epsilon}_{ijj}\overset{2}{\epsilon}_{ikk} + a_2 \overset{2}{\epsilon}_{iik}\overset{2}{\epsilon}_{kjj} \\ & + a_3 \overset{2}{\epsilon}_{iik}\overset{2}{\epsilon}_{jjk} + a_4 \overset{2}{\epsilon}_{ijk}\overset{2}{\epsilon}_{ijk} + a_5 \overset{2}{\epsilon}_{ijk}\overset{2}{\epsilon}_{kji} \\ & + b_1 \overset{3}{\epsilon}_{iijj}\overset{3}{\epsilon}_{kkll} + b_2 \overset{3}{\epsilon}_{ijkk}\overset{3}{\epsilon}_{ijll} + b_3 \overset{3}{\epsilon}_{iijk}\overset{3}{\epsilon}_{jkll} \\ & + b_4 \overset{3}{\epsilon}_{iijk}\overset{3}{\epsilon}_{llkj} + b_5 \overset{3}{\epsilon}_{iijk}\overset{3}{\epsilon}_{lljk} + b_6 \overset{3}{\epsilon}_{ijkl}\overset{3}{\epsilon}_{ijkl} \\ & + b_7 \overset{3}{\epsilon}_{ijkl}\overset{3}{\epsilon}_{jkli} + c_1 \overset{1}{\epsilon}_{ii}\overset{3}{\epsilon}_{jjkk} + c_2 \overset{1}{\epsilon}_{ij}\overset{3}{\epsilon}_{ijkk} \\ & + c_3 \overset{1}{\epsilon}_{ij}\overset{3}{\epsilon}_{kkij} + b_0 \overset{3}{\epsilon}_{iijj} \end{aligned} \tag{4.11}$$

The constants λ and μ are the usual Lamé constants and the five a_n are the additional constants of the linear version of Toupin's strain-gradient theory for isotropic centrosymmetric materials. The additional eleven constants are all in terms containing $\overset{3}{\boldsymbol{\varepsilon}}$ and, consequently, are nonexistent in the lower-order theories. In Eq. (4.11) all terms higher than quadratic in derivatives of $\mathbf{u}$ have been assumed to be negligible because only the linear theory is of interest. Indeed, all terms save the last are quadratic in derivatives of $\mathbf{u}$. The last term, which is linear in derivatives of $\mathbf{u}$, has been retained because the unstrained state has not been assumed to be triple-stress $\overset{3}{\boldsymbol{\tau}}$, free, although it has been assumed to be stress, $\overset{1}{\boldsymbol{\tau}}$, and double-stress, $\overset{2}{\boldsymbol{\tau}}$, free. For a more complete discussion of the order of magnitude of the different quantities in this linear approximation see Ref. [M85]. From Eqs. (4.5) and (4.11) follow the constitutive equations.

$$
\begin{aligned}
\overset{1}{\tau}_{pq} &= \lambda \overset{1}{\epsilon}_{ii}\delta_{pq} + 2\mu \overset{1}{\epsilon}_{pq} + c_1 \overset{3}{\epsilon}_{iijj}\delta_{pq} + c_2 \overset{3}{\epsilon}_{pqii} + \tfrac{1}{2}c_3(\overset{3}{\epsilon}_{iipq} + \overset{3}{\epsilon}_{iiqp}) \\
\overset{2}{\tau}_{pqr} &= a_1(\overset{2}{\epsilon}_{pii}\delta_{qr} + \overset{2}{\epsilon}_{qii}\delta_{pr}) + \tfrac{1}{2}a_2(\overset{2}{\epsilon}_{iip}\delta_{qr} + 2\overset{2}{\epsilon}_{rii}\delta_{pq} + \overset{2}{\epsilon}_{iiq}\delta_{pr}) \\
&\quad + 2a_3 \overset{2}{\epsilon}_{iir}\delta_{pq} + 2a_4 \overset{2}{\epsilon}_{pqr} + a_5(\overset{2}{\epsilon}_{rqp} + \overset{2}{\epsilon}_{rpq}) \\
\overset{3}{\tau}_{pqrs} &= \tfrac{2}{3}b_1 \overset{3}{\epsilon}_{iijj}\delta_{pqrs} + \tfrac{2}{3}b_2 \overset{3}{\epsilon}_{jkii}\delta_{jkpqrs} \\
&\quad + \tfrac{1}{6}b_3[(\overset{3}{\epsilon}_{iijk} + \overset{3}{\epsilon}_{iikj})\delta_{jkpqrs} + 2\overset{3}{\epsilon}_{jsii}\delta_{jpqr}] \\
&\quad + \tfrac{2}{3}b_4 \overset{3}{\epsilon}_{iisj}\delta_{jpqr} + \tfrac{2}{3}b_5 \overset{3}{\epsilon}_{iijs}\delta_{jpqr} + 2b_6 \overset{3}{\epsilon}_{pqrs} \\
&\quad + \tfrac{2}{3}b_7(\overset{3}{\epsilon}_{qrsp} + \overset{3}{\epsilon}_{rspq} + \overset{3}{\epsilon}_{spqr}) + \tfrac{1}{3}c_1 \overset{1}{\epsilon}_{ii}\delta_{pqrs} \\
&\quad + \tfrac{1}{3}c_2 \overset{1}{\epsilon}_{ij}\delta_{ijpqrs} + \tfrac{1}{3}c_3 \overset{1}{\epsilon}_{is}\delta_{ipqr} + \tfrac{1}{3}b_0\delta_{pqrs}
\end{aligned}
\tag{4.12}
$$

where

$$
\begin{aligned}
\delta_{ijkl} &= \delta_{ij}\delta_{kl} + \delta_{ik}\delta_{jl} + \delta_{il}\delta_{jk} \\
\delta_{ijklmn} &= \delta_{ik}\delta_{jl}\delta_{mn} + \delta_{ik}\delta_{jm}\delta_{ln} + \delta_{il}\delta_{jm}\delta_{kn}
\end{aligned}
$$

Mindlin [M85] has shown that the term in the energy density which is linear in derivatives of **u** and produces the homogeneous self-equilibrating components of triple force $(b_0/3)\delta_{pqrs}$ can account for surface tension in an isotropic elastic solid without postulating the existence of a vanishingly thin surface membrane. Mindlin [M85] obtains a result that motivates him to take

$$T = \tfrac{1}{2}b_0 n_l u_{k,kl} \tag{4.13}$$

as the definition of the punctual surface tension, or surface energy per unit area, in a centrosymmetric, isotropic elastic solid. Mindlin [M85] solves the problem of a free plane surface of a semi-infinite elastic solid and obtains an expression for the surface tension along with the accompanying rapidly decaying displacement field. Mindlin also considers the problem within the framework of lattice theory and compares the results with those of the continuum approach. He finds no additional information regarding the character of the decay of the displacement field from the lattice approach.

Mindlin [M85] obtains a complete representation of the isotropic displacement equations of equilibrium in terms of stress functions analogous to those discussed in Section 3 for the couple-stress case. He uses the results to obtain the solution for a concentrated force in an infinite elastic solid according to the second gradient of strain theory.

Mindlin [M85] also considers an elastic liquid as a special case of an elastic solid in which the stored energy U is a function of the dilatation $\Delta = \nabla \cdot \mathbf{u}$ and its first and second gradients only. Within the framework of this description, Mindlin solves the problems of a plane surface and a spherical surface of an elastic liquid. In both cases he solves for the surface tension T and the accompanying rapidly decaying dilatation field.

Before his work on the second gradient of strain, Mindlin [M81] obtained a generalization of the linear theory of elasticity by defining an elastic continuum with microstructure. The basic idea consists of defining *internal* micro-coordinates x_i' in addition to the usual macro-coordinates x_i and expanding the total displacement field u_j^t of a (micro) point as a power series in the micro-coordinates and retaining the first two terms only to obtain

$$u_j^t(x_i', x_i, t) = u_j(x_i, t) + x_k'\psi_{kj}(x_i, t) \tag{4.14}$$

where x_k' is measured from the mass center of the micro-volume V', u_j is the ordinary macro-displacement and

$$u_j' = x_k'\psi_{kj}(x_i, t) \tag{4.15}$$

can be interpreted as the micro-displacement. The micro- as well as the macro-displacement gradients are assumed to be infinitesimal; and, hence, products of derivatives of u_j and/or ψ_{kj} are negligible. The symmetric part of the micro-displacement gradient $\partial u_j'/\partial x_i' \equiv \psi_{ij}$ is

$$\psi_{(ij)} = \tfrac{1}{2}(\psi_{ij} + \psi_{ji}) \tag{4.16}$$

which is interpreted as the micro-strain; and the antisymmetric part of ψ_{ij} is

$$\psi_{[ij]} = \tfrac{1}{2}(\psi_{ij} - \psi_{ji}) \tag{4.17}$$

which is interpreted as the micro-rotation. The usual (now the macro) strain is defined by

$$\epsilon_{ij} = \tfrac{1}{2}(u_{j,i} + u_{i,j}) \tag{4.18}$$

and the relative deformation, i.e., the difference between the macro- and micro-displacement gradients is defined by

$$\gamma_{ij} = u_{j,i} - \psi_{ij} \tag{4.19}$$

and the macro-gradient of the micro-displacement gradient

$$\kappa_{ijk} = \psi_{jk,i} \tag{4.20}$$

is called the micro-deformation gradient. Since u_j and ψ_{ij} are assumed to be single-valued functions of the x_i, there result the compatibility equations

$$\begin{gathered} e_{mik}e_{nlj}\epsilon_{kl,ij} = 0 \\ e_{mij}\kappa_{jkl,i} = 0 \\ (\epsilon_{jk} + \omega_{jk} - \gamma_{jk})_{,i} = \kappa_{ijk} \end{gathered} \tag{4.21}$$

where the macro-rotation

$$\omega_{ij} = \tfrac{1}{2}(u_{j,i} - u_{i,j}) \tag{4.22}$$

The kinetic-energy density, i.e., the kinetic energy per unit of macro-volume, may be defined by

$$T = \frac{1}{2V'}\int_{V'} \rho' \dot{u}_j{}^t \dot{u}_j{}^t dV' \tag{4.23}$$

where ρ' is the micro-mass density and V' is the volume of the appropriately defined micro-medium. Substituting from Eq. (4.14) into Eq. (4.23), we obtain

$$T = \tfrac{1}{2}\rho \dot{u}_j \dot{u}_j + \tfrac{1}{2} I_{kl} \dot{\psi}_{kj} \dot{\psi}_{lj} \tag{4.24}$$

where we have employed the definitions

$$\rho = \frac{1}{V'}\int_{V'} \rho' dV', \qquad I_{kl} = \frac{1}{V'}\int_{V'} \rho' x_k' x_l' dV' \tag{4.25}$$

and used the condition

$$\int_{V'} \rho' x_k' dV' = 0 \tag{4.26}$$

which holds because the x_k' are measured from the center of mass of the micro-volume. Mindlin [M81] assumes that the stored energy function U is of the form

$$U = U(\epsilon_{ij}, \gamma_{ij}, \kappa_{ijk}) \tag{4.27}$$

and takes Hamilton's principle in the form

$$\delta \int_{t_0}^{t_1} (\mathfrak{T} - \mathfrak{U})dt + \int_{t_0}^{t_1} \delta W dt = 0 \tag{4.28}$$

for independent variations δu_j and $\delta \psi_{jk}$ which vanish at t and t_0, and where

$$\mathfrak{T} = \int_V T dV, \qquad \mathfrak{U} = \int_V U dV \tag{4.29}$$

and δW is the virtual work done by the generalized external forces, and is assumed to be of the form

$$\delta W = \int_V [f_j \delta u_j + \Phi_{jk} \delta \psi_{jk}] dV + \int_S [t_j \delta u_j + T_{jk} \delta \psi_{jk}] dS \tag{4.30}$$

The result of substituting from Eqs. (4.24), (4.27), (4.29), and (4.30) into Eq. (4.28), taking the variations in the usual manner, integrating by parts with respect to time, applying the divergence theorem and omitting the time integral is

$$\int_V [(\tau_{ij,i} + \sigma_{ij,i} + f_j - \rho \ddot{u}_j)\delta u_j + (\mu_{ijk,i} + \sigma_{jk} + \Phi_{jk} - I_{jl}\ddot{\psi}_{lk})\delta\psi_{jk}]dV$$

$$+ \int_S [(t_j - n_i\tau_{ij} - n_i\sigma_{ij})\delta u_j + (T_{jk} - n_i\mu_{ijk})\delta\psi_{jk}]dS = 0 \quad (4.31)$$

where

$$\tau_{ij} = \frac{\partial U}{\partial \epsilon_{ij}}, \quad \sigma_{ij} = \frac{\partial U}{\partial \gamma_{ij}}, \quad \mu_{ijk} = \frac{\partial U}{\partial \kappa_{ijk}} \quad (4.32)$$

Since the variations δu_j and $\delta\psi_{jk}$ are arbitrary, there result the twelve stress equations of motion

$$\tau_{ij,i} + \sigma_{ij,i} + f_j = \rho \ddot{u}_j$$
$$\mu_{ijk,i} + \sigma_{jk} + \Phi_{jk} = I_{lj}\ddot{\psi}_{lk} \quad (4.33)$$

and the twelve traction boundary conditions

$$n_i(\tau_{ij} + \sigma_{ij}) = t_j, \qquad n_i\mu_{ijk} = T_{jk} \quad (4.34)$$

On account of linearity the stored energy function U must be a homogeneous quadratic function of its variables, and may be written in the form

$$U = \tfrac{1}{2}c_{ijkl}\epsilon_{ij}\epsilon_{kl} + \tfrac{1}{2}b_{ijkl}\gamma_{ij}\gamma_{kl} + \tfrac{1}{2}a_{ijklmn}\kappa_{ijk}\kappa_{lmn}$$
$$+ d_{ijklm}\gamma_{ij}\kappa_{klm} + f_{ijklm}\kappa_{ijk}\epsilon_{lm} + g_{ijkl}\gamma_{ij}\epsilon_{kl} \quad (4.35)$$

Mindlin [M81] notes that only 903 of the 1764 coefficients in Eq. (4.35) are independent. From Eqs. (4.32) and (4.35), the constitutive equations take the form

$$\tau_{pq} = c_{pqij}\epsilon_{ij} + g_{ijpq}\gamma_{ij} + f_{ijkpq}\kappa_{ijk}$$
$$\sigma_{pq} = g_{pqij}\epsilon_{ij} + b_{ijpq}\gamma_{ij} + d_{pqijk}\kappa_{ijk}$$
$$\mu_{pqr} = f_{pqrij}\epsilon_{ij} + d_{ijpqr}\gamma_{ij} + a_{pqrijk}\kappa_{ijk} \quad (4.36)$$

For a centrosymmetric, isotropic material, the constitutive equations take the reduced form

$$\tau_{pq} = \lambda\delta_{pq}\epsilon_{ii} + 2\mu\epsilon_{pq} + g_1\delta_{pq}\gamma_{ii} + g_2(\gamma_{pq} + \gamma_{qp})$$
$$\sigma_{pq} = g_1\delta_{pq}\epsilon_{ii} + 2g_2\epsilon_{pq} + b_1\delta_{pq}\gamma_{ii} + b_2\gamma_{pq} + b_3\gamma_{qp}$$
$$\mu_{pqr} = a_1(\kappa_{iip}\delta_{qr} + \kappa_{rii}\delta_{pq}) + a_2(\kappa_{iiq}\delta_{pr} + \kappa_{iri}\delta_{pq})$$
$$+ a_3\kappa_{iir}\delta_{pq} + a_4\kappa_{pii}\delta_{qr} + a_5(\kappa_{qii}\delta_{pr} + \kappa_{ipi}\delta_{qr})$$
$$+ a_8\kappa_{iqi}\delta_{pr} + a_{10}\kappa_{pqr} + a_{11}(\kappa_{rpq} + \kappa_{qrp})$$
$$+ a_{13}\kappa_{prq} + a_{14}\kappa_{qpr} + a_{15}\kappa_{rqp} \quad (4.37)$$

which contain a total of eighteen independent coefficients. The substitution of Eq. (4.37) into Eq. (4.33) yields the twelve displacement equations of motion in the twelve dependent variables u_i and ψ_{ij}.

Mindlin [M81] has obtained plane-wave solutions of the displacement equations of motion and shown that they account for optical as well as acoustic modes. In Ref. [M81] Mindlin obtains some low-frequency, very long wavelength approximations of the linear microstructure equations, and shows the relation of the resulting equations to the linear version of Toupin's strain-gradient theory and couple-stress theory. Mindlin [M81] obtains a complete representation of the isotropic displacement equations of equilibrium of the strain-gradient theory in terms of stress functions analogous to those discussed in Section 3 for couple-stress theory, and uses the results to obtain the solution for a concentrated force in an infinite solid. Mindlin [M82] obtains a potential representation of the linear, isotropic, dynamic equations of elastic materials with microstructure analogous to the wave equation representation of Lamé in classical elasticity, but he does not establish the completeness of the representation. Furthermore, Mindlin [M84] notes that the linear equations of equilibrium for a Cosserat continuum may be obtained from the foregoing more general linear equations for an elastic continuum with microstructure if ψ_{ij} is replaced by $\psi_{[ij]}$, i.e., if the symmetric part of ψ_{ij} is set equal to zero. He further shows that setting $\gamma_{[ij]} = 0$ reduces the Cosserat equations to the couple-stress equations. Mindlin [M84] obtains a complete representation of the linear, isotropic, displacement equations of equilibrium for a Cosserat continuum in terms of stress functions analogous to those discussed in Section 3 for the couple-stress equations, and uses the results to obtain solutions for the concentrated force and couple in the infinite solid.

Acknowledgments—The work of one of the authors (H. F. T.) was supported in part by the Office of Naval Research under Contract No. N00014-67-A-0117-0007 and the National Science Foundation under Grant No. GK-11195.

References

1. R. A. Toupin, The elastic dielectric, *J. Rat. Mech. Anal.* **5**, 849 (1956). Sec. 4.
2. C. Truesdell and R. A. Toupin, The classical field theories, in *Encyclopedia of Physics*, edited by S. Flügge, Springer-Verlag, Berlin, 1960, Vol. III/1. Eq. $(303.5)_2$.
3. Reference [2], Eqs. (205.10) and (241.4) with $m^{pqr} = 0$, $h^p = 0$, and $q = 0$. Since an applied point couple is exactly balanced by the antisymmetric part of the stress tensor in such a way that there is no power term associated with the spin tensor, there can be no rate of storage of elastic energy due to the spin or antisymmetric part of the stress tensor.
4. Reference [1], Secs. 5 and 10, Ref. [3].
5. A. C. Eringen, *Nonlinear Theory of Continuous Media*, McGraw-Hill, New York, 1962. Art. 44.3.

6. A. E. H. Love, *The Mathematical Theory of Elasticity*, Cambridge University Press, 1927. (Also Dover Publications, Inc., New York, 1944.) 4th edition, Chap. III.
7. K. S. Viswanathan, The theory of elasticity and of wave propagation in crystals from the atomistic standpoint, *Proc. Indian Acad. Sci.* **A39,** 196 (1954).
K. S. Viswanathan, The theory of elasticity of crystals, *Proc. Indian Acad. Sci.* **A41,** 98 (1955).
C. V. Raman and K. S. Viswanathan, On the theory of the elasticity of crystals, *Proc. Indian Acad. Sci.* **A42,** 1 (1955).
C. V. Raman and D. Krishnamurti, Evaluation of the four elastic constants of some cubic crystals, *Proc. Indian Acad. Sci.* **A42,** 111 (1955).
8. J. Laval, Élasticite des Cristaux, *Compt. Rend.* **232,** 1947 (1951).
Y. LeCorre, Détermination des sept constants elastique dynamics du phosphate monoammonique, *Compt. Rend.* **236,** 1903 (1953).
J. Laval, Théorie atomique de 1 elasticité cristalline excluant les forces centrales, *Compt. Rend.* **238,** 1773 (1954).
9. W. Voigt, Theoretische Studien über die Elasticitätsverhaltnisse der Krystalle, *Abb. Ges. Wiss. Gottingen* **34** (1887).
W. Voigt, Über Medien ohne innere Kräfte und eine durch sie gelieferte mechanische Deutung der Maxwell-Hertzschen Gleichungen, *Gött Abh.* **1894,** 72 (1895).
Reference [2], Sec. 200. Reference [6], Sec. 42.
10. Reference [2], referred to as the polar case.
11. E. et F. Cosserat, Théorie des corps déformables, A. Hermann et Cie, Paris, 1909.
12. R. A. Toupin, Elastic materials with couple-stresses, *Arch. Rat. Mech. Anal.* **11,** 385 (1962).
13. A. E. Green and R. S. Rivlin, Simple force and stress multipoles, *Arch. Rat. Mech. Anal.* **16,** 325 (1964).
14. A. E. Green and P. M. Naghdi, A note on simple dipolar stresses, *J. de Mech.* **7,** 465 (1968).
15. R. Muki and E. Sternberg, The influence of couple-stresses on singular stress concentrations in elastic solids, *Z. angew. Math. Phys.* **16,** 611 (1965).
D. B. Bogy and E. Sternberg, The effect of couple-stresses on singularities due to discontinuous loadings, *Int. J. Solids Struct.* **3,** 757 (1967).
16. D. B. Bogy and E. Sternberg, The effect of couple-stresses on the corner singularity due to an asymmetric shear loading, *Int. J. Solids Struct.* **4,** 159 (1968).
17. E. Sternberg and R. Muki, The effect of couple-stresses on the stress concentration around a crack, *Int. J. Solids Struct.* **3,** 69 (1967).
18. See, e.g., some papers in *Proc. Fifth U.S. National Congress of Appl. Mech.*, 1966. Also see the *J. appl. Mech.* in the years 1964–1967 and *Appl. Math. Mech.* (*Prikl. Math. Mech.*) in the years 1964–1968. The authors and publications are too numerous to detail in this brief coverage.
19. R. A. Toupin, Theories of elasticity with couple-stress, *Arch. Rat. Mech. Anal.* **17,** 85 (1964).
20. J. L. Ericksen and C. Truesdell, Exact theory of stress and strain in rods and shells, *Arch. Rat. Mech. Anal.* **1,** 295 (1958).
21. A. C. Eringen and E. S. Suhubi, Nonlinear theory of simple microelastic solids I, *Intl. J. Engng. Sci.* **2,** 189 (1964).

A. C. Eringen and E. S. Suhubi, Nonlinear theory of microelastic solids II, *Intl. J. Engr. Sci.* **2,** 389 (1964).
22. A. E. Green and R. S. Rivlin, Multipolar continuum mechanics, *Arch. Rat. Mech. Anal.* **17,** 113 (1964).
23. A. E. Green, P. M. Naghdi, and R. S. Rivlin, Directors and multipolar displacements in continuum mechanics, *Int. J. Engng. Sci.* **2,** 611 (1965).
24. A. E. Green, Micro-materials and multipolar continuum mechanics, *Int. J. Engng. Sci.* **3,** 533 (1965).
25. J. L. Bleustein, Effects of micro-structure on the stress concentration at a spherical cavity, *Int. J. Solids Struct.* **2,** 83 (1966).
26. G. Herrmann, Some applications of micromechanics, *Experimental Mechanics* **12,** 235 (1972).
27. G. Herrmann and J. D. Achenbach, On dynamic theories of fiber reinforced composites, *Proc. AISS/ASME, Eighth Structures, Structural Dynamics and Materials Conference,* New York, AIAA, 112 (1967).
C. T. Sun, J. D. Achenbach and G. Herrmann, Time harmonic waves in a stratified medium propagating in the direction of the layering, *J. appl. Mech.* **35,** 408 (1968), and many subsequent publications.
28. Reference [2], Secs. 13, 15, and 16.
29. H. Jeffreys, *Cartesian Tensors,* Cambridge University Press, New York, 1931 (reprinted 1952), p. 12.
30. A. E. Green and R. S. Rivlin, On Cauchy's equations of motion, *Z. angew. Math. Phys.* **15,** 290 (1964).
31. A. L. Cauchy, Mémoire sur les systémes isotropes de points matériels, *Mem. Acad. Sci.* **XXII,** 615 (1850) = Oevres (1) 2, 351.
32. L. Brand, *Vector and Tensor Analysis,* Wiley, New York, 1947, p. 222.
33. E. L. Aero and E. V. Kuvshinskii, Fundamental equations of the theory of elastic media with rotationally interacting particles, *Fizika, Tverdogo Tela* **2,** 1399 (1960); *translation: Soviet Physics Solid State* **2,** 1272 (1961).
34. Reference [6], Sec. 124.
35. R. R. Huilgol, On the concentrated force problem for two-dimensional elasticity with couple-stresses, *Int. J. Engng. Sci.* **5,** 81 (1967).
36. N. I. Muskhelishvili, *Some Basic Problems of the Mathematical Theory of Elasticity,* 4th edition, translated by J. R. M. Radok, Noordhoff, 1963.
37. E. Reisner, Note on the theorem of the symmetry of the stress tensor, *J. Math. and Phys.* **23,** 192 (1944).

Bodies in Contact with Applications to Granular Media†

H. DERESIEWICZ

Department of Mechanical Engineering, Columbia University, New York, N.Y.

Introduction

During the 1940s, R. L. Wegel at the Bell Telephone Laboratories in Murray Hill was studying the propagation of stress waves in granular aggregates. His interest in the subject was motivated by the pressing demand in the telephone industry for a rational method of designing the carbon microphone, to the present day the most widely used type. The essential component of this device is an assemblage of packed, electrically conducting carbon granules in which variation of acoustic pressure, caused by means of a diaphragm, creates changes in contact area between the individual grains. These changes, in turn, alter the electrical resistance of the system and hence provide variations in current flow.

Such analyses as existed were all based on the assumption that relative displacement of the grains occurred in the direction normal to the contact. Wegel was able to demonstrate by means of experiments on pairs of rubber balls that the tangential compliance of the contact was of the same order of magnitude as its normal compliance and argued that it must therefore be taken into account in any analysis of the aggregate.

Sometime in 1946 or 1947 R. D. Mindlin, then serving as a consultant to the Laboratories, visited Wegel, was shown the results of the experiments and was urged to look into the question of predicting the tangential compliance mathematically. The fruit of Mindlin's initial investigation in this field [M29]‡ was the germinal paper in a long series on frictional contact, by himself and his students, and led to a number of subsequent studies on the statical and dynamical response of granular arrays.

The aim of this article is twofold: to present the results of recent work in contact theory and to discuss the application of these results in the construction of a theory of mechanics of granular materials. Accordingly, the subject

† Support by the National Science Foundation under Grant GK-16361 is gratefully acknowledged.

‡ Numbers preceded by the letter M in brackets refer to the Publications of R. D. Mindlin listed after the Preface to this volume.

matter separates naturally into two parts and is therefore dealt with in that manner.

Thus Part I is devoted to various generalizations of the classical Hertz theory of elastic contact, with emphasis on contact due to force systems other than normal. Here frictional effects make their appearance and give rise to load-displacement (or torque-twist) relations which are not only nonlinear but inelastic as well. It follows that the deformation at a contact depends, in general, on the past history of loading. Specifically, a discussion is given of contact of spherical as well as nonspherical solids due to tangential forces, oblique forces and twisting moments, in loading and in unloading. It will be recalled that in Hertz's case assumptions made in the course of the analysis had to be verified by subsequent experiments before the validity of his theory could be accepted. Similarly, assumptions are made in the cases of generalizations involving frictional phenomena, and, accordingly, experiments delving into the validity of the ensuing results are discussed here. Other generalizations of the Hertz theory are touched upon, particularly higher-order contact and contact of anelastic bodies, but space limitations preclude discussion of the related problems of rolling contact.

Part II gives an account of attempts to construct a mechanics of granular aggregates based on considerations of local deformation at each of the contacts between constituent granules. It turns out that, even in the case of models consisting of simple, regular arrays of spherical grains, external loads, transmitted through the medium via the several contacts, in general give rise at these contacts to forces which are oblique to the initial tangent plane between the grains and also to torsional couples. Thus the apparatus of the extended theory of contact, discussed in Part I, must be brought into play in the construction of any viable theory. We get here the first hint of the kind of difficulties involved, for such a theory will contain the nonlinearities and dissipation inherent in the behavior of a single contact. It is because of these difficulties that the theory is at a stage at which it must be viewed as far more tentative than definitive. Yet, despite this, a number of results of interest have emerged from the analysis, and these are presented herein together with results of experiments, both static and dynamic.

The list of references with which the article concludes is intended to be representative, not exhaustive. It should be noted that a number of papers have appeared in the past decade in the Russian literature (in particular in *Prikladnaia Matematika i Mekhanika*) dealing primarily with various punch problems. More often than not, however, the investigation does not proceed much further than the formulation of integral equations and an indication of how an approximate solution might be extracted. Rarely are numerical results obtained, and not once has this writer come across a report of experimental work.

Part I Contact Theory

A. *Normal Forces*

We begin with a brief review of some of the results of the classical theory. This theory, due to Hertz [1], deals with a pair of homogenous, isotropic, elastic solids in contact due to forces normal to their initial common tangent plane. Several assumptions are made in the course of the analysis, and it is useful to enumerate them here since it is the removal of one or more of these assumptions which constitutes the various generalizations which we will discuss. Thus: (1) the warping of the surface of contact due to unequal geometric and elastic properties of the two bodies is neglected and the contact surface is taken to be plane. (2) The linear dimensions of this plane region of contact are assumed small compared with the principal radii of curvature of either body at the initial point of contact. To a first approximation, therefore, the points on either body which come in contact as a result of compression, as well as neighboring points, are taken to lie, before compression, on a surface of the second degree, the coordinates of each of the two surfaces being referred to an origin located at the point of initial contact. It should be noted that singular points of the surfaces at the point of initial contact are thereby excluded. As a consequence of assumptions (1) and (2) the problem is considered as one of the half-space. (3) Unless both bodies have the same material constants or both are incompressible, points on opposite sides of the interface equidistant from the line normal to the initial tangent plane will displace laterally relative to one another. This interfacial slip, which gives rise to frictional forces tending to oppose it, is supposed in the Hertz theory to occur unimpeded, i.e., the tangential components of traction at the contact are neglected.

We consider a pair of bodies of general shape, having elastic constants E_1, ν_1 and E_2, ν_2, respectively, compressed by a force N normal to their common initial tangent plane. The compression gives rise to a distributed pressure, σ. Then, within the framework of the assumptions listed above, Hertz's contact problem may be shown to reduce to the solution of the integral equation

$$(\vartheta_1 + \vartheta_2) \iint \sigma(\xi, \eta) r_1^{-1} d\xi d\eta = \alpha - z(x, y) \tag{1}$$

where z, the "distance function," is quadratic, i.e., $z(x, y) = Ax^2 + By^2$. Further,

$$r_1 = [(x - \xi)^2 + (y - \eta)^2 + z^2]^{1/2}, \qquad \vartheta_i = \frac{1 - \nu_i^2}{\pi E_i}$$

and A, B are known constants descriptive of the geometry of the two surfaces

[1, 2]. The solution of Eq. (1) yields a normal pressure of magnitude

$$\sigma = \frac{3N}{2\pi ab}\left[1 - \left(\frac{x}{a}\right)^2 - \left(\frac{y}{b}\right)^2\right]^{1/2} \tag{2}$$

distributed over a plane contact bounded by an ellipse of semi-axes a and b. These semi-axes, given by the expressions

$$\frac{B}{A} = \frac{\mathbf{E} - (1 - e^2)\mathbf{K}}{(1 - e^2)(\mathbf{K} - \mathbf{E})}$$

$$2e^2 A a^3 = 3N(\vartheta_1 + \vartheta_2)(\mathbf{K} - \mathbf{E}) \tag{3}$$

where $e = (1 - b^2/a^2)^{1/2}$ denotes the eccentricity of the ellipse and $\mathbf{K}(e)$, $\mathbf{E}(e)$ are complete elliptic integrals of the first and second kind, respectively, can, in general, be evaluated only numerically. Once a and b are known, the relative approach of the bodies, α, (parallel to the load N) is obtained from the expression

$$2a\alpha = 3N(\vartheta_1 + \vartheta_2)\mathbf{K}(e) \tag{4}$$

The values of these geometric quantities may be expressed by the formulas

$$a = \mu\left[\frac{N(k_1 + k_2)}{A + B}\right]^{1/3}$$

$$b = \nu\left[\frac{N(k_1 + k_2)}{A + B}\right]^{1/3}$$

$$\alpha = \lambda[N^2(k_1 + k_2)^2(A + B)]^{1/3} \tag{5}$$

Here $k_i = 3\pi\vartheta_i/4$, $A + B$ is one-half the sum of the principal curvatures of both bodies, and the value of the coefficients μ, ν, λ is given [3] in Fig. 1 as function of $\tau = \arccos[(B - A)/(B + A)]$. It is seen that the dimensions of the ellipse are proportional to $N^{1/3}$ while the approach varies as $N^{2/3}$, the latter relation being known as Hertz's contact law. Finally, the normal compliance of the contact is [M29]

$$C = \frac{d\alpha}{dN} = \frac{(\vartheta_1 + \vartheta_2)\mathbf{K}(e)}{a} \tag{6}$$

The simplest special case is that of like spheres, each of radius R. In Eq. (1) $\vartheta_1 = \vartheta_2 = \vartheta$ and $z = z(\rho) = \rho^2/R$, where ρ denotes the radial distance from the center of contact. The solution may be obtained either from the axisymmetric integral equation, or extracted from the general solution by not-

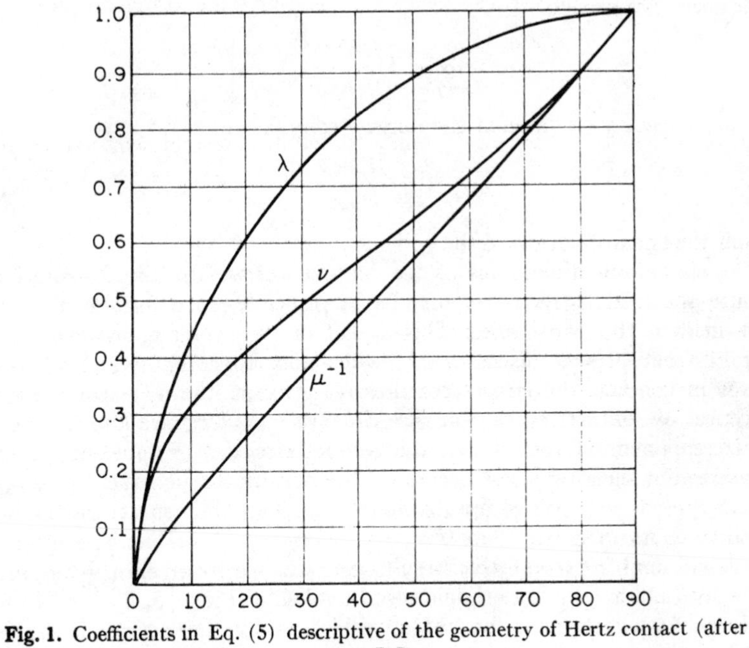

Fig. 1. Coefficients in Eq. (5) descriptive of the geometry of Hertz contact (after Ref. [3]).

ing that, for $e \ll 1$, Eqs. (3) and (4) become

$$\frac{B}{A} \doteq 1 + \frac{3e^2}{4} + \frac{21e^4}{32} + \frac{309e^6}{512}$$

$$4Aa^3 \doteq 3\pi\vartheta N\left(1 + \frac{3e^2}{8} + \frac{15e^4}{64} + \frac{175e^6}{1024}\right)$$

$$\alpha \doteq \left[\frac{A(3\pi\vartheta N)^2}{2}\right]^{1/3}\left(1 + \frac{e^2}{8} + \frac{e^4}{16} + \frac{127e^6}{3072}\right)$$

This time the theory predicts a circular contact region of radius

$$a = (kRN)^{1/3} \tag{7}$$

a pressure distribution

$$\sigma = \frac{3N}{2\pi a^3}(a^2 - \rho^2)^{1/2} \tag{8}$$

and a normal approach

$$\alpha = 2\left(\frac{k^2N^2}{R}\right)^{1/3} \tag{9}$$

The latter leads to a value of the normal compliance

$$C = \frac{1-\nu}{2\mu a} \tag{10}$$

μ being the shear modulus of the spheres.

The shape and dimensions of the contact known and the distribution of pressure specified there, it is possible, in principle, to determine the entire stress field in the two bodies. This aspect of the problem, however, is not central to our present discussion. It will suffice to note that, for a pair of spheres in contact, the largest compressive stress is the axial component at the center of contact, $\sigma_{zz} \equiv -\sigma_0 = -3N/2\pi a^2$; the largest tensile stress, the radial component at the edge of the contact circle, $\sigma_{rr} = (1-2\nu)\sigma_0/3$; and the maximum shearing stress occurs on the normal through the center of the contact area at a depth of approximately one-half the contact radius below the surface, its magnitude, for $\nu = 0.3$, being $\tau_{\max} \doteq 0.31\sigma_0$. The interested reader will find representative results and further references given in the review by Lubkin [4]; note should also be taken of Refs. [5, 6].

Numerous experiments, beginning with those by Hertz himself, have confirmed the essential results of his theory, specifically, the contact law, the shape and size of the contact region and aspects of the resulting stress field. Specific references may once again be found in Ref. [4].

We now come to the first of several generalizations of the Hertz theory, namely the removal, or rather weakening, of the assumption that the surfaces in the vicinity of contact are adequately represented by quadric surfaces. Such a generalization is desirable, for example, in treating the case of contact of a sphere in a spherical seat. The contact may now be expected to occur over a region whose dimension is no longer very small compared with the radius of curvature at the contact.

This problem, for the axisymmetric case, was investigated by Cattaneo [7] who solved the integral equation given in Eq. (1) but with $z = A_0\rho^2 + B_0\rho^4$. His analysis leads to a circular contact area over which the pressure distribution is of the form

$$\sigma = \frac{4}{\pi^2(\vartheta_1+\vartheta_2)}\left[\left(A_0+\frac{8B_0a^2}{3}\right)(a^2-\rho^2)^{1/2} - \frac{16B_0}{9}(a^2-\rho^2)^{3/2}\right] \tag{11}$$

and to a normal approach

$$\alpha = 2A_0a^2 + \frac{8B_0a^4}{3} \tag{12}$$

the radius of the contact being given by the (positive) root of

$$64B_0a^5 + 40A_0a^3 - 15\pi(\vartheta_1 + \vartheta_2)N = 0 \tag{13}$$

It should be noted that $A_0 > 0$ and $A_0 + 6B_0\rho^2 > 0$; for a pair of spheres, of radii R_1 and R_2, respectively, $A_0 = (R_1^{-1} + R_2^{-1})/2$, $B_0 = (R_1^{-3} + R_2^{-3})/8$.

The root of Eq. (12) can, evidently, be evaluated only numerically for a given configuration and contact force. However, an approximate, closed-form expression can be found, valid for problems for which the Hertz theory yields insufficiently accurate results but for the solution of which small corrections to the Hertz predictions are acceptable. Proceeding in this vein Deresiewicz [8] showed that, providing $8a^2|B_0| \ll 5A_0$,

$$a \doteq a_0(1 - \lambda a_0^2 + 4\lambda^2 a_0^4) \tag{14}$$

and

$$\alpha \doteq \alpha_0\left(1 - \frac{\lambda\alpha_0}{4A_0} - \frac{\lambda^2\alpha_0^2}{4A_0^2}\right) \tag{15}$$

where $\lambda = 8B_0/15A_0$, and a_0, α_0 are the Hertzian values of the contact radius and normal approach, respectively. Further the maximum contact pressure is

$$\sigma_m \doteq \sigma_{0m}\left(1 + \frac{2\lambda a_0^2}{3} - \lambda^2 a_0^4\right) \tag{16}$$

where $\sigma_{0m} = 4A_0a_0/\pi^2(\vartheta_1 + \vartheta_2)$ is the value given by the Hertz theory.

It may be noted that an extension of Cattaneo's "second approximation" may be made to account for terms of any order. Thus, it is easily shown that, for a distance function

$$z(\rho) = \sum_{m=1}^{n} A_m\rho^{2m},$$

the integral equation, Eq. (1), yields a pressure distribution of the form

$$\sigma = \sum_{m=1}^{n} k_m(a^2 - \rho^2)^{(2m-1)/2} \tag{17}$$

The coefficients k_m in Eq. (17) are related by virtue of the expression for the total load,

$$N = 2\pi\int_0^a \sigma\rho d\rho \tag{18}$$

and the radius of contact is a root of an algebraic equation of degree $2n + 1$.

A similar computation was carried out by Steuermann [9] for the case of solids of revolution in osculating contact of any order, i.e., configurations for which $[d^{2n-1}z/d\rho^{2n-1}]_{\rho=0} = 0$. The distance function now is given by $z = A\rho^{2n}$

and the analysis shows the contact law to be

$$\alpha = kN^{2n/(2n+1)} \tag{19}$$

Further, the radius of contact is proportional to the $(2n + 1)$th root of the normal force and the pressure distribution is the Hertzian one multiplied by a polynomial in ρ of degree $2(n - 1)$ with odd-power terms absent. Figure 2 shows the pressure profile for various values of n. We note that only if $n = 1$, corresponding to Hertzian contact, does the maximum pressure occur at the center of contact. For $n \geq 2$ the maximum is seen to be displaced toward the edge of the contact region and approaches it ever more closely with increasing n.

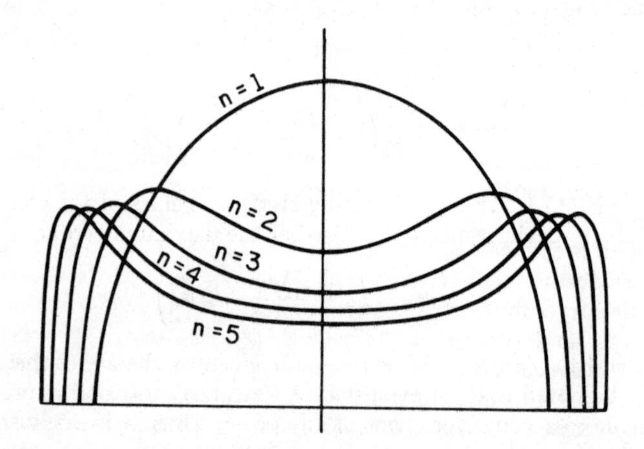

Fig. 2. Pressure distribution in osculating contact (after Ref. [9]).

In a subsequent paper [10] Steuermann formulated the problem of higher-order contact for nonspherical bodies, i.e., such that the three second derivatives of the distance function, $z(x, y)$, vanish at the origin. Korovchinskii [11] discussed the case of a polynomial $z(x, y)$ of the fourth degree and showed that under certain circumstances the contact region is approximately elliptic, the form of the ellipse depending, however, on the total load. Lundberg [12] gave a solution in quadrature of the integral equation, Eq. (1), for an arbitrary axisymmetric distance function and also treated the problem of contact of conical bodies. Moreover, it was shown by Galin [13, Chap. II, Art. 8] that a polynomial distance function in Eq. (1) of degree n will result in a contact pressure of the form

$$\sigma(x, y) = P(x, y)\left(1 - \frac{x^2}{a^2} - \frac{y^2}{b^2}\right)^{-1/2} \tag{20}$$

where $P(x, y)$ is also a polynomial of degree n.

Steuermann, in another paper [14], discussed several problems of contact in which the initial point of contact is a singular point for the second derivative of the distance function. Among others, he examined the case of axisymmetric surfaces whose second derivative at the origin is unbounded. This situation is exemplified by $z(\rho) = A\rho^{3/2}$ for which the radius of contact is proportional to the $\frac{2}{5}$ power of the normal force and the pressure distribution has a cusp at the center.

The Russian literature is replete with other examples of generalization, such as the effect on the contact problem of surface forces outside the region of contact and the interaction due to multiple contacts. The bulk of the solutions concern plane problems. These being outside our scope of interest, the reader is referred to Galin's monograph [13] for particulars and references.

All of the results described above were obtained on the basis of the Hertzian assumption which treats the problem as one of the plane. The case of the sphere in a closely fitting spherical cavity was formulated as a problem of a loaded spherical cap and an approximate solution deduced by Goodman and Keer [15]. The results show, and experiments tend to confirm, a stiffer load-deflection relationship than is given by the Hertz theory, but no comparison is made with predictions of Steuermann's or Cattaneo's second-order solution. It is shown further that if the radius of contact is large compared with the depth of penetration, it may be varied independently of this depth. In a recent study of the same problem, compression occurring due to the weight of the sphere, it was concluded that, when the region of contact is large, the maximum pressure predicted by the Hertz theory is 20% too low [16].

An important extension of Hertz's theory concerns the effect of shearing components of traction on the contact surface resulting from a relative lateral displacement which occurs whenever the contacting bodies are elastically dissimilar. This question, for a pair of spheres, was examined by Goodman [17] on the assumption that normal displacements due to lateral shears may be neglected in comparison with such displacements due to normal tractions. This is equivalent to assuming that Hertz's contact law, the size of the contact, and the normal pressure are unaffected by the presence of the frictional effects, experimental evidence being cited by the author in justification. His solution of the problem predicts a distribution of radial shearing stress (not expressible by elementary functions) depicted by curve (b) in Fig. 3, its maximum never greater than 28% of the normal stress (the extreme value occurring for the case $\nu_1 = 0$, $\nu_2 = \frac{1}{2}$, $\mu_1/\mu_2 \ll 1$). On the other hand, as may be seen from the curve (c) in the same figure, the ratio of the shearing stress to the normal pressure at a given point may exceed the coefficient of friction, and then slip may be expected to occur. In fact, this ratio is unbounded at the edge of the contact circle and, hence, slip will surely take place over an annular portion of the contact. The width of this annulus is, however, judged to be extremely small for contact between unlubricated, oxide-free metals.

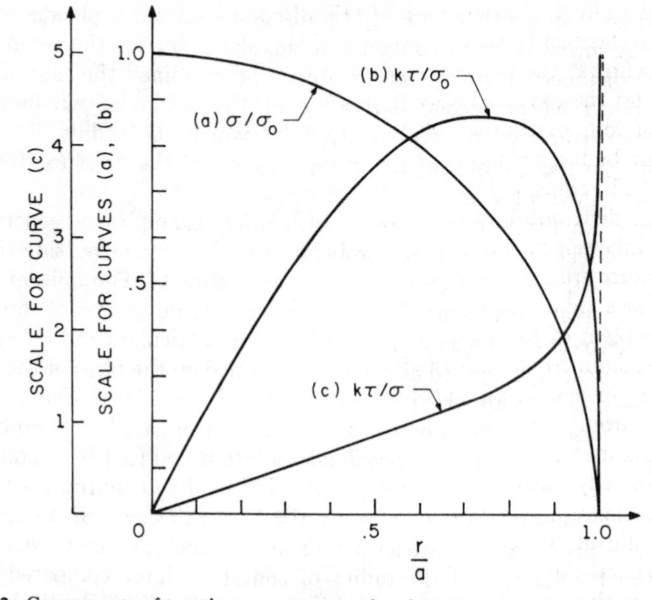

Fig. 3. Components of traction on contact surface between normally loaded spheres in the presence of friction (after Ref. [17]); (a) (Hertzian) pressure, (b) shear, (c) ratio of (b) to (a) showing singularity at the edge of contact.

$$k = \pi\left(\frac{1-\nu_1}{\mu_1}+\frac{1-\nu_2}{\mu_2}\right)\Big/\left(\frac{1-2\nu_1}{\mu_1}-\frac{1-2\nu_2}{\mu_2}\right)$$

An earlier approximate solution of the plane problem of contact of cylinders in the presence of friction also predicted the existence of an annulus of slip [18]. Goodman's solution has been employed to calculate displacements due to the frictional forces generated at the contact. It was shown [19] that the normal deflection of the contact surface due to these shearing tractions cannot exceed $12\frac{1}{2}\%$ of its value due to the normal pressure; on the other hand, the lateral (i.e., radial) component of displacement in the region of contact can be of the same order of magnitude as that generated by the Hertzian pressure distribution. In a related study Conway [20] showed that, for a thin layer compressed and pulled between two stationary rigid cylinders, the effect of friction is to increase slightly the pressure near the trailing end and decrease it near the leading end; however, even in this case of relatively large shearing stresses, the effect on the normal pressure is quite small.

Other generalizations of the classical theory of contact under normal loading have been made to elastic materials which exhibit anisotropy and inhomogeneity, and to materials displaying rate or time effects.

Thus the effect of anisotropy was discussed in a paper by Conway [21] treating the classical Hertz contact of spheres of transversely isotropic material; it was shown that the pressure distribution remains given by Eq. (8) and the expression for the contact radius now contains "equivalent" moduli expressible in terms of four of the moduli of each material. More recently the contact problem for bodies displaying general aeolotropic behavior has been studied by Willis [22] using a Fourier transform method. He found that here, too, the pressure distribution remains Hertzian.

For an inhomogeneous half-space whose shear modulus increases as a power of the distance from the boundary the solution was given for circular contact [23] and for a rigid cylindrical punch of arbitrary shape [24].

The problem of indentation by a rigid sphere of a viscoelastic half-space has been examined in several papers, beginning with the one by Lee and Radok [25] in which the solution valid for loading up to a single maximum is deduced from the elastic solution. For indentation occurring at constant velocity in a material with Maxwell response in shear the behavior was found to be essentially elastic at times short compared with the relaxation time, but at the relaxation time the pressure distribution shows flattening and subsequently a dip at the center of contact. The solution was extended by Yang [26] to general linear viscoelastic materials and arbitrary quadric surfaces in contact. For the special case of a rigid sphere pressed by its own weight, W, into a three-parameter, incompressible half-space, the indentation was shown to be $\alpha(t) = R^{-1/3}[(3W/16E)(2 - e^{-t/2\tau})]^{2/3}$, E denoting the instantaneous elastic modulus and τ the relaxation time.

Another extension, by Hunter [27], was made to the case where the contact area possesses a single maximum, this problem being of significance in the interpretation of scleroscopic tests on polymers. The formulation leads to a pair of dual integral equations with kernels related by hereditary integrals; these equations were transformed to a pair of integral equations in the complex plane and the pressure distribution was obtained in quadrature involving the stress relaxation function. Numerical and experimental results on the problem of impact were reported by Calvit [28]. Hunter's solution was further extended by Graham [29] to the indentation of materials of arbitrary linear viscoelastic behavior by a punch of arbitrary profile (i.e., resulting in a distance function quadratic in x and y). Finally, a method was given by Ting [30] for obtaining the contact stresses arising from the indentation of a viscoelastic half-space by a spherical punch for any number of loadings and unloadings.

The distribution of normal traction at a contact resulting from the reduction of the external force in the presence of adhesion has been discussed in Refs. [31, 32]. This is but one aspect of the more general question of the effect of surface forces between solids. To account for the recently observed phenomenon that, under certain circumstances, contact at low load levels occurs over an area larger than predicted by the Hertz formula, Johnson,

Kendall, and Roberts [33] proposed a modification of this formula which predicts, for spherical bodies, a circular contact region of radius given by

$$a^3 = (k_1 + k_2)\tilde{R}N(1 + \eta + \sqrt{2\eta + \eta^2})$$

where $\tilde{R} = R_1R_2/(R_1 + R_2)$ and $\eta = 3\pi\gamma\tilde{R}/N$, γ denoting the energy of adhesion of both surfaces per unit contact area. It is seen that for light loads the contact region is larger than the Hertzian and, indeed, remains nonzero for vanishingly small load, i.e., $a_{N=0}{}^3 = 6\pi\gamma\tilde{R}^2(k_1 + k_2)$. The (tensile) load required to effect separation is $N_{a=0} = -3\pi\gamma\tilde{R}/2$, independent of the elastic moduli. For large loads the contribution of the surface energy terms becomes small and the Hertzian result holds.

The predictions of the foregoing analysis were borne out in experiments carried out on carefully prepared optically smooth rubber and gelatine spheres; these materials were chosen for their low elastic modulus, which property tends to minimize the effect of surface asperities and to insure intimate contact. The value of the adhesive energy of rubber spheres (2.2 cm radius) was computed at $\gamma = 35$ erg/cm for each surface when dry, but the presence of liquids on the interface reduced this value drastically, e.g., to $\gamma = 3.4$ erg/cm per surface when the surfaces were immersed in water and to $\gamma < 0.05$ erg/cm for the contact when immersed in the highly wetting solution of sodium dodecyl sulphate. For materials with a high elastic modulus, such as metals and glass, the deformation produced by the surface forces is too small to be separated experimentally from the effects of surface roughness.

Of other extensions of Hertz's theory which fall within the scope of contact due to normal forces we note Keer's discussion [34] of the indentation by an elastic sphere of an elastic layer; he finds that, for a given load, the contact radius and normal approach are smaller than for the Hertz case. Further solutions deal with the compression of a plate between two spheres, like, Refs. [35, 36], and unlike, Ref. [37], and studies of the contact of a hollow sphere with a plate, by means of a finite element approach [38] and within the scope of shell theory [39]. On the technologically important subject of lubricated contact the reader's attention is called to a recent paper by Christensen [40] and to the references cited there. Finally, note may be taken of numerical methods, based on quadratic programming, applicable to any geometry of elastic bodies in contact [41, 42].

B. *Tangential Forces*

Unless otherwise specified the discussion will concern two like spheres. These, initially compressed by constant normal forces, N, are supposed subjected to additional loading which results in a monotonically increasing force, T, on the region of contact. Owing to symmetry the distribution of normal pressure remains unchanged during the tangential loading.

In what follows a distinction will be made between "slip," denoting relative displacement of contiguous points on a portion of the contact surface, and "sliding," a term reserved for relative displacement over the entire contact.

Now, if it is assumed that no slip occurs, considerations of symmetry lead to the conclusion that the tangential component of displacement of the contact surface is that of a rigid body. The situation is described by a mixed boundary-value problem for the elastic half-space: the normal component of traction is zero and the tangential component of displacement, parallel to T, is constant on the entire area of contact, and all three components of traction vanish on the remainder of the boundary. The solution, due to Cattaneo [43] and Mindlin [M29], yields a tangential traction on the contact parallel to the applied force, its magnitude axially symmetric and unbounded at the edge of the contact circle; specifically, the magnitude is given by

$$\tau = \frac{T}{2\pi a}(a^2 - \rho^2)^{-1/2}, \qquad \rho < a \tag{20}$$

the distribution being illustrated in Fig. 4. The relative displacement, δ, of points on one sphere remote from the contact with respect to similarly located points on the other sphere is proportional to the applied force, i.e.,

$$\delta = \frac{(2 - \nu)T}{4\mu a} \tag{21}$$

this relation being shown in Fig. 5.

As infinite traction is required in the absence of slip, it is expected that slip does accompany tangential forces and that this occurs regardless of how small these forces may be. Also, it is reasonable to suppose that slip will start at the edge of contact where the singularity in traction occurs and progress radially inward; further, since, in the absence of slip, the traction distribution is symmetric, slip will take place over an annular region. Finally, it is assumed that the tangential traction on the annulus of slip is in the direction of applied force and that at each point where slip occurs Coulomb's law of friction holds locally, i.e., $\tau = f\sigma$, where f is a (constant) coefficient of friction. Considerations of symmetry may once again be invoked to show that on the portion of the contact area over which no slip has taken place, called the "adhered" region, the tangential component of displacement is constant. Thus the situation is reduced to another mixed boundary-value problem: the tangential displacement (constant) and normal traction (zero) are given on the adhered portion and the traction is prescribed over the remainder of the boundary (normal component zero, tangential component proportional to Hertz pressure on the annulus; all components zero outside the contact region). Solution of this problem [M29], [43] yields the inner radius of the adhered portion,

$$c = a\left(1 - \frac{T}{fN}\right)^{1/3} \tag{22}$$

the distribution of tangential traction on the contact area,

$$\tau = \begin{cases} \dfrac{3fN}{2\pi a^3}\,(a^2 - \rho^2)^{1/2}, & c \leq \rho \leq a \\[2ex] \dfrac{3fN}{2\pi a^3}\,[(a^2 - \rho^2)^{1/2} - (c^2 - \rho^2)^{1/2}], & \rho \leq c \end{cases} \tag{23}$$

(see Fig. 4), and the relative tangential displacement of points in the two spheres far away from the contact,

$$\delta = \frac{3(2 - \nu)fN}{8\mu a}\left[1 - \left(1 - \frac{T}{fN}\right)^{2/3}\right] \tag{24}$$

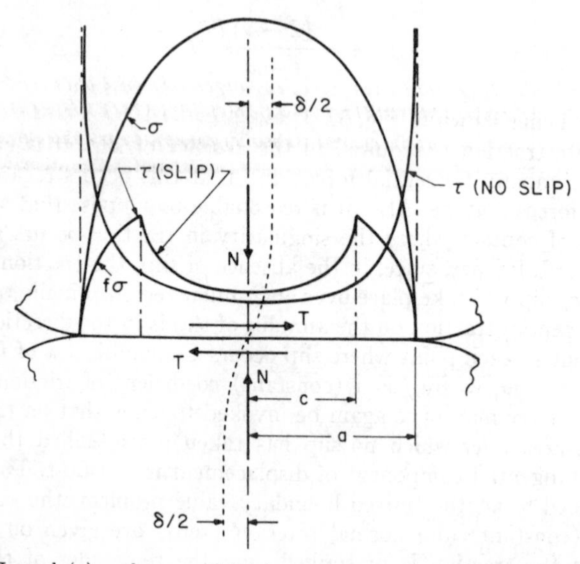

Fig. 4. Normal (σ) and tangential (τ) components of traction on contact region between two spheres subjected to normal force followed by monotonic tangential force [M49].

This load-displacement relationship is illustrated in Fig. 5 along with experimental results obtained by Johnson [44] which appear to confirm its validity. The tangential compliance of the contact is

$$S = \frac{d\delta}{dT} = \frac{2-\nu}{4\mu a}\left(1 - \frac{T}{fN}\right)^{-1/3} \tag{25}$$

and represents the reciprocal of the slope of the curve in Fig. 5. As the applied force T approaches fN, i.e., as the maximum tangential force compatible with Coulomb's law is developed, the radius of the adhered portion tends to zero and the tangential compliance grows without bound; when $T > fN$ the displacement δ becomes indeterminate. This, then, corresponds to rigid-body sliding.

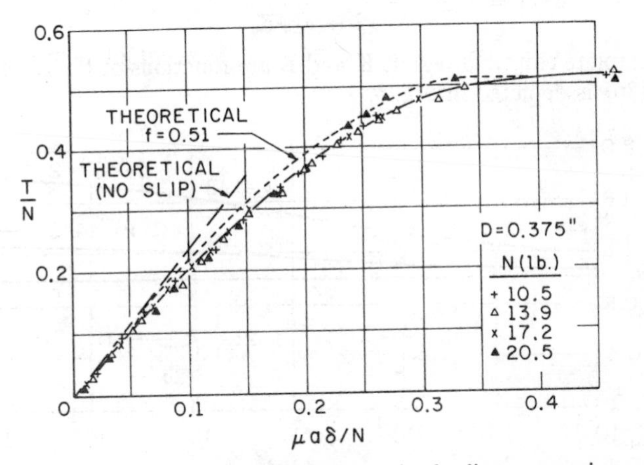

Fig. 5. Tangential force-displacement relation for loading: comparison of Eqs. (21) and (24) with results of static tests (after Ref. [44]).

The initial value of the tangential compliance, obtained by setting $T = 0$ in Eq. (25), is the same as obtained from the no-slip solution, Eq. (21). Moreover, it is seen to be of the same order as the normal compliance of Hertz contact, Eq. (10), the ratio of the initial tangential to the normal compliance being $\mathcal{R} = (2-\nu)/2(1-\nu)$, i.e., $\mathcal{R}$ ranges from unity (for $\nu = 0$) to $\frac{3}{2}$ (for $\nu = \frac{1}{2}$).

When the elastic properties of the two bodies are different a normal component of traction on the contact is in general required in order to insure continuity of the normal displacement. This is analogous to the tangential components which are neglected in the Hertz theory, and is ignored in the present theory. It may be noted that, for the (plane strain) contact of two cylinders in the presence of tangential traction, the Hertz pressure is modified slightly whenever the ratio $\mu/(1 - 2\nu)$ is unequal for the bodies. The pressure distribution is asymmetric and the width of the contact region is decreased [45].

For a pair of nonspherical bodies the problem of tangential loading in the presence of constant normal load leads to results similar to those discussed above. The distribution of traction on the elliptical contact, analogous to Eq. (23), is given in Ref. [43], while the ratio of the initial tangential to the normal compliance for bodies having the same elastic properties is [M29]

$$\mathcal{R} = \frac{2 - \nu}{2(1 - \nu)} \psi(e) \tag{26}$$

where

$$\psi(e) = 1 \pm \frac{\nu[(2 - e^2)\mathbf{K} - 2\mathbf{E}]}{(2 - \nu)e^2\mathbf{K}}, \qquad a \gtrless b$$

and the complete elliptic integrals **K** and **E** are functions of the eccentricity, e. Relation (26) is depicted in Fig. 6.

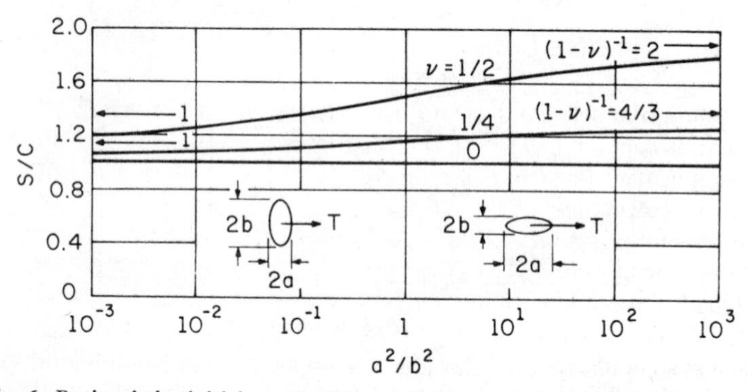

Fig. 6. Ratio of the initial tangential compliance to the normal compliance for contact of nonspherical bodies of like elastic properties (after Ref. [M29]).

For the plane problem of cylinders it was shown by Smith and Liu [46] that the maximum shearing stress in the presence of tangential forces may occur on the contact surface rather than below it as in the case of normal contact. Computations by Hamilton and Goodman [47] of the stress field generated when a pair of spheres is at the point of sliding (i.e., when the contact is loaded by the limiting tangential force, $T = fN$) show that, except when the coefficient of friction is small, the region where failure is most likely to occur, in ductile as well as in brittle materials, is the front edge of the contact circle. For nonspherical bodies subjected to limiting tangential loading the stresses and displacements on the contact region are given in Ref. [48].

The solution of tangential contact of two bodies of revolution has been given by Cattaneo [49] for second-order Hertz contact, i.e., when the distance function is of the fourth degree. It is shown there that the radius of the adhered portion, c, is given by the root of the quintic (13) in which N is replaced by $N - T/f$ and a by c, and that the distribution of the tangential traction on the contact region is

$$\tau = \begin{cases} f\sigma, & c \leq \rho \leq a \\ f(\sigma - \sigma'), & \rho \leq c \end{cases} \tag{27}$$

where σ is the corrected Hertz distribution, Eq. (11), and σ' is obtained from Eq. (11) by replacing a by c. If the tangential loading is small, i.e., $T \ll fN$, then an approximation to this solution may be obtained valid under the same circumstances which led to Eqs. (14)–(16). Thus the radius of contact is given by Eq. (14) and the radius of the adhered portion by the same expression with a_0 replaced by c_0, this being its value, Eq. (22), in the first-order theory. Accordingly,

$$\frac{c}{a} = \frac{c_0}{a_0}\left[1 + \lambda a_0^2\left(1 - \frac{c_0^2}{a_0^2}\right)\right] \tag{28}$$

A second-order solution has also been attempted for two asymmetric bodies in tangential contact [50], but the assumption of an elliptic contact region and a simple, two-term normal pressure distribution does not inspire confidence in the correctness of the results.

The unloading problem has been dealt with by Mindlin, Mason, Osmer, and Deresiewicz [M40]. Here, the system of two like spheres under constant normal load N has had its tangential load reduced from a peak value $T^* < fN$ to $T < T^*$. The situation may be conceived of as resulting from the imposition, to a system of forces N and T^*, of a tangential force of magnitude $\Delta T = T^* - T$ in the direction opposite to that of T^*. If slip were prevented, the tangential traction would, as in the case of initial tangential loading, become singular on the edge of the contact circle, except that its algebraic sign would be reversed. Therefore slip may once again be presumed to occur, but this time in the direction opposite to that of the initial slip. Thus an annulus of counter-slip will be formed and grow radially inward as the tangential force continues to be reduced, its inner radius given by

$$b = a\left[1 - \frac{T^* - T}{2fN}\right]^{1/3} \tag{29}$$

The resultant tangential traction is depicted by curve $ABCa$ in Fig. 7. The corresponding relative displacement of distant points of the two spheres during

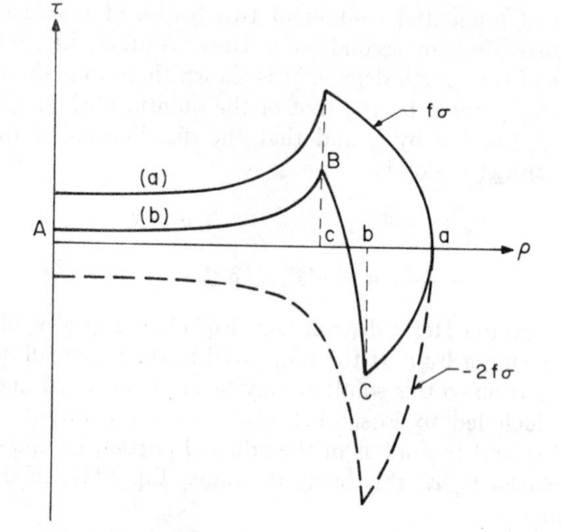

Fig. 7. Distribution of tangential traction on contact surface between two spheres subjected to a constant normal force followed by a tangential force; (a) loading, (b) unloading (after Ref. [M40]).

unloading is

$$\delta_u = \frac{3(2-\nu)fN}{8\mu a}\left[2\left(1-\frac{T^*-T}{2fN}\right)^{2/3} - \left(1-\frac{T^*}{fN}\right)^{2/3} - 1\right] \tag{30}$$

and is illustrated by curve PRS in Fig. 8. When the tangential load is removed an annulus of counter-slip remains, its inner radius given by Eq. (29) with $T = 0$; the tangential traction, though self-equilibrating, is not identically zero. A permanent set appears, its value given by Eq. (30) with $T = 0$, corresponding to point R in Fig. 8. The only way this set can be removed is by applying a tangential force in the reverse direction. For $T = -T^*$, i.e., when the loading has been fully reversed, the counter-slip will have penetrated to the depth of the original slip, as may be seen from a comparison of Eq. (22) at $T = T^*$ and Eq. (29) at $T = -T^*$. Except for reversal of sign the tangential traction at this stage is identical with that at $T = T^*$. Thus points S and P in Fig. 8 represent identical situations except for reversal of sign.

During the unloading stage the tangential compliance of the contact is

$$S = \frac{2-\nu}{4\mu a}\left(1-\frac{T^*-T}{2fN}\right)^{-1/3} \tag{31}$$

its initial value on unloading (i.e., for $T = T^*$) being the same as that of the initial compliance on loading [$T = 0$ in Eq. (25)].

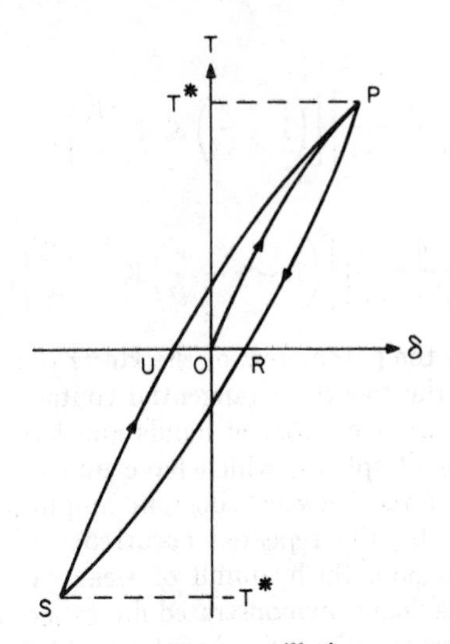

Fig. 8. Theoretical hysteresis loop due to oscillating tangential force at constant normal force [M40].

A subsequent increase of T from $-T^*$ to T^* gives rise to the same sequence of events as occurred in the course of the decrease of T from T^* to $-T^*$ except for reversal of sign. The displacement during this portion of the loading cycle is $\delta_l = -\delta_u(-T)$ and is illustrated by curve SUP in Fig. 8; the compliance is given by Eq. (31) with the sign of T reversed. The load-displacement curve of Fig. 8 is seen to form a closed loop, $PRSUP$, this path being traversed during subsequent variation of T between the limits $\pm T^*$ as long as the normal load is kept unchanged. The area enclosed in the loop represents the energy dissipated due to friction in each loading cycle:

$$E = \frac{9(2-\nu)f^2N^2}{10\mu a}\left\{1 - \left(1 - \frac{T^*}{fN}\right)^{5/3} - \frac{5T^*}{6fN}\left[1 + \left(1 - \frac{T^*}{fN}\right)^{2/3}\right]\right\} \tag{32}$$

When $T^*/fN \ll 1$, i.e., for small amplitudes of loading, Eq. (32) reduces to

$$E \doteq \frac{(2-\nu)\,T^{*3}}{36\mu afN} \tag{33}$$

i.e., the energy loss per cycle is proportional to the cube of the amplitude of the tangential force.

The solution for oscillating tangential forces has been extended by Deresiewicz [51] to the contact of nonspherical bodies. He showed this problem to be identical, qualitatively, with that of two spheres, the expressions for the displacement, compliance and energy loss being obtainable, respectively, from Eqs. (24) and (30), (31), and (32) simply by multiplying these by the constant factor

$$\Phi = \begin{cases} \left[\dfrac{4a}{\pi b(2-\nu)}\right]\left[\left(1-\dfrac{\nu}{e^2}\right)\mathbf{K}+\dfrac{\nu\mathbf{E}}{e^2}\right], & a<b \\[2ex] \left[\dfrac{4}{\pi(2-\nu)}\right]\left[\left(1-\nu+\dfrac{\nu}{e^2}\right)\mathbf{K}-\dfrac{\nu\mathbf{E}}{e^2}\right], & a>b \end{cases} \tag{34}$$

which depends only on the parameters of the Hertz problem.

The predictions of the theory of tangential contact [M40] have been the subject of experiments by a number of hands and have now been verified in all essentials. In a pair of spheres which have undergone a large number of reversals of tangential force between constant amplitudes surface wear may be expected, produced by the repeated occurrence of slip and counter-slip over a fixed annular region. Such annuli of wear were observed and force-displacement hysteresis loops demonstrated in experiments on crown glass lenses [M40]; moreover, the dimensions of the annuli were in good agreement with values determined from theory. More elaborate experiments, static as well as dynamic, were performed by Johnson [44] using steel balls in contact with a steel plane. The dynamic tests were carried out in such a way that the period of stress reversals was large compared with travel time of stress waves through the spheres. As may be seen from Figs. 9 and 10, the agreement between theory and experiment, insofar as the load-displacement relations are concerned, is very good indeed. A typical annulus of wear [52] is shown in Fig. 11.

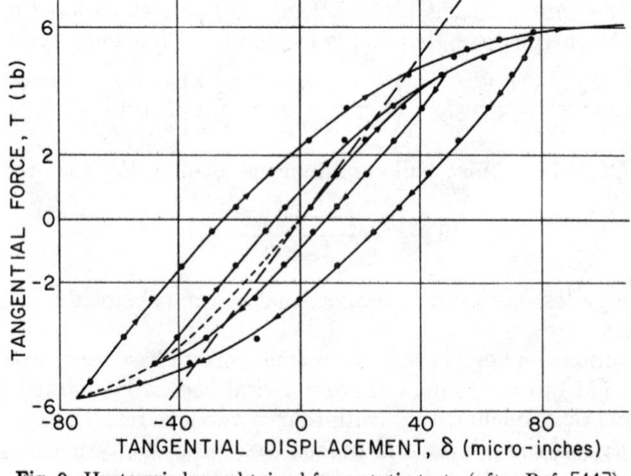

Fig. 9. Hysteresis loop obtained from static tests (after Ref. [44]).

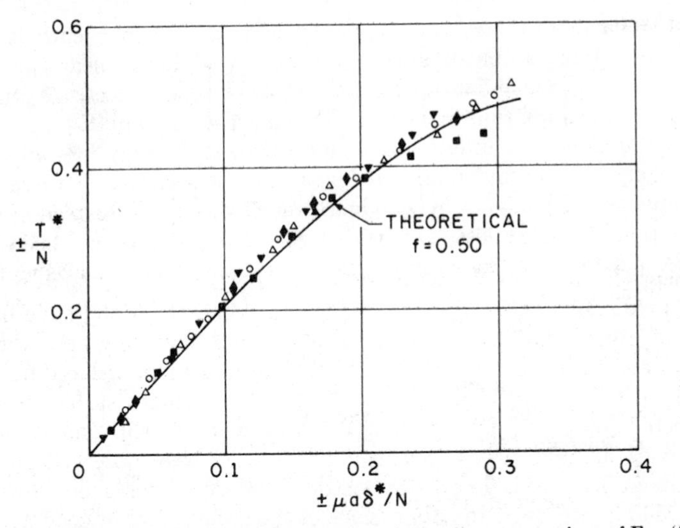

Fig. 10. Tangential force-displacement relation for loading: comparison of Eq. (24) with results of dynamic tests [44].

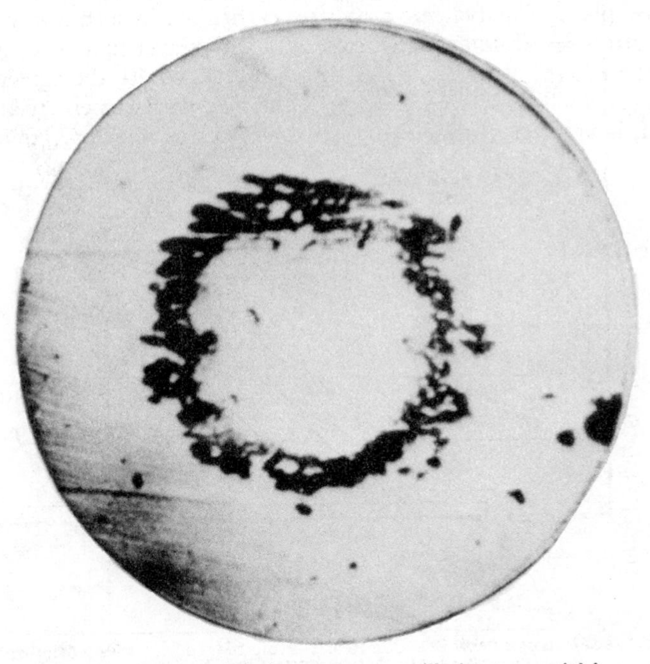

Fig. 11. Example of an annulus of wear due to oscillating tangential force at constant normal force [52].

Measurements of the energy loss per cycle, while in agreement with theory at the larger force amplitudes (i.e., for tangential forces near the limiting value at which gross sliding occurs), showed, for small forces, a variation as the square of the amplitude rather than the cube predicted by Eq. (33) [M40], [44, 53]. Moreover, at intermediate amplitudes the energy loss appeared to be noticeably affected by the sphere size [44]. The reasons for these discrepancies have been elucidated by Goodman and Brown [54]. Instead of plotting the variation of energy loss, ΔE, with force amplitude, as had been done in all earlier papers, they constructed curves of ΔE versus displacement amplitude. Using as units of energy and displacement $T_{\max}\delta_{\max}$ and $\delta_{\max}$, respectively, where $T_{\max} = fN$ is the limiting tangential force and $\delta_{\max} = 3(2 - \nu)fN/8\mu a$ the corresponding displacement, obtained from Eq. (24), they showed that the resulting curve is independent of the value of the coefficient of friction. A comparison between their experimental data (obtained from the loading of a sphere between two flat plates) and theory reveals effects neither of variation of size nor of normal load; this may be seen from Fig. 12. It would appear that the size effect encountered in earlier work was the result of using

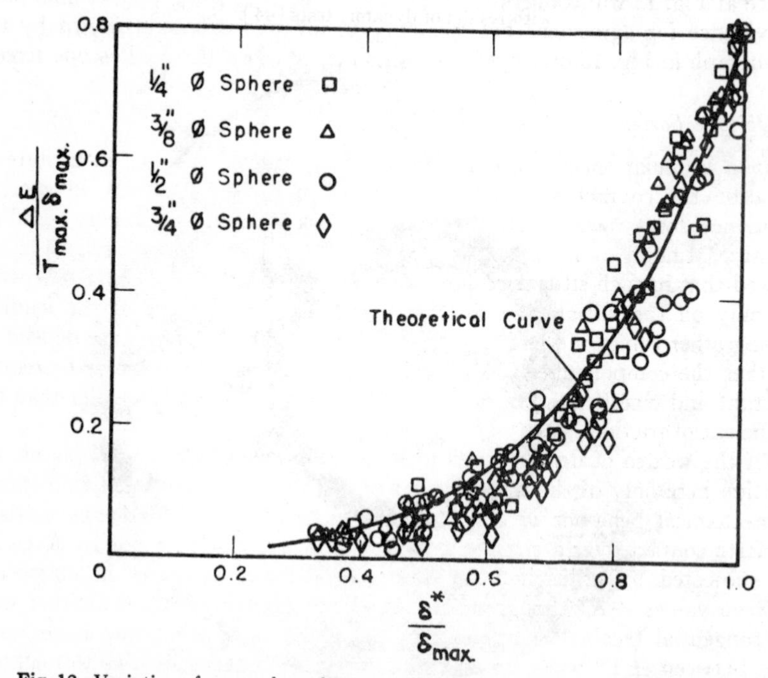

Fig. 12. Variation of energy loss, due to oscillating tangential forces, with displacement amplitude: comparison of theory with experiment [54].

an average value of the coefficient of friction, the same in all cases. Measurements indicate, however, that there is a sizable variation in the value of this coefficient. The energy loss at small amplitudes has been attributed to internal hysteresis, this effect being significant only at these amplitudes; an experiment on purely normal oscillating contact, reported by Johnson in the discussion of Ref. [54], appears to confirm this view. More recently, on the other hand, White [55], using an intuitive argument, has shown that energy expended to overcome static friction (i.e., sticking) just prior to slip is proportional to the square of the applied force, so that the energy dissipated per cycle is amplitude independent and hence does not vanish even at very low force levels. This energy loss, incidentally, varies as the square of the difference between the coefficients of static and sliding friction.

Johnson has also suggested that the reason why measured values of dissipation tend to be lower than theoretical ones for moderate values of force amplitude is the rise in the coefficient of friction with progressive surface damage, stemming from rupture of oxide films and the development of metallic cold-welding, and, indeed, its variation over the contact region. Goodman and Brown's use of stainless steel spheres minimized this effect, though a glance at Fig. 12 will confirm that it did not eliminate it entirely. A final piece of evidence for agreement between theory and experiment is shown by the photograph in Fig. 13 of a hysteresis loop displayed on the oscilloscope screen.

C. *Oblique Forces*

In a granular mass subjected to varying external load or in a state of vibration the contact surfaces between individual granules will, in general, experience forces whose normal and tangential components vary simultaneously. Mindlin and Deresiewicz, in a long and exhaustive study [M42], showed that in such situations the tangential compliance of a contact depends not only on the initial state but on the entire past history of the loading. Among other things it was found that the results in any given case depend on whether the components of force at the contact increase, decrease or remain constant and whether the ratio of their changes exceeds or is smaller than the coefficient of friction.

Of the wealth of details given in Ref. [M42] space limitations permit the mention here only of the following set of circumstances pertinent in a theory of mechanical behavior of a granular aggregate. Two like spheres, initially in Hertz contact over a circular region of radius a_0 due to a normal force N_0, are subjected to additional forces whose resultant oscillates in magnitude between values $\pm R^*$ though retaining a fixed direction. Stated another way, the tangential (scalar) component of the resultant of additional forces oscillates between $\pm T^*$ while its normal component changes such as to maintain constant the ratio $\beta \equiv dT/dN$. It is shown that when $\beta \geq f$ the tangential

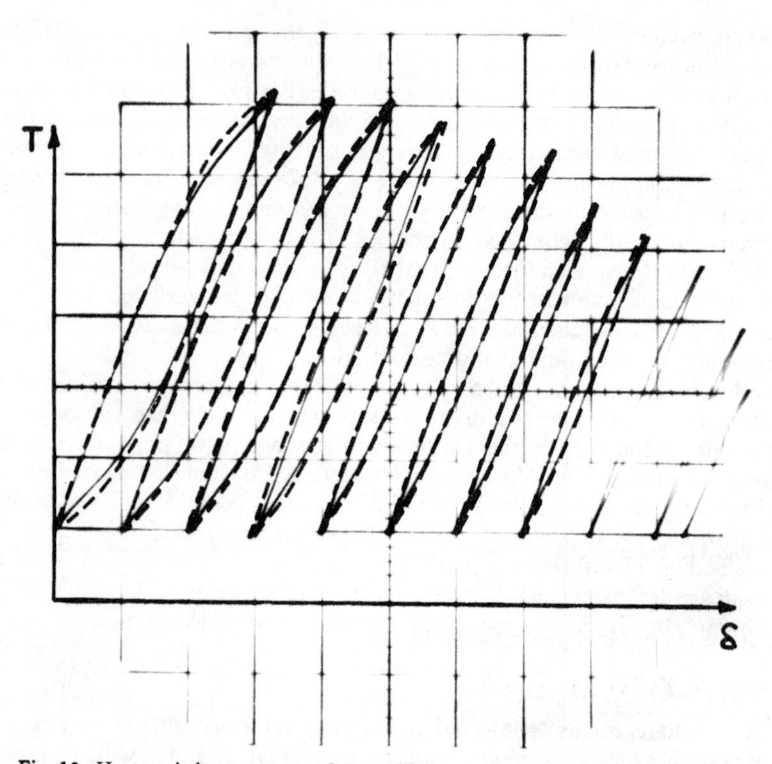

Fig. 13. Hysteresis loops generated on oscilloscope screen by oscillating tangential forces; superposed dashed lines represent theoretical curves [54].

force-displacement relation, after traversing the path $OPRS$, stabilizes along the loop $SUVWS$, shown in Fig. 14. Here $L = T/T_0^*$ and $\Delta = \delta/\delta_{\max}°$ are the force and displacement measured, respectively, in units of $T_0^* = fN_0$ and $\delta_{\max}° = 3(2-\nu)fN_0/8\mu a_0$, denoting the limiting magnitudes of these quantities which would occur if the normal force remained constant at its initial value. The tangential compliance of the contact in the stabilized cycle during loading (curve SUV) is

$$S = \frac{2-\nu}{4\mu a}\left\{\theta + (1-\theta)\left[1-(1+\theta)\frac{L^*+L}{2(1+\theta L)}\right]^{-1/3}\right\} \tag{35}$$

where $\theta = f/\beta \leq 1$; the compliance during unloading (curve VWS) is given by the same expression except that the signs θ and L are now reversed. The

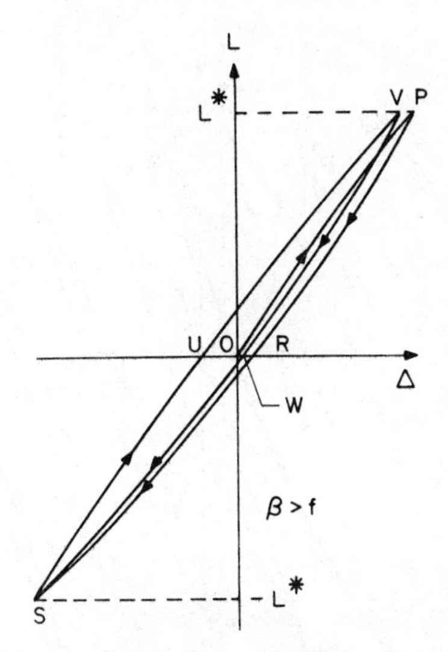

Fig. 14. Theoretical hysteresis loop due to oscillating force of constant obliquity $\beta > f$ (after Ref. [M42]).

frictional energy loss per cycle turns out to be

$$E = \frac{9(2-\nu)(fN_0)^2}{10\mu a_0}\left\{\frac{1}{4\theta}\left[\frac{1+\theta}{1-\theta}(1-\theta L^*)^{5/3} - \frac{1-\theta}{1+\theta}(1+\theta L^*)^{5/3}\right]\right.$$

$$\left. - \frac{1}{1-\theta^2}(1-L^*)^{2/3}\left[1 - \frac{1+5\theta^2}{6}L^*\right]\right\}, \qquad 0 \le \theta \le 1 \quad (36)$$

this expression, for small values of L^*, reducing to

$$E \doteq \frac{(2-\nu)T^{*3}}{36\mu a_0 fN_0}(1-\theta^2) \quad (37)$$

In all these expressions there is imposed on T^* an upper limit which may be written in the form $0 \le L^* \le (1+\theta)^{-1}$. For $\theta = 0$, corresponding to $N =$ constant, Eq. (36) may be shown to reduce to Eq. (32), Eq. (37) to Eq. (33), and Eq. (35) to Eq. (31) with the sign of T reversed.

The situation is quite different for the case $\beta \le f$ (so that $1 \le \theta \le \infty$). This time the load-displacement relation has the appearance shown in Fig. 15, *OPOS* representing the initial loading and unloading paths and *SUVUS* the stabilized path. No loop is formed and no frictional energy loss occurs. The

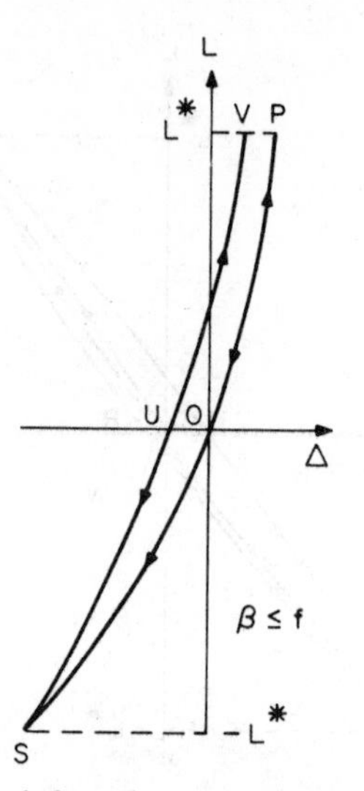

Fig. 15. Theoretical hysteresis loop due to oscillating force of constant obliquity $\beta \leq f$ (after Ref. [M42]).

tangential compliance in the stabilized cycle is given by

$$S = \frac{2 - \nu}{4\mu a} \tag{38}$$

It is noteworthy that expressions (36) and (37) vanish for $\theta = 1$, indicating that dT/dN must exceed the coefficient of friction in order that slip take place and energy be dissipated.

Results of experiments obtained by Johnson [56] bear out the foregoing theory. It is found that for angles of obliquity of the oscillating force smaller than the angle of friction of the surfaces the energy dissipation hardly differs from the small loss which occurs when the oscillating force is purely normal. As was noted in the discussion in the preceding section this loss is attributable to internal dissipation in the material. At angles of obliquity greater than the angle of friction the experiments revealed energy dissipation and surface damage which increased rapidly with increasing obliquity, reaching a maximum when the additional force was tangential. Figure 16 illustrates the observed fretting, the angle of friction being arc $\tan f \doteq 29°$.

D. *Twisting Moments*

A situation closely analogous to that involving tangential contact forces arises when the bodies in contact are subjected to couples tending to cause relative rotation about the normal to the initial tangent plane. Such local

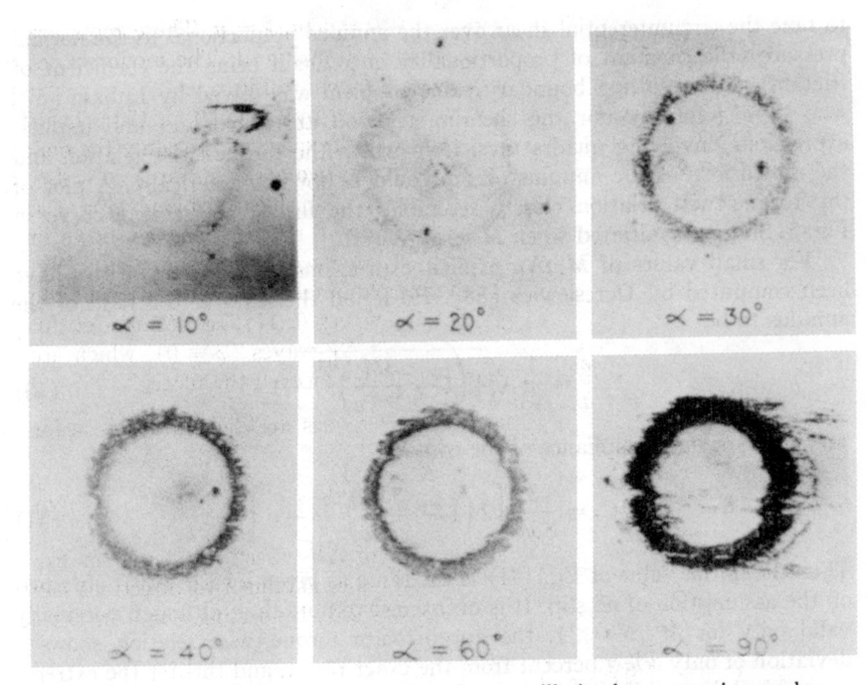

Fig. 16. Fretting damage in a steel sphere due to oscillating forces at various angles of obliquity [56].

force systems will arise in a granular assemblage when it is loaded in a general manner.

A pair of like spheres in Hertz contact due to a constant force N is subjected to a couple of magnitude M about the axis of N. On the assumption of no slip, Mindlin [M29] found that the entire contact surface rotates as a rigid body about the line of centers, the torsional compliance of the pair of spheres being

$$C_\tau = \frac{3}{8\mu a^3} \tag{39}$$

and that the (shearing) component of traction in the circumferential direction rises to infinity at the edge of the contact.

The appearance of this singularity in traction once again indicates the occurrence of slip, again, because of symmetry, over an annular region, but this time in the circumferential direction. The adhered portion, occupying the region of contact interior to the annulus, undergoes rigid-body rotation. Recalling Mindlin's treatment of the tangential contact problem, it is reasonable to take the circumferential shear over the annulus proportional to the normal pressure (the constant of proportionality once again being the coefficient of friction). The resulting boundary-value problem was solved by Lubkin [57] who gave a formula for the shearing traction at the contact and implicit expressions, involving quadratures, from which the torque-twist relation and the dimensions of the annulus of slip could be found numerically. A plot of the torque-twist relation closely resembles the force-displacement curve in Fig. 5. Sliding is initiated when $M = 3\pi fNa/16$.

For small values of M/fNa explicit expressions for these quantities have been computed by Deresiewicz [58]. He found the ratio of the radii of the annulus to be

$$\frac{c}{a} = \frac{1}{\sqrt{3}} \sqrt{4\left(1 - \frac{3M}{2fNa}\right)^{1/2} - 1} \tag{40}$$

and the torsional compliance of the contact,

$$C_\tau = \frac{3}{8\mu a^3}\left[2\left(1 - \frac{3M}{2fNa}\right)^{-1/2} - \right] \tag{41}$$

Thus the initial value of Eq. (41) is the same as given by Eq. (39), obtained on the assumption of no slip. It is of interest to note that, although nominally valid only for $M/fNa \ll 1$, the approximate torque-twist relation shows a deviation of only a few percent from the exact result, and this for the extreme case of incipient sliding.

The analysis carried out in Ref. [M40] for oscillating tangential contact may now be extended to oscillating torsional contact. It turns out [58] that slip and counter-slip succeed one another over the same region as the applied couple is reversed in direction with constant amplitude, M^*. A stable torque-twist cycle is established whose appearance is qualitatively the same as that shown in Fig. 8, the torsional compliance being given by

$$C_\tau = \frac{3}{8\mu a^3}\left[2\left(1 - \frac{3}{2}\frac{M^* + M}{2fNa}\right)^{-1/2} - 1\right] \tag{42}$$

during loading and by the same expression with the sign of M reversed during unloading. The energy loss per cycle is

$$E = \frac{2f^2N^2}{\mu a}\left\{\frac{8}{9}\left[1 - \left(1 - \frac{3M^*}{2fNa}\right)^{3/2}\right] - \frac{M^*}{fNa}\left[1 + \left(1 - \frac{3M^*}{2fNa}\right)^{1/2}\right]\right\}, \tag{43}$$

which reduces to

$$E = \frac{3M^{*3}}{16\mu fNa^4} \tag{44}$$

for $M/fNa \ll 1$.

An experimental verification was obtained by Hetényi and McDonald [59] of the validity of the assumption that within the region over which slip has occurred the circumferential shear is proportional to the Hertz pressure. This was done by means of photoelastic tests, applying the three-dimensional stress-freezing technique to the case of incipient sliding. In addition, the entire stress and displacement fields were determined analytically for this limiting situation. The stress field for the other extreme situation, where complete adhesion is assumed, is given in Ref. [60].

Torsional contact of nonspherical bodies was investigated by several authors beginning with Mindlin [M29] who determined the compliance of the contact in the absence of slip (or, what is equivalent, the initial compliance in the presence of slip):

$$C_\tau = \frac{3}{8\mu b^3} \cdot \frac{8(\mathbf{BD} - \nu\mathbf{CE})}{\pi[\mathbf{E} - 4\nu(1 - e^2)\mathbf{C}]} \tag{45}$$

The functions appearing in Eq. (45) are expressible in terms of the elliptic integrals defined immediately below Eq. (3), i.e., $\mathbf{D} = (\mathbf{K} - \mathbf{E})/e^2$, $\mathbf{B} = [\mathbf{E} - (1 - e^2)\mathbf{K}]/e^2$ and $\mathbf{C} = [(2 - e^2)\mathbf{K} - 2\mathbf{E}]/e^4$. Relation (45) is shown graphically in Fig. 17. The same problem, in the presence of slip, was discussed by Cattaneo [61] and Pacelli [62]. The former devised a method of successive approximations for the determination of the contact shearing stress; the latter, guided by the appearance of these approximations and of Lubkin's exact solution for spheres, succeeded in constructing an exact solution for the general case.

Related problems studied include the determination of the shearing stress generated at the contact: due to a twisting moment applied to a rigid circular

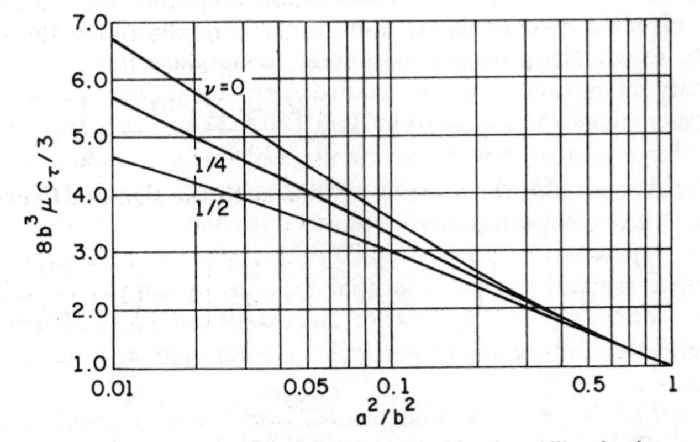

Fig. 17. Initial torsional compliance for nonspherical bodies of like elastic properties [M29].

disk fastened to an elastic layer which, in turn, is fixed on a rigid foundation [63]; in the presence of an arbitrary pressure distribution acting over a (circular) annular region (equivalent to the assumption of an arbitrary law of friction) [64]; and between a plate and a pair of like spheres in torsion [65].

Part II Granular Materials

The mechanical response, primarily static, of granular aggregates has classically been the concern of soil mechanics. Much of the work done within this framework has been experimental and empirical; the analytical has invariably been based on models of the material as a continuum subjected to special constraint conditions. The interested reader will find these matters, many of them of rather specialized interest, adequately discussed in any modern text on soil mechanics. A good deal of experimental work has also been done on the dynamics of granular aggregates, consolidated as well as unconsolidated, particularly on wave propagation through marine sediments and sedimentary rocks; a representative bibliography is given in Ref. [66].

Our present interest is in predictions of behavior of granular materials based on the analysis of models consisting of arrays of discrete granules. As noted in the remarks prefatory to this article, it was R. D. Mindlin who first succeeded in incorporating the idea of tangential compliances of two contacting bodies into such an analysis and who also performed experiments to test the results.

Of course, the analysis of behavior of particulate materials through discrete models is an old idea dating back almost a century to Reynolds [67], who assumed the constituent particles to be rigid. A model consisting of rigid granules being evidently unsuitable for problems of wave propagation, a number of investigations have been carried out which treat the granular aggregate as an array of elastic spheres in normal (Hertzian) contact. An early study along these lines, concerned with elucidating the behavior of the carbon microphone, was made by Hara [68]. For disturbances whose wavelength is large compared with the diameter of the spheres he found the phase velocity of compressional waves in such an array to vary as the sixth root of the contact force and the cube root of an elastic modulus, a relation essentially verified in experiments by Iida [69, 70]. Among other work on wave propagation through regular arrays of like spheres, based on the assumption of purely Hertzian contact, we may cite that of Takahashi and Satô [71] and Gassmann [72], the latter author being particularly interested in a simulation of the earth's crust.

Brandt [73, 74] was the first to deal with irregular packings of spherical particles (still basing his theory on Hertzian contact). His model consisted of spheres of several sizes but of the same material. Spheres of largest radius are packed randomly with a porosity β. Spheres of the next smaller size are assumed packed, with the same porosity β, in the interstices of the primary array. This process is continued with subsequent smaller spheres until m sets of spheres are contained in the model. Boundary effects on each interstitial set as well as on the primary array are neglected. Loading on the array is taken to consist of external pressure p_0 transmitted through flexible confining walls and interstitial pressure p_L supplied by means of a liquid admitted through a tube. The model is thus capable of simulating experiments in which the material is either jacketed and pressurized or else exposed to liquid under pressure. An ingenious analysis leads to equations from which the relationship between dilatational wave speed and pressure may be obtained (Fig. 18). For a dry medium the proportionality between the speed of a pressure wave and the one-sixth power of the product of the external pressure and the square of a modulus is once again predicted, a result in reasonable agreement with experimental data at low pressures [75, 76]. Among other measurements of wave velocities we note those reported in Ref. [77], obtained in experiments with dry and water-filled sand and also glycerine-filled sand. Brandt's model has also been used in an investigation of longitudinal wave velocities in sand, clay, and sandstone formations represented by aggregates consisting of arbitrarily shaped elastic grains, partly cemented to one another, the interstices being filled with liquid [78].

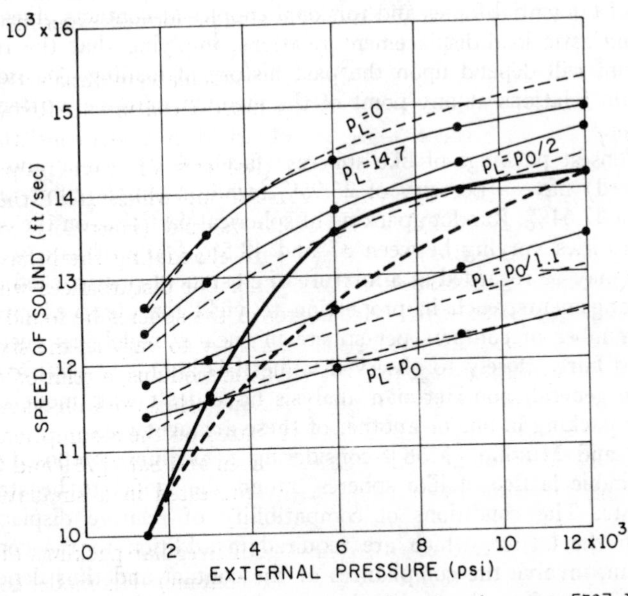

Fig. 18. Speed of sound through sandstone as a function of pressure [73]. Dashed and solid curves refer, respectively, to theoretical and experimental results; heavy and light lines pertain to dry and oil-saturated samples, respectively.

It must be stressed that the brief notice which has here been accorded to these researches is by no means a reflection of their lack of importance but, rather, the result of severe space limitation. For the same reason the balance of this section is not treated with the detail it deserves. The unbalance between the lengthy Part I and the abbreviated Part II of this article is, however, mitigated by two facts: (1) the contents of the latter have been dealt with extensively in an earlier review by the present author [79]; and (2) in the intervening decade and a half a great deal of research has been done in the various aspects of the contact problem, in contrast with little work on its application to granular assemblages.

Although theories based on normal contact have been successful in their qualitative predictions of mechanical behavior of granular aggregates, quantitative agreement has generally lagged behind. The linear variation, for example, of wave speed with the cube root of an elastic modulus and the sixth root of applied isotropic pressure has been verified in experiments, but the predicted magnitudes have invariably been too low.

The source of this discrepancy may be sought in the omission of effects of non-Hertzian loads to which individual granules are, in general, subjected. It was seen in Part I that compliances due to tangential forces are of the same order of magnitude as normal compliances and therefore corrections stemming from the inclusion of their effects may be expected to be significant. The presence of tangential forces and torsional couples at contacts gives rise, however, to inelastic load-displacement relations, implying that the response of the medium will depend upon the past history of loading. Accordingly, the stress-strain relations at any point of the medium must be written in differential form.

The densest packings of like spheres (face-centered cubic and hexagonal close-packed) have a porosity of 25.95%, while a simple cubic array has a porosity of 47.64%. Random packing of spherical particles, on the other hand, yields porosities varying between 37 and 42.5% [80]. Thus a well-agitated aggregate may be regarded as a mixture of clusters of close-packed and simple cubic arrangements, each in proportion to yield the observed porosity; the average number of contacts per sphere in such a model has been found to correspond fairly closely to observed values [81]. This is why first attempts at a more general, non-Hertzian analysis have dealt with models consisting of regular packing in one or another of these arrangements.

Duffy and Mindlin [M58], considering a medium composed of a face-centered cubic lattice of like spheres, found the model to be statically indeterminate. The conditions of compatibility of relative displacements of spheres in the lattice, which are required in addition to the conditions of equilibrium, involve the compliances at the contact and thus depend on the contact forces and on the loading history. An exact solution was obtained for the case of an initial isotropic compressive stress σ_0 followed by an arbitrary incremental stress. The incremental stress-strain relations have a form corresponding to a cubic crystal, the moduli being

$$c_{11} = 2c_{44} = \frac{(4 - 3\nu)c_0}{2 - \nu}$$

$$c_{12} = \frac{\nu c_0}{2(2 - \nu)}$$

$$c_0 = \left[\frac{3\mu^2\sigma_0}{2(1 - \nu)^2}\right]^{1/3} \tag{46}$$

where μ, ν denote the shear modulus and Poisson's ratio, respectively, of the spheres.

A test of the theory was provided by measuring the speed of a longitudinal wave in a "granular" bar similar to the one shown in Fig. 19, constructed of like steel balls arranged in a face-centered cubic array within a thin rubber jacket and held in place by an external isotropic pressure obtained by a partial evacuation of the container. Here, as required by a theory using differential stress-strain relations, the variation of stress due to the passage of the wave is small compared with the initial pressure. The wave speed in the bars, arranged with the longitudinal axis parallel either to the {100} or to the {110} direction of the lattice in order to avoid coupling with flexural waves, and of small cross-sectional dimensions compared with the wavelength of the disturbance, is given by

$$\rho v_{100}^2 = \frac{2(8 - 7\nu)c_0}{8 - 5\nu}$$

$$\rho v_{110}^2 = \frac{2(4 - 3\nu)(8 - 7\nu)c_0}{(4 - 3\nu)^2 + (2 - \nu)(8 - 7\nu)} \tag{47}$$

where ρ is the density of the medium. When the effect of tangential contact

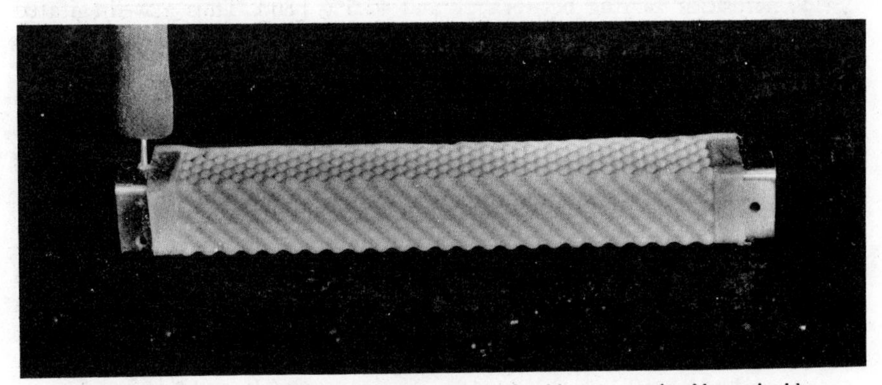

Fig. 19. Bar made of steel balls in a face-centered cubic array; axis of bar coincides with the {100} direction of the packing (courtesy of C. W. Thurston).

forces is neglected the corresponding expressions are

$$\frac{3\rho v_{100}^2}{2} = \rho v_{110}^2 = c_0 \tag{48}$$

Thus both theories yield the same qualitative result, namely the familiar proportionality of the wave speed to the sixth root of the pressure and the cube root of the elastic modulus. The magnitudes, however, do not coincide. For example, for numerical values appropriate to steel, Eq. (47) gives $v_{100} = 1330\sigma_0^{1/6}$ and $v_{110} = 1350\sigma_0^{1/6}$, while Eq. (48) yields $v_{100} = 800\sigma_0^{1/6}$ and $v_{110} = 980\sigma_0^{1/6}$, the units being feet per second when σ_0 is given in pounds per square inch. A comparison of theoretical with experimental results, shown in Fig. 20, indicates better agreement resulting from the inclusion of tangential forces in the theory. Agreement, moreover, is seen to improve with the use of high tolerance balls since these minimize the occurrence of unequal initial contact forces. For longitudinal as well as shear waves in such a granular bar White [55] computed the attenuation due to energy expended in overcoming static friction (cf. the discussion in Part I, Section B). He found the attenuation per wavelength to be independent of the initial pressure, the frequency, and sphere size.

In Ref. [M58] are also examined the effects of loading resulting in initial as well as subsequent incremental stress consisting of an isotropic pressure σ_0 and a uniaxial pressure σ_a parallel to one edge of the lattice. The incremental stress-strain relations contain six independent elastic moduli appropriate to a tetragonal crystal. For arbitrary initial loading the incremental relations have

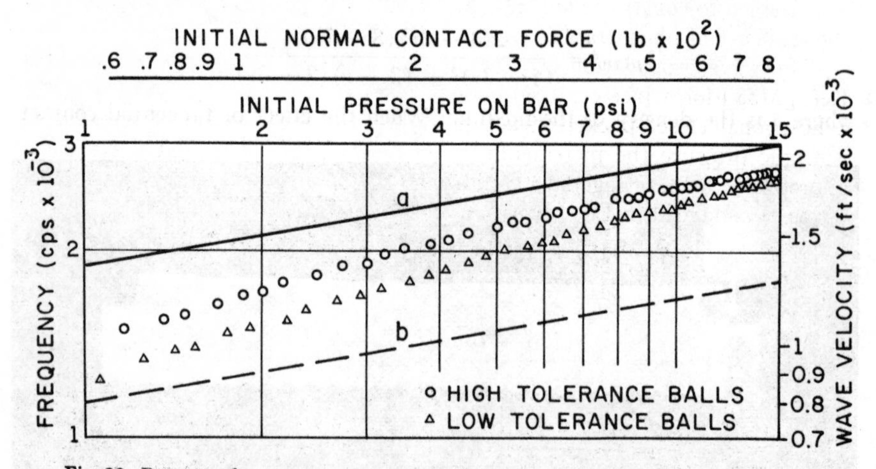

Fig. 20. Resonant frequency and wave velocity of the first compressional mode of a granular bar (constructed as in Fig. 19) as a function of the initial pressure [M58]: (a) prediction of theory which takes into account tangential contact forces; (b) prediction of theory based on normal contact forces only.

been shown to exhibit twenty-one independent moduli [82]. In each case, however, the moduli are functions of the loading history, and the contact forces themselves, in turn, are shown to depend on the contact compliances. Thus the formulation of finite, or total, stress-strain relations requires, for a given variation of applied loading, the solution of a system of simultaneous, nonlinear, integro-differential equations.

Such finite stress-strain relations are needed in assessing the response of the medium to stresses whose variation is of the order of the initial stress, as in static loading tests. An inverse procedure employed by Thurston and Deresiewicz [83] leads to a prescribed variation of σ_0 and σ_a which obviates the need to solve integro-differential equations. Subject to certain conditions, the loading, superimposed on an initial isotropic pressure, turns out to consist of a uniaxial compression imposed simultaneously with an isotropic pressure, the magnitudes of the two systems being related in a complicated but almost linear manner. The resulting stress-strain relations exhibit stiffening behavior and correspond closely to those determined experimentally.

When a face-centered cubic array of spheres, under constant initial isotropic pressure, is subjected to a monotonically increasing compressive force parallel to a principal axis of the array, a limiting magnitude of this force is reached beyond which rearrangement of the spheres within the lattice takes place. It was shown in Ref. [83] that a series of shearing displacements of individual layers of spheres is initiated in a manner analogous to twinning in crystals. The displacements in this process being several orders of magnitude larger than the elastic displacements due to static loading, the onset of twinning was deemed to constitute failure in the aggregate. The process is illustrated by the sequence of photographs in Fig. 21.

The wave propagation experiments and much of the analysis reported in Ref. [M58] for a face-centered cubic array were repeated by Duffy [84] for a hexagonal closed-packed structure. When the initial state of stress is an isotropic pressure the incremental stress-strain relations exhibit a symmetry appropriate to a tetragonal crystal with six independent elastic moduli. In terms of c_0 defined in Eq. (46) these moduli are given by

$$\frac{c_{11}}{c_0} = \frac{1152 - 1848\nu + 725\nu^2}{24(2-\nu)(12-11\nu)}$$

$$\frac{c_{12}}{c_0} = \frac{\nu(120 - 109\nu)}{24(2-\nu)(12-11\nu)}, \qquad \frac{c_{13}}{c_0} = \frac{\nu}{3(2-\nu)}$$

$$\frac{c_{33}}{c_0} = \frac{4(3-2\nu)}{3(2-\nu)}, \qquad \frac{c_{44}}{c_0} = \frac{c_{55}}{c_0} = \frac{6-5\nu}{3(2-\nu)}$$

$$\frac{c_{66}}{c_0} = \frac{576 - 948\nu + 417\nu^2}{24(2-\nu)(12-11\nu)} \tag{49}$$

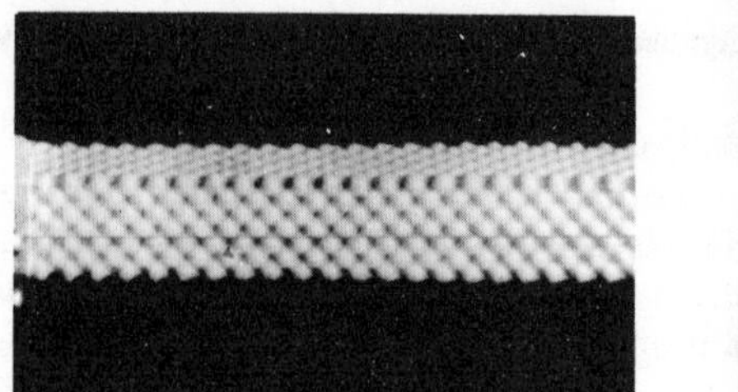

3 frames (0.00379 sec.)
before onset of twinning

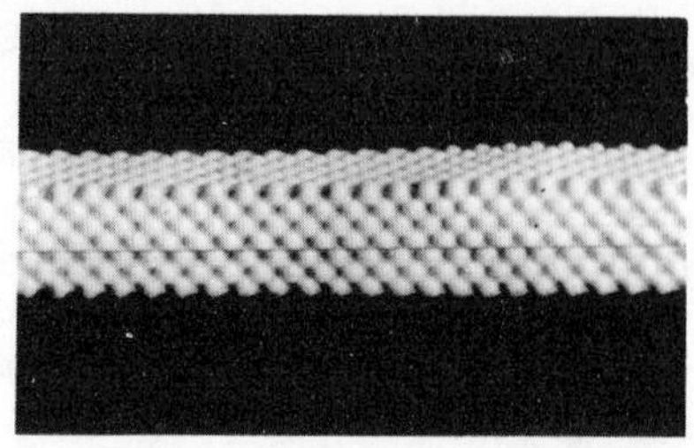

Onset of twinning

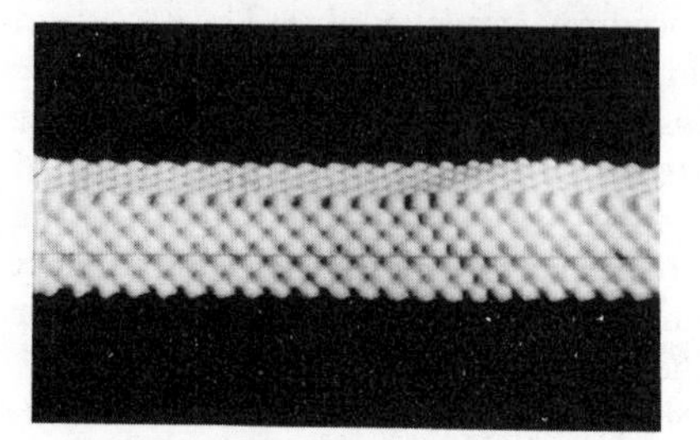

1 frame (0.00126 sec.)
after onset of twinning

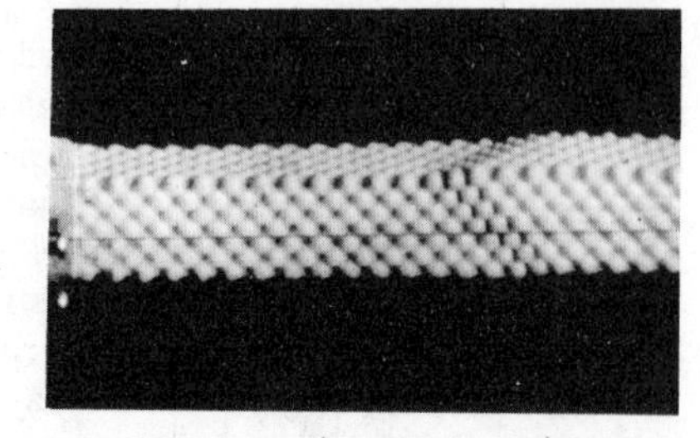

2 frames (0.00253 sec.)
after onset of twinning

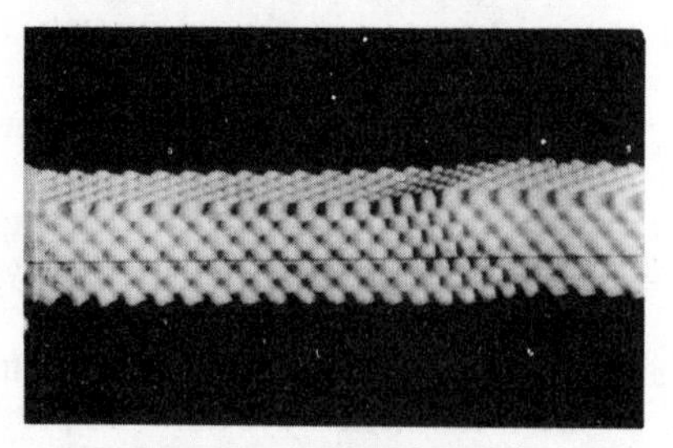

4 frames (0.00506 sec.)
after onset of twinning

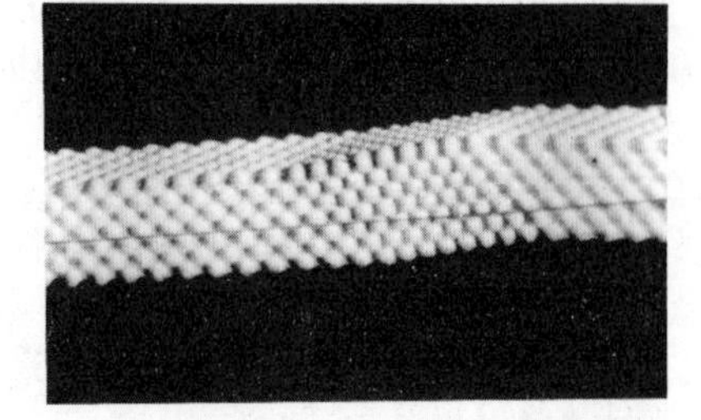

8 frames (0.01012 sec.)
after onset of twinning

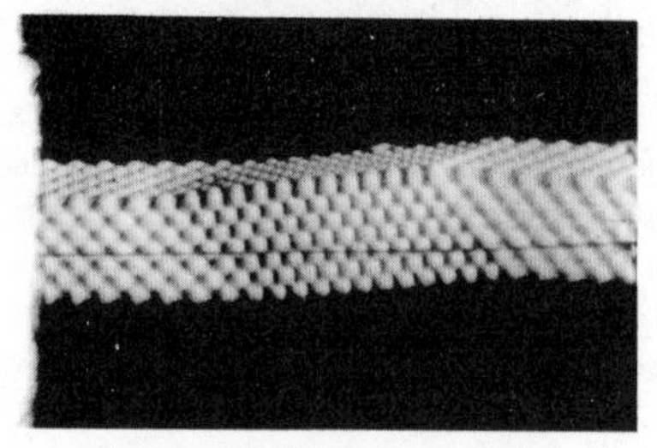

14 frames (0.01771 sec.)
after onset of twinning

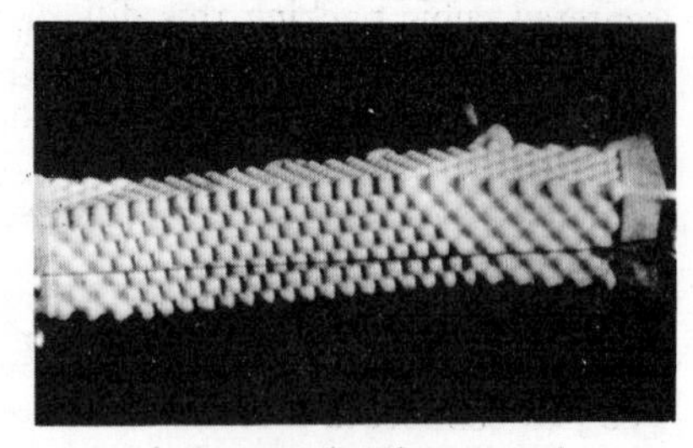

65 frames (0.08222 sec.)
after onset of twinning

Fig. 21. Failure by twinning in the granular bar of Fig. 19 due to axial compression [83].

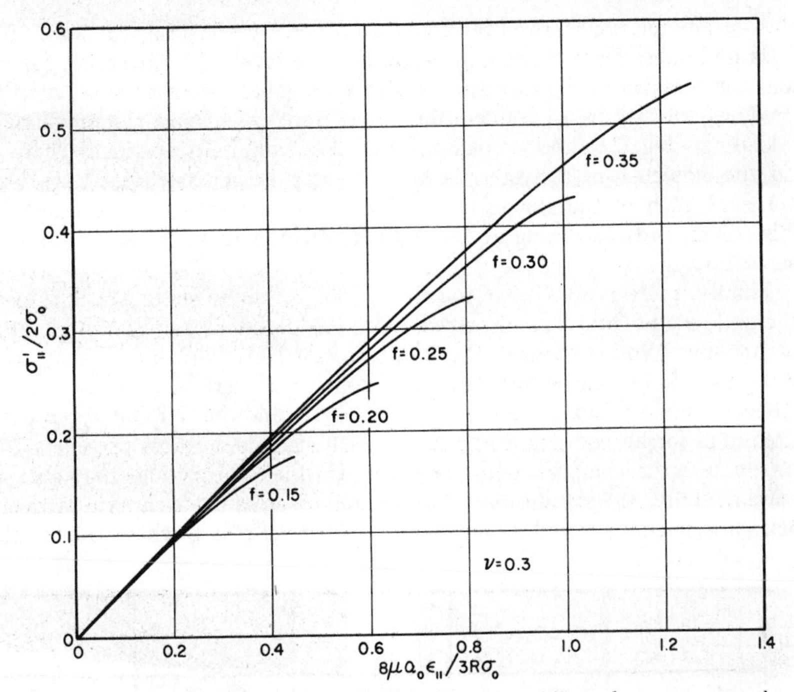

Fig. 22. Stress-strain relation for a simple cubic array of like spheres compressed in the {110} direction: loading portion of cycle [85].

It may be noted that, when $\nu = 0$, the moduli in Eq. (49) correspond to those in an isotropic material. The anisotropy is most pronounced when $\nu = \frac{1}{2}$, but even in this case the speed of the fastest dilatational wave in the packing is only 1.9% greater than that of the slowest wave. (In the face-centered cubic packing this difference is 3.3%.) The fastest wave is the one propagating along the hexagonal axis of the lattice; its speed, for a packing of steel balls, is $v_{100} = 1345\sigma_0^{1/6}$, as opposed to $v_{100} = 980\sigma_0^{1/6}$ obtained by using the Hertz theory. A comparison of the theoretical and experimental relations of wave speed versus pressure results in a diagram all but indistinguishable from that in Fig. 20.

In contrast with the two close-packed arrays the simple cubic lattice of like spheres under initial isotropic compression has been shown by Deresiewicz [85] to constitute a statically determinate structure. The loading history in this case does not enter into the calculation of the contact forces, though it does affect the value of the tangential compliance at each contact. As a result, for homothetic but otherwise arbitrary loading, i.e., loading such that the stress quadrics are similar and similarly oriented, integration of the incremental stress-strain relations reduces to a set of quadratures. Explicit total stress-strain relations are obtained in Ref. [85] for loading to finite stress levels and subsequent unloading between finite levels. In particular, for uniaxial compression in the direction of a face diagonal the relation between the stress, σ_{11}, and strain, ϵ_{11}, (above the initial state of isotropic compression, σ_0) in loading (Fig. 22) and unloading (Fig. 23) exhibits a nonlinear character and the existence of a hysteresis loop. It may be noted that, at $\sigma_{11}/2\sigma_0 = f/(1-f)$ each of the curves in Fig. 22 terminates with a horizontal slope. This corresponds to sliding at the individual contacts, constituting failure in the assemblage.

Finally, more recently Ko and Scott [86] reexamined the static response of simple cubic and face-centered cubic models under initial hydrostatic compression. Working within the framework of Hertzian contact, they based their analysis on the assumption that spheres at certain locations in each lattice are slightly smaller than the other spheres and that a minimum pressure is required for all contacts to be made. Although the authors present a comparison between compressibility predicted by their theory and that obtained experimentally, the significance of this comparison is not clear. In particular, their experiments show the existence of a hysteresis loop which they explain

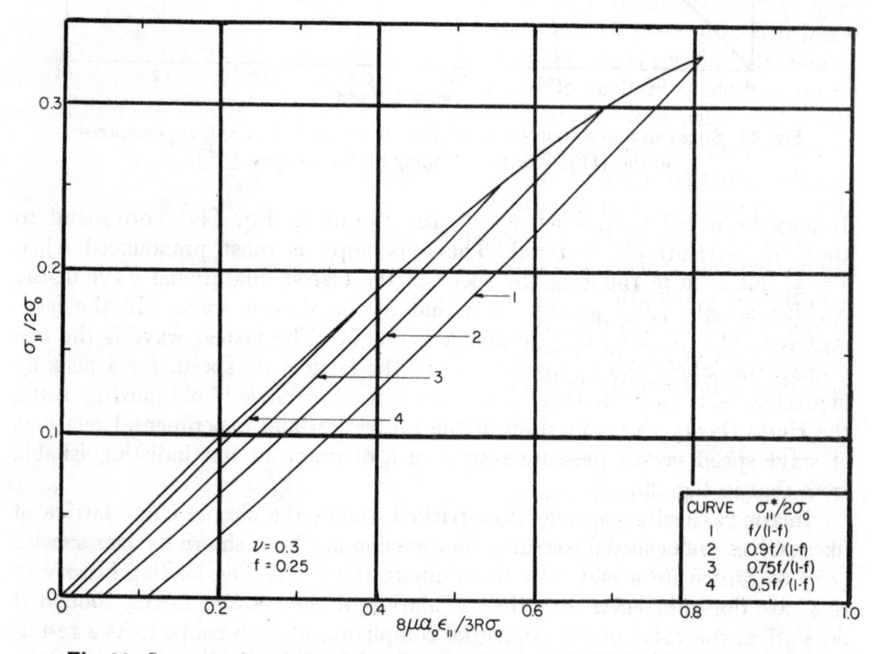

Fig. 23. Stress-strain relation for a simple cubic array of like spheres compressed in the {110} direction: loading-unloading cycle [85].

away as being due to various causes such as membrane penetration, wall bulging and friction between membrane and grains, but not due to possible intergranular slip.

Closure

It is instructive to conclude this review with a comparison of the state of the art in the field today and at the inception of Mindlin's activity in it just about twenty-five years ago.

Contact theory had by that time already taken on the aspect of a classical field offering few new directions for fruitful inquiry. In fact, except for Cattaneo's 1938 paper on tangential contact, his two 1947 papers on second-order contact and some work in Germany (particularly on rolling contact) published about the same time, one finds little activity in this area, at least outside the Soviet Union. Mindlin's 1949 paper on tangential and torsional contact started a trickle of research which rapidly broadened to include not only his subsequent work but also that of his students and even of their students, all this activity stimulating researches by others as well. In the list of references which follows, about two dozen items dealing with contact theory may be said to have their direct antecedents in the initial effort by Mindlin. Moreover, it is of interest to note that, of the six open problems of contact theory enumerated fifteen years ago in Ref. [79], four have been solved and problems of anisotropic as well as of inelastic contact which had not been mentioned there have also been treated. The remaining unsolved questions deal with contact under arbitrarily varying oblique forces.

The situation with a theory of granular media is somewhat different. Although here, too, the effect of Mindlin's work has been pronounced, leading to significant improvement over then existing theory, the research which it stimulated can be said to have made only a first step on a long road toward a satisfactory analytical edifice. Thus none of the four open questions listed in Ref. [79] has been discussed in the subsequent literature, this fact being a reflection not of lack of importance of the problems or of interest in them but rather an indication of the formidable difficulties which lie in store.

References

1. H. Hertz, Ueber die Berührung fester elastischer Körper, *J. reine angew. Math.* **92**, 156–171 (1882). The date is frequently, though erroneously, given as 1881.
2. D. C. Gazis, Graphical investigation of geometric aspects of the Hertz problem, *J. appl. Mech.* **27**, 735–737 (1960).
3. D. H. Cooper, Hertzian contact-stress deformation coefficients, *J. appl. Mech.* **36**, 296–303 (1969).
4. J. L. Lubkin, Contact problems, *Handbook of Engineering Mechanics*, edited by W. Flügge, McGraw-Hill, New York, 1962, Chap. 42.

5. N. Usunoff, Über den Spannungszustand im Halbraum bei halbkugelförmiger Druckverteilung, *Z. angew. Math. Mech.* **22,** 262–269 (1942).
6. C. Weber, Halbraum mit Halbkugelbelastung, *Z. angew. Math. Mech.* **22,** 318–321 (1942).
7. C. Cattaneo, Teoria del contatto elastico in seconda approssimazione, *Rc. Mat. Applic.* **6,** 504–512 (1947).
8. H. Deresiewicz, A note on second-order Hertz contact, *J. appl. Mech.* **28,** 141–142 (1961).
9. E. Steuermann, To Hertz's theory of local deformations in compressed elastic bodies, *C.R. Acad. Sci. URSS* **25,** 359–361 (1939).
10. E. Steuermann, On the question of local deformations in elastic bodies pressed against one another, *C.R. Acad. Sci. URSS* **31,** 738–741 (1941).
11. M. V. Korovchinskii, Local contact of closely conforming bodies (in Russian), *Mashinovedenie,* No. 2, March–April, 71–79 (1970). Reported in *Appl. Mech. Rev.* 856 (1972).
12. G. Lundberg, Elastische Berührung zweier Halbräume, *Forsch. Geb. Ingenieurw.* **10,** 201–211 (1939).
13. L. A. Galin, *Contact Problems in the Theory of Elasticity* (English translation), Applied Mathematics Research Group, North Carolina State College, Raleigh, N.C., 1961.
14. E. Steuermann, Some special cases of the contact problem, *C.R. Acad. Sci. URSS* **38,** 197–200 (1943).
15. L. E. Goodman and L. M. Keer, The contact stress problem for an elastic sphere indenting an elastic cavity, *Int. J. Solids Struct.* **1,** 407–415 (1965).
16. V. F. Bondareva, Contact problems for an elastic sphere, *PMM* **35,** 37–45 (1971).
17. L. E. Goodman, Contact stress analysis of normally loaded rough spheres, *J. appl. Mech.* **29,** 515–522 (1962).
18. A. Schäfer, Die Reibung beim Walzendruck bei verschiedenen Elastizitätsmoduln, *Arch. Eisenhüttenw.* **23,** 253–256 (1952).
19. G. C. Feng, Discussion of Ref. [17], *J. appl. Mech.* **30,** 317 (1963).
20. H. D. Conway, The effects of friction on normal contact stresses, *J. appl. Mech.* **38,** 1094–1095 (1971).
21. H. D. Conway, The pressure distribution between two elastic bodies in contact, *Z. angew. Math. Phys.* **7,** 460–465 (1956).
22. J. R. Willis, Hertzian contact of anisotropic bodies, *J. Mech. Phys. Solids* **14,** 163–176 (1966).
23. G. Ya. Popov and V. V. Savchuk, A contact problem for theory of elasticity for a circular contact region including consideration of surface structure of the contacting bodies (in Russian), *Izv. Akad. Nauk SSSR, Mekh. Tverd. Tela,* No. 3, May–June, 80–88 (1971). Reported in *Appl. Mech. Rev.,* No. 6089 (1972).
24. M. K. Kassir, Boussinesq problems for nonhomogeneous solid, *J. Eng. Mech. Div., Proc. ASCE* **98,** 457–470 (1972).
25. E. H. Lee and J. R. M. Radok, The contact problem for viscoelastic bodies, *J. appl. Mech.* **27,** 438–444 (1960).
26. W. H. Yang, The contact problem for viscoelastic bodies, *J. appl. Mech.* **33,** 395–401 (1966).
27. S. C. Hunter, The Hertz problem for a rigid spherical indenter and a viscoelastic half-space, *J. Mech. Phys. Solids* **8,** 219–234 (1960).
28. H. H. Calvit, Numerical solution of the problem of impact of a rigid sphere onto a linear viscoelastic half-space and comparison with experiment, *Int. J. Solids Struct.* **3,** 951–966 (1967).

29. G. A. C. Graham, The contact problem in the linear theory of viscoelasticity, *Int. J. Engng. Sci.* **3,** 27–46 (1965).
30. T. C. T. Ting, The contact stresses between a rigid indenter and a viscoelastic half-space, *J. appl. Mech.* **33,** 845–854 (1966).
31. K. L. Johnson, A note on the adhesion of elastic solids, *Br. J. appl. Phys.* **9,** 199–200 (1958).
32. R. C. Drutowski, Hertzian contact and adhesion of elastomers, *J. Lubr. Technol., Trans. ASME,* Ser F, **91,** 732–737 (1969).
33. K. L. Johnson, K. Kendall, and A. D. Roberts, Surface energy and the contact of elastic solids, *Proc. Roy. Soc.,* **A324,** 301–313 (1971).
34. L. M. Keer, The contact stress problem for an elastic sphere indenting an elastic layer, *J. appl. Mech.* **31,** 143–145 (1964).
35. G. Ya. Popov, The contact problem of the theory of elasticity for the case of a circular area of contact, *PMM* **26,** 207–225 (1962).
36. Y. O. Tu, A numerical solution for an axially symmetric contact problem, *J. appl. Mech.* **34,** 283–286 (1967).
37. Y. O. Tu and D. C. Gazis, The contact problem of a plate pressed between two spheres, *J. appl. Mech.* **31,** 659–666 (1964).
38. J. H. Rumbarger, R. C. Herrick, and P. R. Eklund, Analysis of the elastic contact of a hollow ball with a flat plate, *J. Lubr. Technol., Trans. ASME,* Ser F, **92,** 138–144 (1970).
39. D. P. Updike and A. Kalnins, Contact pressure between an elastic spherical shell and a rigid plate, *J. appl. Mech.* **39,** 1110–1114 (1972).
40. H. Christensen, Elastohydrodynamic theory of spherical bodies in normal approach, *J. Lubr. Technol., Trans. ASME,* Ser. F, **92,** 145–154 (1970).
41. T. F. Conry and A. Seireg, A mathematical programming method for design of elastic bodies in contact, *J. appl. Mech.* **38,** 387–392 (1971).
42. J. J. Kalker and Y. van Randen, A minimum principle for frictionless elastic contact with application to non-Hertzian half-space contact problems, *J. engng. Math.* **6,** 193–206 (1972).
43. C. Cattaneo, Sul contatto di due corpi elastici, *Atti Accad. naz. Lincei Rc.,* Ser. 6, **27,** 342–348, 434–436, 474–478 (1938).
44. K. L. Johnson, Surface interaction between elastically loaded bodies under tangential forces, *Proc. Roy. Soc.* **A230,** 531–548 (1955).
45. H. Poritsky, Stresses and deflections of cylindrical bodies in contact with application to contact of gears and of locomotive wheels, *J. appl. Mech.* **17,** 191–201, 466–468 (author's closure) (1950).
46. J. O. Smith and C. K. Liu, Stresses due to tangential and normal loads on an elastic solid with application to some contact stress problems, *J. appl. Mech.* **20,** 157–166 (1953).
47. G. M. Hamilton and L. E. Goodman, The stress field created by a circular sliding contact, *J. appl. Mech.* **33,** 371–376 (1966).
48. P. J. Vermeulen and K. L. Johnson, Contact of nonspherical elastic bodies transmitting tangential forces, *J. appl. Mech.* **31,** 338–340 (1964).
49. C. Cattaneo, Teoria del contatto elastico in seconda approssimazione: compressione obliqua, *Rc. Sem. Fac. Sci. Univ. Cagliari* **17,** 13–28 (1947).
50. V. C. Mow, P. L. Chow, and F. F. Ling, Microslips between contacting paraboloids, *J. appl. Mech.* **34,** 321–328 (1967).

51. H. Deresiewicz, Oblique contact of nonspherical elastic bodies, *J. appl. Mech.* **24,** 623–624 (1957).
52. K. L. Johnson, Recent developments in the theory of elastic contact stresses: their significance in the study of surface breakdown, *Proc. Conf. Lubric. and Wear, Instn. Mech. Engrs.*, 620–627 (1957).
53. R. V. Klint, Oscillating tangential forces on cylindrical specimens in contact at displacements within the regions of no gross slip, *Amer. Soc. Lubr. Eng. Trans.* **3,** 255–264 (1960).
54. L. E. Goodman and C. B. Brown, Energy dissipation in contact friction: constant normal and cyclic tangential loading, *J. appl. Mech.* **29,** 17–22, 763–764 (discussion and authors' closure) (1962).
55. J. E. White, Static friction as a source of seismic attenuation, *Geophysics* **31,** 333–339 (1966).
56. K. L. Johnson, Energy dissipation at spherical surfaces in contact transmitting oscillating forces, *J. Mech. Eng. Sci.* **3,** 362–368 (1961).
57. J. L. Lubkin, The torsion of elastic spheres in contact, *J. appl. Mech.* **18,** 183–187 (1951).
58. H. Deresiewicz, Contact of elastic spheres under an oscillating torsional couple, *J. appl. Mech.* **21,** 52–56 (1954).
59. M. Hetényi and P. H. McDonald, Jr., Contact stresses under combined pressure and twist, *J. appl. Mech.* **25,** 396–401 (1958); **26,** 305–306 (discussion) (1959).
60. N. A. Rostovstev, On the problem of torsion of an elastic half-space (in Russian), *Prikl. Mat. Mekh.* **19,** 55–60 (1955).
61. C. Cattaneo, Compressione e torsione nel contatto tra corpi elastici di forma qualunque, *Ann. Scuola Norm. Sup. Pisa*, Ser. III, **9,** 23–42 (1955).
62. M. Pacelli, Contatto con attrito tra due corpi elastici di forma qualunque: compressione e torsione, *Ann. Scuola Norm. Sup. Pisa*, Ser. III, **10,** 155–184 (1956).
63. A. L. Florence, Two contact problems for an elastic layer, *Quart. J. Mech. appl. Math.* **14,** 453–459 (1961).
64. L. M. Keer, The torsion of a rigid punch in contact with an elastic layer where the friction law is arbitrary, *J. appl. Mech.* **31,** 430–434 (1964).
65. J. J. O'Connor, Compliance under a small torsional couple of an elastic plate pressed between two identical elastic spheres, *J. appl. Mech.* **33,** 377–383 (1966).
66. H. Deresiewicz, On the mechanics of granular media, *Applied Mechanics Surveys*, Spartan Books, Washington, 1966, pp. 277–284.
67. O. Reynolds, On the dilatancy of media composed of rigid particles in contact, *Philosophical Magazine*, Ser. 5, **20,** 469–481 (1885).
68. G. Hara, Theorie der akustischen Schwingungsausbreitung in gekörnten Substanzen und experimentelle Untersuchungen an Kohlepulver, *Elektrische Nachrichten-Technik* **12,** 191–200 (1935).
69. K. Iida, The velocity of elastic waves in sand, *Bull. Earthq. Res. Inst. Tokyo Univ.* **16,** 131–144 (1938).
70. K. Iida, Velocity of elastic waves in a granular substance, *Bull. Earthq. Res. Inst. Tokyo Univ.* **17,** 783–808 (1939).
71. T. Takahashi and Y. Satô, On the theory of elastic waves in granular substance, *Bull. Earthq. Res. Inst. Tokyo Univ.* **27,** 11–16 (1949); **28,** 37–43 (1950).
72. F. Gassmann, Elastic waves through a packing of spheres, *Geophysics* **16,** 673–685 (1951); **18,** 269 (1953).

73. H. Brandt, A study of the speed of sound in porous granular media, *J. appl. Mech.* **22,** 479–486 (1955).
74. H. Brandt, Factors affecting compressional wave velocity in unconsolidated marine sand sediments, *J. Acoust. Soc. Amer.* **32,** 171–179 (1960).
75. N. Nasu, Studies on the propagation of an artificial earthquake wave through superficial soil or sand layers, and the elasticity of soil and sand (in Japanese), *Bull. Earthq. Res. Inst. Tokyo Univ.* **18,** 289–304 (1940).
76. D. S. Hughes and J. L. Kelly, Variation of elastic wave velocity with saturation in sandstone, *Geophysics* **17,** 739–752 (1952).
77. N. R. Paterson, Seismic wave propagation in porous granular media, *Geophysics* **21,** 691–714 (1956).
78. E. A. Kozlov, Longitudinal wave velocities in terrigenous deposits, *Bull. Acad. Sci. USSR, Geophys. Ser.*, 647–656 (1962).
79. H. Deresiewicz, Mechanics of granular matter, *Advances in Applied Mechanics*, Academic Press, New York, 1958, Vol. 5, pp. 233–306.
80. A. E. R. Westman and H. R. Hugill, The packing of particles, *J. Amer. Ceram. Soc.* **13,** 767–779 (1930).
81. W. O. Smith, P. D. Foote and P. F. Busang, Packing of homogenous spheres, *Phys. Rev.* **34,** 1271–1274 (1929).
82. C. W. Thurston, Discussion of Ref. [M58], *J. appl. Mech.* **25,** 310–311 (1958).
83. C. W. Thurston and H. Deresiewicz, Analysis of a compression test of a model of a granular medium, *J. appl. Mech.* **26,** 251–258 (1959).
84. J. Duffy, A differential stress-strain relation for the hexagonal close-packed array of elastic spheres, *J. appl. Mech.* **26,** 88–94 (1959).
85. H. Deresiewicz, Stress-strain relations for a simple model of a granular medium, *J. appl. Mech.* **25,** 402–406 (1958).
86. H. Y. Ko and R. F. Scott, Deformation of sand in hydrostatic compression, *J. Soil Mech. Found. Div., Proc. ASCE* **93,** 137–156 (1967).

Added in proof

We list here several papers which would have been noticed in the text had they come to our attention earlier or had been completed earlier.

Y. M. Tsai, Stress distribution in elastic and viscoelastic plates subjected to symmetrical rigid indentations, *Quart. appl. Math.* **27,** 371–380 (1969).

Y. M. Tsai, Thickness dependence of the indentation hardness of glass plates, *Int. J. Fracture Mech.* **5,** 157–165 (1969).

Y. M. Tsai, Dynamic contact stresses produced by the impact of an axisymmetric projectile on an elastic half-space, *Int. J. Solids Struct.* **7,** 543–558 (1971).

C. Hardy, C. N. Baronet and G. V. Tordion, The elasto-plastic indentation of a half-space by a rigid sphere, *Int. J. Numer. Methods in Engng.* **3,** 451–462 (1971).

L. K. Agbezuge and H. Deresiewicz, On the indentation of a consolidating half-space, presented at *13th Int. Congr. Theor. Appl. Mech.*, Moscow, 1972.

R. A. Davis, A discrete probabilistic model for mechanical response of a granular medium, Ph.D. dissert., Columbia Univ., 1973.

* * *

Figures used with permission.

Waves and Vibrations in Isotropic and Anisotropic Plates

YIH-HSING PAO
Cornell University, Ithaca, New York 14850

and

RAJ K. KAUL
State University of New York, Buffalo, New York 14214

1 Introduction

Since Chladni demonstrated the nodal patterns of a vibrating plate in 1787, the subject of vibration of plates has fascinated many scientist and engineers. In a prize-winning memoir, offered for the third time by the Academy of Science of the French Institute in 1815, Sophie Germain [1] obtained the equation for an isotropic plate in bending motion. At about the same time, the subject also engaged the interest of many other distinguished scholars including Lagrange, Legendre, Fourier, Poisson, Cauchy, and Euler. The equation which is listed as Eq. (2.1) in this article is now referred to as the Germain–Lagrange equation because of certain contributions which Lagrange made when the paper was submitted for the prize.

The publication of Germain's memoir also started a prolonged controversy on the proper boundary conditions of a thin plate in bending. Saint-Venant, Thomson, Tait, Mathieu, Boussinesq, and Kirchhoff all made contributions to this subject. The paper by Kirchhoff [2] merits particular attention as it contained a complete derivation of the plate equation and the correct boundary conditions based on a variational method. It also contained a theorem of the uniqueness of the solutions and a comparison of theoretical results with Chladni's experiments.

At the same period of time, the in-plane motion of a plate was also investigated. The governing equations for a thin isotropic plate in extensional motion were derived by Poisson [3], and they are listed as Eq. (2.3). However, in contrast to flexural motion, the subject of extensional vibrations of a thin plate attracted much less attention until very recently.

In 1888, Rayleigh [4] published a paper which marked a turning point in the history of the theory of plates. In this paper, Rayleigh determined the vibrations of an homogeneous, isotropic plate of arbitrary thickness from the linear theory of elasticity. He showed also that his results were reducible to those deduced from the thin-plate theories in the limit of zero-plate thickness, and to those of the surface waves in half-space (the Rayleigh surface waves) when the wavelength approached zero. Based on Rayleigh's formulation, Lamb [5] investigated in detail the propagation of waves in a thick plate (the Rayleigh–Lamb waves).

As shall be shown later, the characteristic frequencies of Rayleigh–Lamb waves are well over 10 kilohertz for an ordinary structural member. Thus it is not surprising that the research of Rayleigh and Lamb drew little attention at that time. In fact, the subject was not even discussed in Rayleigh's own treatise, *The Theory of Sound*, because "the subject belongs rather to the theory of elasticity," whereas only a reference to that subject was cited in Love's *Treatise on the Mathematical Theory of Elasticity*.

The history of research on the waves and vibrations in anisotropic plate dates as far back as that of isotropic plates. Equations for extensional and flexural vibrations of thin crystal plates were derived by Cauchy [6] in 1829. Toward the end of last century, crystal physics was an actively pursued subject. Among many other new findings, the discovery of piezoelectricity by Curie in 1880 had a great impact on the modern technology of telecommunications, and gave an impetus toward further development of plate theories.

During the Second World War, quartz crystals, a piezoelectric material, were used in large quantities as oscillators and frequency standards in radio equipments. It then required the precise determination of the natural modes and frequencies of quartz crystal plates, all in high kilocycle or megacycle range. The successful development of quartz and other materials as transducers, filters, resonators, and delay lines depended to a large extent on the understanding of the vibration and wave propagation in plates (or rods) made of these materials.

Through his association with Bell Telephone Laboratories (the major developer of quartz crystals during the war), Mindlin became interested in the subject of waves and high-frequency vibrations of elastic plates and rods. From 1951 onward, he and his associates, most of them graduate students; published more than forty papers on the theory of plates. The main goal of all these researches was to study the dispersion of waves in a plate of arbitrary thickness, and to analyze the vibrational modes and frequencies of a bounded plate.

The basic difficulty in the analysis is the inability to find exact solutions for the boundary-value problems of plates in motion, based on the linear theory of elasticity. Rayleigh and Lamb were successful only in finding exact solutions of an infinitely extended plate with the major plane surfaces free of stresses. For a plate with additional traction-free boundary conditions imposed on the edge surface of the plate (a plate bounded by traction-free surfaces on all sides), the general solution has not yet been found. Hence analytically, one is forced either to find an approximate solution of the exact theory of elasticity as applied to a bounded plate, or to develop an approximate theory for the plate so that closed form solutions are obtainable.

Following Cauchy and Poisson, Mindlin adopted the latter approach which in view of the historical development of the subject seemed to be the natural one to follow. In 1951 he presented an approximate theory of plate [M35]† by including the effects of rotatory inertia and shear on the transverse vibration of an isotropic plate. The basic equation of the theory, which is analogous to the Timoshenko equation of beams, is now commonly referred to as the Mindlin equation of plates. It is an extension of the Germain–Lagrange equation of thin plates and can be applied to a higher range of frequencies. The equations were found extremely useful in predicting the high-frequency flexural vibrations of a circular disk which is a component of a mechanical filter [7]. Analogous theories were developed for anisotropic plates, which have been successfully applied to analyze the vibration of certain piezoelectric crystal plates [M36, M69].

To cover an even wider range of frequencies, higher-order plate theories were developed. Following Cauchy's work on crystal plates, Mindlin developed a systematic method of approximation by expanding the displacement components of a plate into a power series (or a series of orthogonal polynomials) of the thickness coordinate. The coefficients of the series, which are functions of time, and the remaining two space coordinates are designated as the ordered displacements of the plate. By properly truncating the series and by using the variational method as developed by Kirchhoff, various order-plate theories have been obtained, the classical theory being the zeroth-order one. The applicable frequency range increases with the order of approximation. Since the governing equations are two-dimensional, exact solutions for a bounded plate with traction-free surfaces can often be found.

One test of the validity of various approximate theories is to compare the dispersion relation based on them to that based on the exact theory for an infinitely extended plate. In the case of isotropic plates, the exact solution is the Rayleigh–Lamb solution. For anisotropic plates, the solution was formulated by Ekstein [8]. However, because of the complexity of the frequency equations in both cases, no detailed information on the dispersion relations were available at that time. Such results are important in deciding what to include in various orders of approximate plate theories. The exact dispersion relation also supplies the data necessary for the adjustment of the constitutive equations which are obtained after truncation. It is for these reasons that major efforts were launched by Mindlin to investigate fully the Rayleigh–Lamb and Ekstein frequency spectra.

This article is an attempt to summarize a part of the research by Mindlin

† Numbers preceded by the letter M in brackets refer to the Publications of R. D. Mindlin listed after the Preface to this volume.

and his associates on the theory of plates and to pay tribute to his monumental contributions in this area.

2 Waves in an Isotropic Plate

Let x_1, x_2, x_3 be a system of right-handed rectangular Cartesian coordinate system and let t be the time. Consider a homogeneous, isotropic, elastic plate of infinite extent and bounded by two parallel planes $x_2 = \pm b$. Let u_i $(i = 1, 2, 3)$ be the Cartesian components of the displacement vector in the direction x_i. If the plate is extremely thin and the transverse (flexural) motion is predominant, then the waves and vibrations, in the absence of body force, can be analyzed from a study of the classical Germain–Lagrange equation [1]

$$D\nabla_1^2\nabla_1^2 u_2 + 2\rho b \frac{\partial^2 u_2}{\partial t^2} = 0 \tag{2.1}$$

In the above

$$\nabla_1^2 \equiv \frac{\partial^2}{\partial x_1^2} + \frac{\partial^2}{\partial x_2^2} \tag{2.2}$$

is the two-dimensional Laplace operator, and $D = 4\mu b^3/3(1 - \nu)$, is the flexural rigidity of the thin plate, where μ is Lamé's modulus of shear rigidity, ν is Poisson's ratio, and ρ is the mass density.

When the in-plane (extensional) motion is predominant, the appropriate equations of motion in the absence of in-plane body forces, are the Poisson extensional equations [3]

$$(\bar{\lambda} + \mu) \frac{\partial}{\partial x_1}\left(\frac{\partial u_1}{\partial x_1} + \frac{\partial u_3}{\partial x_3}\right) + \mu\nabla_1^2 u_1 = \rho \frac{\partial^2 u_1}{\partial t^2}$$

$$(\bar{\lambda} + \mu) \frac{\partial}{\partial x_3}\left(\frac{\partial u_1}{\partial x_1} + \frac{\partial u_3}{\partial x_3}\right) + \mu\nabla_1^2 u_3 = \rho \frac{\partial^2 u_3}{\partial t^2} \tag{2.3}$$

where

$$\bar{\lambda} = \frac{2\lambda\mu}{\lambda + \mu} \tag{2.4}$$

and λ is Lamé's constant.

In Eqs. (2.1) and (2.3) displacement components are independent of the thickness coordinate, i.e., $u_i \equiv u_i(x_1, x_3, t)$, (see Fig. 1). These two-dimensional thin-plate equations are valid only for waves with wavelengths much larger than the thickness of the plate and at low frequencies.

For short wavelengths and high frequencies, the thin-plate theories are no longer valid because in general the u_i are functions of the thickness coordinate

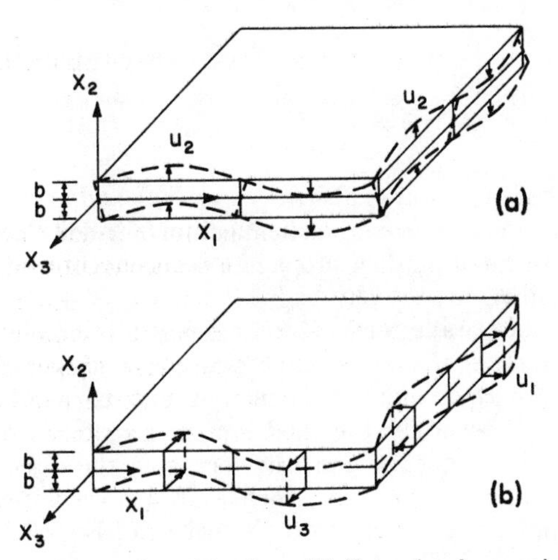

Fig. 1. (a) Flexural waves in a thin plate. (b) Extensional waves in a thin plate.

x_2, and this variation of displacement across the thickness can no longer be disregarded. As such, study of waves in plates requires analysis of the exact three-dimensional theory of linear elasticity, which in terms of displacement components $u_i = u_i(x_1, x_2, x_3; t)$ are given by

$$(\lambda + \mu)\partial_i\partial_k u_k + \mu\partial_k\partial_k u_i = \rho\ddot{u}_i \tag{2.5}$$

where $\partial_i \equiv \partial/\partial x_i$. Latin indices $i, j, k, \ldots$ take the values from 1 to 3, and repetition of an index implies the usual summation over the range 1 to 3. Partial differentiation with respect to time is indicated by a superposed dot with $\dot{u}_i \equiv \partial u_i/\partial t$.

The stress tensor τ_{ij} is symmetric and in the case of an isotropic material, it is related to the displacement gradients in accordance with the linear constitutive law

$$\tau_{ij} = \lambda\delta_{ij}\partial_k u_k + \mu(\partial_i u_j + \partial_j u_i) \tag{2.6}$$

where δ_{ij} is the Kronecker delta.

Equation (2.5) can be decomposed into two uncoupled system of equations, by using the Poisson–Lamé [3, 9] decomposition formula

$$u_i = \partial_i\varphi + \epsilon_{ijk}\partial_j H_k; \qquad \partial_i H_i = 0 \tag{2.7}$$

where φ is a scalar point function and H_i are the three components of a vector field. The four components of the displacement potentials φ and H_i are particular solutions of the wave equations

$$v_p^2\nabla^2\varphi = \ddot{\varphi}$$
$$v_s^2\nabla^2 H_i = H_i \tag{2.8}$$

where

$$v_p^2 = \frac{\lambda + 2\mu}{\rho}; \qquad v_s^2 = \frac{\mu}{\rho} \tag{2.9}$$

represent the speed of dilatational wave (P-wave) and shear wave (S-wave), respectively. Thus in an isotropic elastic medium of infinite extent, the P-wave and S-wave are uncoupled; they propagate independently of each other with wave speeds v_p and v_s, respectively.

At the two plane parallel surfaces $x_2 = \pm b$ of a thick plate, when the two faces are free of traction, the two waves are coupled and the resulting motion as viewed along the length or width of the plate is quite complicated. The primary interest therefore is to study the modes of propagation and the dispersion relation between the frequency and wavelength in the direction of the wave propagation. In order to carry out such a study, we impose the following traction-free boundary conditions to the solutions of Eq. (2.5),

$$\tau_{2i}(x_1, x_2, x_3; t)\big]_{x_2=\pm b} = 0, \qquad i = 1, 2, 3 \tag{2.10}$$

Later we shall briefly discuss the effect of additional boundary conditions prescribed on the parallel planes $x_1 = \pm a$ and $x_3 = \pm c$. Note that the boundary condition (2.10) is identically satisfied in the case of classical flexural and extensional equations of motion of thin plates.

(A) *SH-Modes*

Consider two horizontally polarized shear waves (SH) with displacement components

$$u_1 = u_2 = 0$$
$$u_3 = E_1 \exp i(\xi x_1 + \beta x_2 - \omega t) + E_2 \exp i(\xi x_1 - \beta x_2 - \omega t) \tag{2.11}$$

where

$$\omega = v_s(\xi^2 + \beta^2)^{1/2} \tag{2.12}$$

These are the displacement components obtained by setting $\varphi = H_1 = H_3 = 0$ in Eq. (2.7) and choosing $H_2 = H_2(x_1, x_2; t)$ in Eqs. (2.7) and (2.8).

For an infinitely extended medium, each term in the third of Eq. (2.11) represents a plane harmonic wave with distinct amplitude and wavelength. However, in the presence of boundaries, the amplitude and wavelength can no longer be arbitrary. It can be verified that for the solution (2.11) to satisfy the boundary conditions (2.10), we must have

$$\text{(i)} \quad E_1 = E_2(\equiv E/2) \qquad \text{and } \sin \beta b = 0$$
$$\text{(ii)} \quad E_1 = -E_2(\equiv F/2i) \qquad \text{and } \cos \beta b = 0$$

and the solution takes the form

$$\text{(i)}\quad u_3 = E\cos\beta x_2 \exp i(\xi x_1 - \omega t); \qquad b\beta = \frac{n\pi}{2}, \quad n = 0, 2, 4, \ldots \tag{2.13}$$

$$\text{(ii)}\quad u_3 = F\sin\beta x_2 \exp i(\xi x_1 - \omega t); \qquad b\beta = \frac{n\pi}{2}, \quad n = 1, 3, 5, \ldots \tag{2.14}$$

where

$$\omega^2 = v_s^2\left[\xi^2 + \left(\frac{n\pi}{2b}\right)^2\right] \tag{2.15}$$

Each one of these solutions represent a harmonic wave traveling in the x_1-direction with wavelength $2\pi/\xi$ and phase velocity $v = \omega/\xi$, and across the thickness of the plate the amplitude varies sinusoidally.

This analysis shows that the *SH*-modes in a plate are the resultant of two horizontally polarized shear waves, which are multiply reflected between the two free major faces $x_2 = \pm b$. The angles of incidence and reflection are shown in Fig. 2 and are given by

$$\theta_3 = \tan^{-1}\frac{\xi}{\beta} \tag{2.16}$$

The *SH*-modes are not contained in either of the thin-plate Eqs. (2.1) and (2.3). From Eq. (2.15) we observe that the phase velocity v is a function of the wavelength $\lambda = 2\pi/\xi$, and therefore in general, the waves are dispersive, except for $n = 0$. The smallest value of ω is $v_s\pi/2b$, which is much higher than the natural frequencies of a thin plate undergoing flexural motion.

(B) *P- and SV-Modes (Rayleigh–Lamb Waves)*

In an unbounded medium, a pressure wave and a vertically polarized shear wave (*SV*-wave) may be represented, respectively, by setting in Eq. (2.7) $H_1 = H_2 = 0$ and choosing

$$\varphi = \varphi_0 \exp i(\xi x_1 \pm \alpha x_2 - \omega t), \qquad \omega^2 = v_p^2(\alpha^2 + \xi^2)$$

$$H_3 = H_0 \exp i(\xi x_1 \pm \beta x_2 - \omega t), \qquad \omega^2 = v_s^2(\beta^2 + \xi^2) \tag{2.17}$$

as the solutions of Eq. (2.8).

The wave-normal of the *P*-wave (scalar potential φ) makes an angle $\theta_1 = \pm\tan^{-1}(\xi/\alpha)$ with the x_2-axis and the wave-normal of the *SV*-wave (vector potential H_3) makes an angle $\theta_2 = \pm\tan^{-1}(\xi/\beta)$ with the same axis. In a bounded medium these two waves cannot propagate independent of each

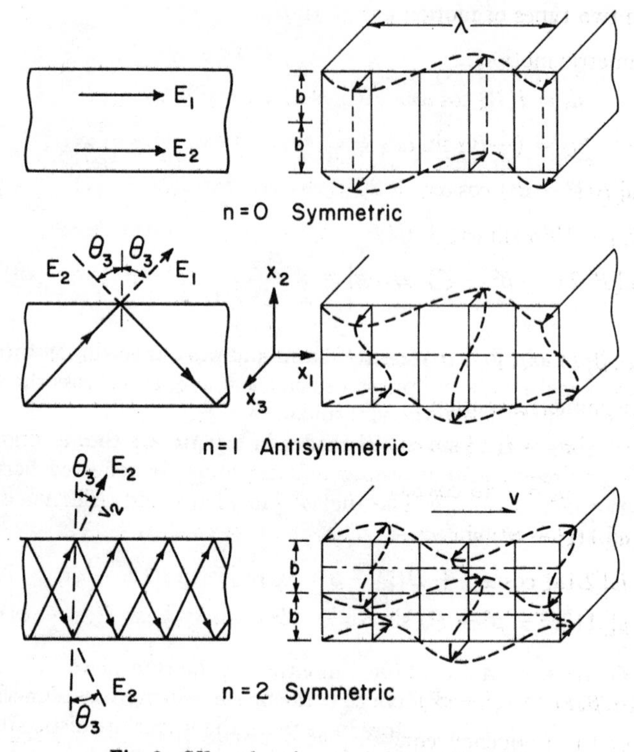

Fig. 2. *SH*-modes of waves in plates ($\lambda = 4b$).

other. Combining four waves of the type $(2.17)_1$ and four of the type $(2.17)_2$ each with properly chosen amplitudes and phase angles, we obtain

$$\varphi = (A \sin \alpha x_2 + B \cos \alpha x_2) \exp i(\xi x_1 - \omega t)$$

$$H_3 = i(C \sin \beta x_2 + D \cos \beta x_2) \exp i(\xi x_1 - \omega t) \tag{2.18}$$

Thus a suitable superposition of eight waves produces a harmonic wave motion along the x_1-axis, with phase velocity $v = \omega/\xi$ and wavelength $2\pi/\xi$. The amplitudes vary sinusoidally along the x_2-axis.

For a thick isotropic plate, with major faces free of traction, the motion separates into two parts: (1) displacements symmetric about the (x_1, x_3)-plane, and (2) displacements antisymmetric about the (x_1, x_3)-plane. With $u_3 = \tau_{23} = \tau_{13} = 0$, the displacements, stresses and the frequency equations, for each of the two types of motion are given by:

1. Symmetric motion

$$u_1 = i(B\xi \cos \alpha x_2 + C\beta \cos \beta x_2) \exp i(\xi x_1 - \omega t)$$
$$u_2 = (-B\alpha \sin \alpha x_2 + C\xi \sin \beta x_2) \exp i(\xi x_1 - \omega t) \quad (2.19)$$

$$\tau_{22} = \mu[B(\xi^2 - \beta^2) \cos \alpha x_2 + 2C\xi\beta \cos \beta x_2] \exp i(\xi x_1 - \omega t)$$
$$\tau_{21} = i\mu[-2B\xi\alpha \sin \alpha x_2 + C(\xi^2 - \beta^2) \sin \beta x_2] \exp i(\xi x_1 - \omega t)$$
$$\tau_{11} = \mu[B(2\alpha^2 - \beta^2 - \xi^2) \cos \alpha x_2 - 2C\xi\beta \cos \beta x_2] \exp i(\xi x_1 - \omega t) \quad (2.20)$$

and

$$F_1(\alpha, \beta, \xi) \equiv (\xi^2 - \beta^2)^2 \cos \alpha b \sin \beta b + 4\alpha\beta\xi^2 \sin \alpha b \cos \beta b = 0 \quad (2.21)$$

2. Antisymmetric motion

$$u_1 = i(A\xi \sin \alpha x_2 - D\beta \sin \beta x_2) \exp i(\xi x_1 - \omega t)$$
$$u_2 = (A\alpha \cos \alpha x_2 + D\xi \cos \beta x_2) \exp i(\xi x_1 - \omega t) \quad (2.22)$$

$$\tau_{22} = \mu[A(\xi^2 - \beta^2) \sin \alpha x_2 - 2D\xi\beta \sin \beta x_2] \exp i(\xi x_1 - \omega t)$$
$$\tau_{21} = i\mu[2A\alpha\xi \cos \alpha x_2 + D(\xi^2 - \beta^2) \cos \beta x_2] \exp i(\xi x_1 - \omega t)$$
$$\tau_{11} = \mu[A(2\alpha^2 - \beta^2 - \xi^2) \sin \alpha x_2 + 2D\xi\beta \sin \beta x_2] \exp i(\xi x_1 - \omega t) \quad (2.23)$$

and

$$F_2(\alpha, \beta, \xi) \equiv (\xi^2 - \beta^2)^2 \sin \alpha b \cos \beta b + 4\alpha\beta\xi^2 \sin \beta b \cos \alpha b = 0 \quad (2.24)$$

Each of the frequency equations is obtained from the surface boundary conditions requiring that the tractions τ_{21} and τ_{22} vanish at the faces $x_2 = \pm b$. These two frequency equations can be written in the convenient form

$$F(\alpha, \beta, \xi) \equiv \frac{\tan \beta b}{\tan \alpha b} + \left[\frac{4\alpha\beta\xi^2}{(\xi^2 - \beta^2)^2}\right]^{\pm 1} = 0 \quad (2.25)$$

where

$$\alpha^2 + \xi^2 = \frac{\omega^2}{v_p^2}; \qquad \beta^2 + \xi^2 = \frac{\omega^2}{v_s^2} \quad (2.26)$$

and the exponent $+1(-1)$ belongs to symmetric (antisymmetric) modes of motion. Elimination of α and β from Eq. (2.25) with the use of Eq. (2.26) leads to a highly complicated transcendental frequency equation relating the variables ω and ξ (or phase velocity v and ξ). For each ξ there is an ω and therefore every pair of (ω, ξ) which satisfies the frequency equation, is a proper mode of wave propagation of the isotropic thick plate, with traction-free faces.

In the (ω, ξ)-plane, for real values of ξ, the frequency Eq. (2.25) can be cast into three different suitable forms depending on whether α and β are real or imaginary. Thus if

(1) $\omega/\xi < v_s$, α and β being imaginary, we replace α by $i\alpha$ and β by $i\beta$ in Eq. (2.25).
(2) $v_s < \omega/\xi < v_p$, α being imaginary and β being real, we replace α by $i\alpha$ in Eq. (2.25).
(3) $\omega/\xi > v_p$, α and β are both real.

The frequency Eq. (2.25) with α and β both imaginary, was first derived by Rayleigh [4] in 1888. In 1917 this frequency equation was analyzed by Lamb [5], and he calculated the roots of this equation for low frequencies. In the limit of long wavelengths and low frequency the dispersion relation obtained from the exact frequency equation for antisymmetric modes of motion agrees with that deduced from the flexural Eq. (2.1), and that for the symmetric modes coincides with the one derived from the extensional Eq. (2.3).

The complete details of the intricate behavior of the modes of propagation of a thick plate are contained in the complicated transcendental Eq. (2.25). Before the advent of electronic digital computers a few efforts were made to explore the nature of these roots, in the region of long wavelengths and low frequencies. Lamb in 1917 studied the lowest symmetric and antisymmetric modes. He also revealed the presence of Lamé [9] modes and the cut-off modes. A search for the roots of the transcendental equation in the high-frequency region was first initiated by Holden [10]. He made use of the bounds α = constant, β = constant to construct a portion of the spectrum for the symmetric modes and real wave numbers. Onoe [11] considerably extended and developed this procedure and his study included antisymmetric modes and imaginary phase velocities. A comprehensive study of these equations for both low and high frequencies, associated with real, imaginary, or complex wave numbers was completed by Mindlin and several of his associates. The research contributions of Mindlin in this area are contained in his two fundamental memoirs [12, 13] and a review article [M66].

(C) *Analysis of Rayleigh–Lamb Frequency Equation*

The transcendental frequency Eq. (2.25) relates the frequency ω and the three wave numbers α, β, and ξ. Eliminating the wave numbers α and β with the help of Eq. (2.26) leads to an implicit functional relationship between the remaining two variables ω and ξ. This implicit relationship which represents dispersion spectrum of the modes of propagation, can be exhibited graphically by a family of curves in an affine space with ω as ordinate and ξ as the abscissa. Each curve of the family of curves is called a "branch" of the frequency spectrum.

Each and every root of the frequency equation is a point on this branch, and each point on every branch represents a mode of wave propagation in the plate. Different modes corresponding to the same branch possess certain common characteristic features, but modes belonging to one branch are distinctly different from the modes of another branch. On a (ω, ξ)-plane, the ratio ω/ξ corresponding to a particular point on a branch is the phase velocity of the particular mode; and the slope of the tangent to the branch at that point is the corresponding group velocity, $v_g \equiv \partial\omega/\partial\xi$. On a phase velocity diagram, representing the relation between phase velocity v and wave number ξ, the group velocity can be calculated from the slope of the tangent and the formula

$$v_g \equiv \frac{\partial \omega}{\partial \xi} = v + \xi \frac{\partial v}{\partial \xi} \tag{2.27}$$

For plotting the dispersion spectrum we may use the frequency ω and the wave number ξ as the ordinate and abscissa, respectively. However, it is useful to first convert them into a dimensionless form according to the relations

$$\Omega = \frac{\omega}{\omega_s}; \qquad \xi = \frac{2b\xi}{\pi}$$

$$\omega_s = \frac{\pi v_s}{2b}; \qquad v_s^2 = \frac{\mu}{\rho}$$

$$V = \frac{v}{v_s}; \qquad k^2 = \frac{v_p^2}{v_s^2} = \frac{2(1-\nu)}{1-2\nu} \tag{2.28}$$

where ω_s is the frequency of the lowest antisymmetric thickness-shear wave. The (ω, ξ)-plane may be divided into three sectors by lines OL $(\alpha = 0)$ and OE $(\beta = 0)$ as shown in Fig. 3. In sector 1, $\Omega < \xi$, $(V < 1)$ and both α and β in Eq. (2.25) are imaginary. In sector 2, $\xi < \Omega < k\xi$, $(1 < V < k)$ α is imaginary and β is real. In sector 3, $\Omega > \xi$, $(V > k)$ and both α and β are real. A division similar to this can also be made in the (V_p, ξ)-plane, as shown in Fig. 4, where $V_p = v/v_p$, and $V_p = 2V$ for $\nu = \frac{1}{3}$.

We now discuss the asymptotic behavior of the branches in the two limiting cases of infinite and zero wavelengths, i.e., as $\xi \to 0$ and $\xi \to \infty$, respectively, and in the simple degenerate case of $\xi = \beta$. There is another limiting case when $\Omega/\xi = v_p/v_s$, $\alpha = 0$, but we do not include it in this discussion.

Surface waves In sector 1 of the spectrum, $\alpha \to i\alpha$, $\beta \to i\beta$, and as $\xi \to \infty$, both parts of Eq. (2.25) reduce to

$$4\alpha\beta\xi^2 - (\xi^2 + \beta^2)^2 = 0 \tag{2.29}$$

or, equivalently

$$V^2\left[V^6 - 8V^4 + 8\left(3 - \frac{2}{k^2}\right)V^2 - 16\left(1 - \frac{1}{k^2}\right)\right] = 0 \tag{2.30}$$

This is the equation governing the speed of Rayleigh's surface waves in an isotropic elastic half-space. Since both the wave numbers α and β across the thickness are imaginary, the displacement components, across the thickness coordinate, exhibit hyperbolic variation, attaining a maximum at the two end points of the thickness. In Fig. 3, this asymptotic behavior is represented by the

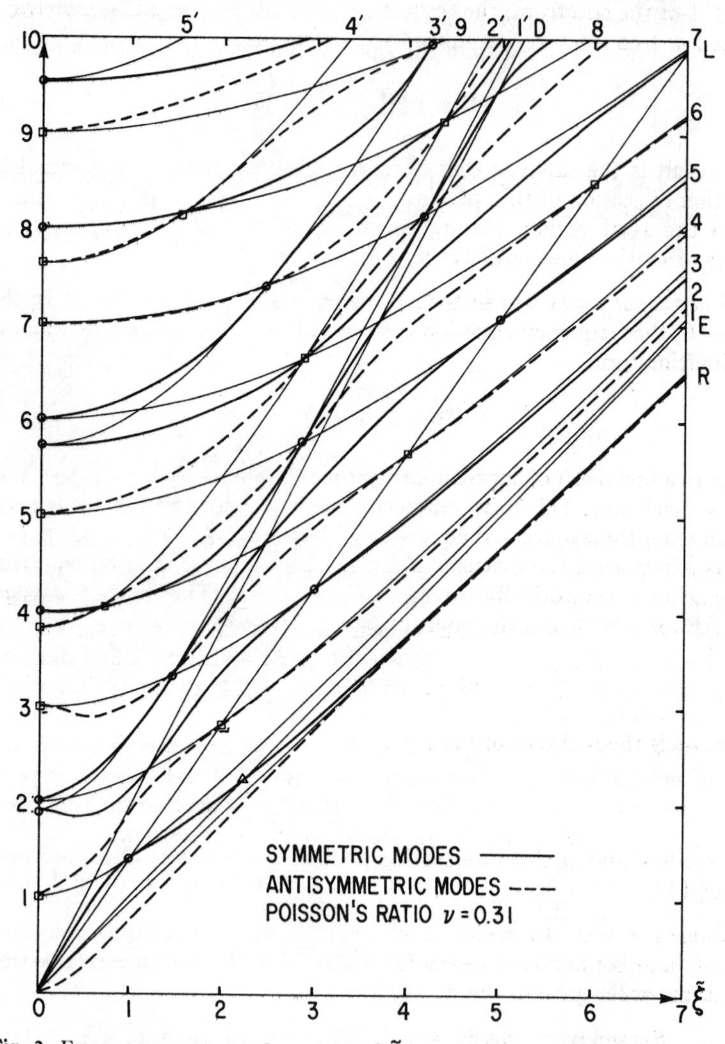

Fig. 3. Frequency spectrum, Ω versus real $\tilde{\xi}$ for an infinite, isotropic plate with traction-free faces. $\nu = 0.31$, $\tilde{\xi} = 2b\xi/\pi$, $\Omega = 2b\omega/\pi\sqrt{\mu/\rho}$.

straight line OR, which is the asymptote to the lowest symmetric and lowest antisymmetric branches of the frequency spectrum. The antisymmetric flexure branch approaches it asymptotically from below and the symmetric extensional branch approaches it from above.

Flexural waves As the dimensionless wave number ξ tends to zero in sector $\Omega/\xi < 1$ of the spectrum, the frequency equation for the antisymmetric modes in the low frequency region takes the limiting form

$$\Omega = \pi\xi^2 \left[\frac{1}{3}\left(1 - \frac{1}{k^2}\right)\right]^{1/2} \tag{2.31}$$

This result is the same as that obtained from the zero-order classical flexural equation of motion of thin plates.

In the same region, the frequency equation corresponding to symmetric modes of motion does not have any real roots.

Extensional waves In sector 2 of the spectrum, α is imaginary. In the limit as $\xi \rightarrow 0$, the frequency equation corresponding to the symmetric modes, takes the limiting form

$$\Omega = 2\left(1 - \frac{1}{k^2}\right)^{1/2} \xi \tag{2.32}$$

Thus, to a first order of approximation, the phase velocity $V = 2(1 - 1/k^2)^{1/2} = v/v_s$ is a constant. This is in complete agreement with the analogous results obtained from the extensional equations of Poisson.

As ξ increases, the extensional branch leaves sector 2 of the spectrum and approaches asymptotically the line OR in sector 1. This branch intersects the line OE for $\beta = 0$, and the point of intersection is given

$$\Omega = \xi = \frac{2b\alpha}{\pi(1 - 1/k^2)^{1/2}} \tag{2.33}$$

where αb is the real root of the equation

$$\tanh \alpha b = \frac{\alpha b}{4(1 - 1/k^2)} \tag{2.34}$$

The mode shape at this intersecting point is analogous to the Goodier–Bishop wave [14].

Lamé's modes In sector 2 of the frequency spectrum, one can easily identify another family of modes for which $\xi = \beta$. In this case the two frequency equations degenerate to the simple forms

$$\begin{aligned} &\text{Symmetric:} \quad \cos \beta b = 0 \qquad (\beta = n\pi/2b, \quad n = 1, 3, 5, \ldots) \\ &\text{Antisymmetric:} \quad \sin \beta b = 0 \qquad (\beta = n\pi/2b, \quad n = 0, 2, 4, \ldots) \end{aligned} \tag{2.35}$$

The frequencies are given by $\omega = \sqrt{2}\beta v_s$, or $\Omega = \sqrt{2}n$, $\xi = n$. These roots are indicated by circles and squares on the line OL in Fig. 3. The mode shapes corresponding to the first two frequencies are shown in Fig. 5.

These families of modes were first investigated by Lamé [9, p. 170]. Their physical significance lies in the fact that at these frequencies, not only the normal and shear stresses (τ_{22} and τ_{21}) vanish on the two major surfaces $x_2 = \pm b$, but also the normal and shear stresses (τ_{11} and τ_{12}) vanish on the planes $x_1 = \pm b$. In fact, the normal and shear stresses vanish on all nodal planes, as shown in Fig. 5. Since in this case the dilatation $(\partial u_1/\partial x_1 + \partial u_2/\partial x_2)$ vanishes identically, these modes of motion are also called equivoluminal modes of motion of the plate. In terms of multiple reflections of the waves (2.17), Lamé's modes can be interpreted as the result of interference of SV-waves ($\varphi \equiv 0$), which are incident and reflected at an angle $\theta_2 = 45°$ to the faces of the plate.

Simple thickness modes Simple thickness modes are defined as those for which the displacement is of the type

$$u_i = u_i(x_2, t), \qquad i = 1, 2, 3 \tag{2.36}$$

The case of $i = 3$, corresponds to the case of SH-wave with $\xi = 0$ in Eq. $(2.11)_2$. Thickness modes corresponding to the remaining two directions can be obtained from Eqs. (2.19), (2.22), and (2.25), by setting $\xi = 0$ in these equations. Thus we obtain

Symmetric motion:

$$\text{(i)}\quad u_1 = 0, \qquad u_2 = B' \sin (m\pi\, x_2/2b), \qquad m = 1, 3, \ldots$$

$$\text{(ii)}\quad u_2 = 0, \qquad u_1 = C' \cos (n\pi\, x_2/2b), \qquad n = 2, 4, \ldots \qquad (2.37)_1$$

Antisymmetric motion:

$$\text{(iii)}\quad u_2 = 0, \qquad u_1 = D' \sin (n\pi\, x_2/2b), \qquad n = 1, 3, \ldots$$

$$\text{(iv)}\quad u_1 = 0, \qquad u_2 = A' \cos (m\pi\, x_2/2b), \qquad m = 2, 4, \ldots \qquad (2.37)_2$$

where the exponential involving time is omitted from these expressions, and

$$\omega = \frac{m\pi v_p}{2b}; \qquad \omega = \frac{n\pi v_s}{2b} \tag{2.38}$$

These roots are indicated by the ordinates

$$\Omega = n; \quad \Omega = mk; \quad n, m = 1, 2, 3, \ldots \tag{2.39}$$

along the Ω-axis, as shown in Fig. 3.

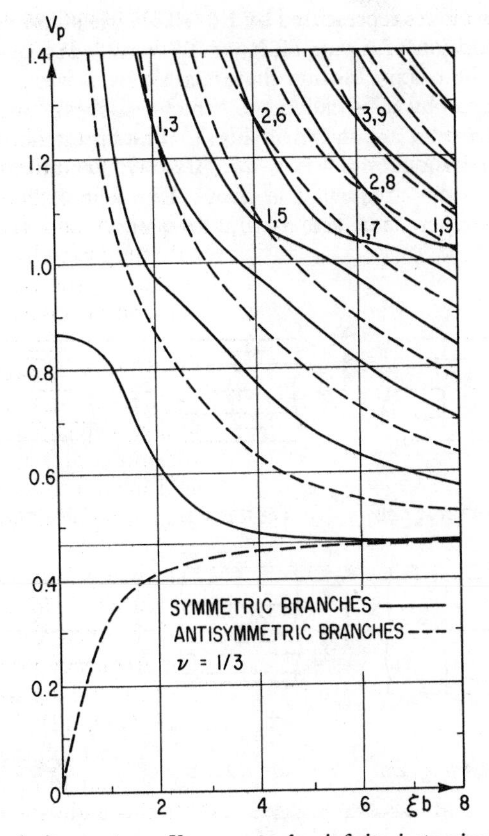

Fig. 4. Phase velocity spectrum, V_p versus ξ, of an infinite, isotropic plate with traction-free faces. $V_p = v/v_p$. $\nu = \frac{1}{3}$.

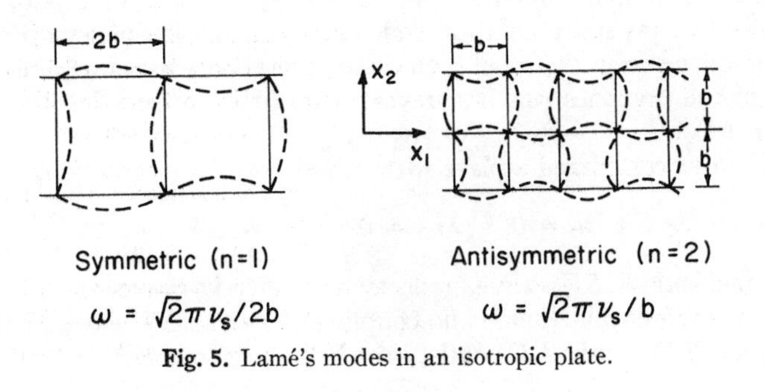

Fig. 5. Lamé's modes in an isotropic plate.

The thickness modes represented by Eq. (2.37) are standing waves resulting from the superposition of a pair of P- or SV-waves, propagating in opposite directions along the x_2-axis. Motion characterized by u_2 is called the "thickness-stretch" mode; that by u_1 is called the "thickness-shear" mode. The first six simple thickness modes are shown in Fig. 6. The lowest one is the thickness-shear mode with frequency $\omega = \pi v_s/2b$, $(\Omega = 1)$. For an aluminum plate of thickness 0.1 cm, this frequency is about 1.5 megahertz ($v_s = 3040$ m/s), which is much higher than the natural frequency of a structural plate in flexural vibration.

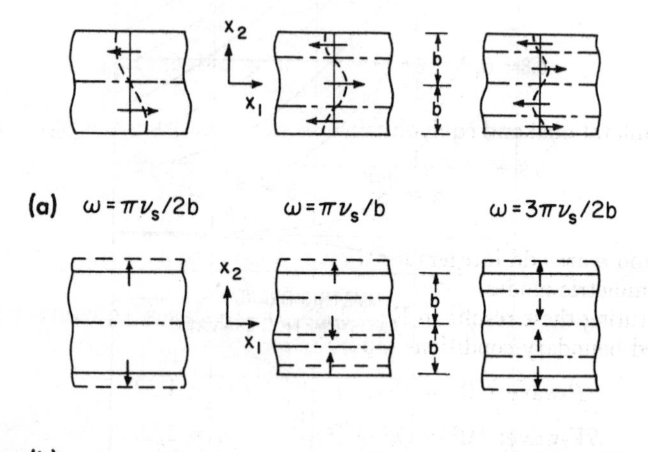

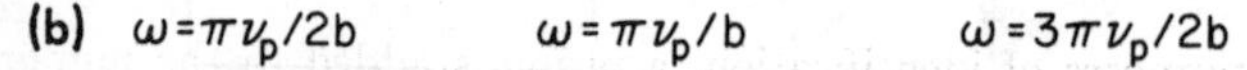

Fig. 6. (a) Simple thickness-shear modes (u_1). (b) Simple thickness-stretch modes (u_2).

Bounds The results (2.30)–(2.34) represent the behavior of the two lowest-order branches of the frequency Eq. (2.25). Lamé's modes, $\Omega = \sqrt{2}n$, $\xi = n$, and the simple thickness modes, $\Omega = n$, $\Omega = mk$ for $\xi = 0$, provide additionally two pivotal points for each branch of the frequency spectrum in the high-frequency region. In between these pivotal critical points, the intricate behavior of the branches was first revealed by Mindlin by using the analysis which now follows.

Mindlin first considered a plate with mixed boundary conditions

$$u_2 = 0, \qquad \tau_{21} = 0 \quad \text{at } x_2 = \pm b \tag{2.40}$$

Since an incident $P(SV)$-wave, reflects as a $P(SV)$-wave at this type of boundary, it follows that there is no coupling of P- and SV-waves. For $\xi \neq 0$, the solutions (2.19) and (2.22) satisfy the boundary conditions (2.40), for the antisymmetric modes, if

$$A = 0, \qquad \beta b = \frac{n\pi}{2}; \qquad n = 1, 3, 5, \ldots$$

$$D = 0, \qquad \alpha b = \frac{m\pi}{2}; \quad m = 1, 3, 5, \ldots \tag{2.41}$$

and, for the symmetric modes, if

$$B = 0, \qquad \beta b = \frac{n\pi}{2}; \qquad n = 2, 4, 6, \ldots$$

$$C = 0, \qquad \alpha b = \frac{m\pi}{2}; \quad m = 2, 4, 6, \ldots \tag{2.42}$$

Thus the dilatational and equivoluminal modes can exist as uncoupled modes if

$$\alpha = \frac{m\pi}{2b}; \qquad \beta = \frac{n\pi}{2b} \tag{2.43}$$

where m and n are odd integers for the antisymmetric modes and even integers for the symmetric modes.

Substituting these results in Eq. (2.26), the frequency equations for a plate with mixed boundary conditions takes the form

$$P\text{-wave:} \quad \Omega^2 = k^2(m^2 + \xi^2), \qquad m = 1', 2', 3', \ldots$$

$$SV\text{-wave:} \quad \Omega^2 = (n^2 + \xi^2), \qquad n = 1, 2, 3, \ldots \tag{2.44}$$

where for purposes of identification, a prime is added to the numerals for m. If we consider the variation of Ω with ξ, we note that as $\xi \to 0$, the limiting frequencies are those of simple thickness modes (2.39). The Eqs. (2.44) represent two families of hyperbolas in the (Ω, ξ)-plane and are represented by thinner lines in Fig. 3 for $\nu = 0.31$. The numerals without a prime represent the family of curves for n = constant and those with a prime identify the curves m = constant.

These family of curves discussed above are not only the frequency spectrum of the mixed boundary condition problem, but, as shown by Mindlin [12], they also serve to form the **bounds** to the branches of the frequency spectrum of a traction-free plate. From here it implies that the branches of the Rayleigh–Lamb spectrum cannot intersect the families of the curves (2.44), except at those points where the two families intersect themselves. The reason is as follows:

Consider the case of a plate with elastically restrained boundary conditions

$$\tau_{22} = \mp e u_2; \quad \tau_{21} = 0; \quad \text{on } x_2 = \pm b \tag{2.45}$$

When $e = 0$, we have the case of traction-free boundary conditions, and when $e = \infty$ we have the condition of mixed boundary conditions. Thus, as $e \to \infty$, the roots (2.44) are the locii of the roots of the frequency equation for the case of the mixed boundary-value problem. As the value of the spring constant e changes, the locii of the roots change gradually as shown in Fig. 7. In the limit when the spring becomes infinitely soft, $e = 0$ and the locii of the roots become the branches of the spectrum for a traction-free plate.

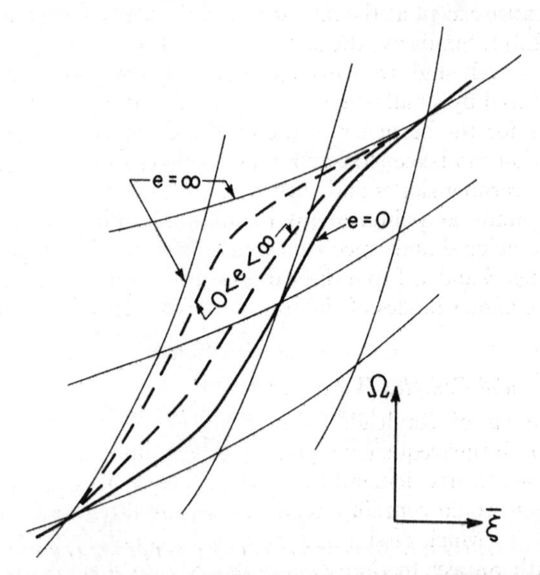

Fig. 7. Bounds ($e = \infty$) and branches for a traction-free plate ($e = 0$).

Another reason is that a typical frequency equation in the case of traction-free boundaries, can be considered as the sum of two parts $f_1(\alpha, \beta, \xi)$ and $f_2(\alpha, \beta, \xi)$ such that

$$F(\alpha, \beta, \xi) \equiv f_1(\alpha, \beta, \xi) + f_2(\alpha, \beta, \xi) = 0 \tag{2.46}$$

Instead of finding the roots of $F(\alpha, \beta, \xi) = 0$, we evaluate the roots of each of the two equations $f_1 = 0$ and $f_2 = 0$, separately. The roots which are common to the two equations, are then the roots of the original equation.

For symmetric modes, the frequency Eq. (2.21) can be decomposed as follows:

$$\begin{aligned} \xi - \beta = 0, &\quad \text{and} \quad \cos \beta b = 0 \\ \xi = 0, &\quad \text{and} \quad \cos \alpha b = 0 \end{aligned} \tag{2.47}$$

$$\begin{aligned} \cos \alpha b = 0, &\quad \text{and} \quad \cos \beta b = 0 \\ \sin \beta b = 0, &\quad \text{and} \quad \sin \alpha b = 0 \end{aligned} \tag{2.48}$$

We may note that the first pair of equations in Eq. (2.47) determines Lamé's modes, and the second pair gives the simple thickness modes. Similarly, study of the first pair of equations in Eq. (2.48) shows that they are contained in Eq. (2.41), the locii of whose roots are mapped as thin lines in Figs. 3 and 4. The points of intersections of the lines $m = 1', 3', 5', \ldots$, and $n = 1, 3, 5, \ldots$, which are the common points of intersections are the roots of the first pair of Eq. (2.48) and are indicated by the small circles in Fig. 3. These lines are also the bounds because except at the intersections, the zeros of each part are not the roots of Eq. (2.21). Similarly, the second pair of Eq. (2.48) are represented by the curves n = even and m = even integers. These points of intersections, which are indicated by small squares are also roots of Eq. (2.21).

The bounds for the branches of the frequency equation representing anti-symmetric motion can be constructed in an analogous manner. With additional information concerning slopes and curvatures of each branch at $\xi = 0$, and the slope of the bounds at points of intersection, Mindlin was able to map all branches of Rayleigh–Lamb spectrum. Such intricate details of the spectrum are shown in Figs. 3 and 4. From these diagrams it is obvious that the frequency spectrum of the higher modes of the free plate has a familiar terrace-like structure.

(D) *Imaginary and Complex Wave Numbers*

In the analysis of Rayleigh–Lamb frequency equation, we have so far assumed that both the frequency ω and the wave number ξ are real. Now physical considerations clearly demand that the frequency be real, for if it was complex, the solution would contain a factor indicating attenuation with time. For an elastic plate, in which dissipative mechanism is not accounted for, such an attenuation with respect to time is not allowed and therefore ω must be real. However, there is no compelling reason to restrict the wave number ξ to real values.

For complex-valued $\xi = \gamma + i\sigma$, an attenuation factor, $\exp(-\sigma x_1)$, would appear in all solutions. This is to be interpreted as a mode of wave propagation which is of significance near the edge $x_1 = 0$. As x_1 increases, the amplitude of the wave decreases, but the energy contained in this wave is not lost, because of the absence of dissipative mechanism. Instead, the energy associated with this mode is transmitted and carried away by waves of a different mode. Experimentally, this type of wave motion has been observed in circular disks [15] of finite dimensions. Such modes, identified by Gazis and Mindlin [M65] with complex wave numbers are called *edge modes*. In the case of rectangular plates, the edge modes were found by Onoe [16].

Although imaginary wave numbers in the solutions of lowest-order flexure equations are well known, they were neither admitted nor evaluated in the Rayleigh–Lamb wave solutions until recently. Their existence was recognized by Aggarwal and Shaw [17] in 1954 and soon after Lyon [18] revealed a family of imaginary branches, based on numerical computations. Complex wave numbers and phase velocities associated with real frequencies, were first discovered by Mindlin and Medick [M61] in an approximate theory of vibrations of plates. The existence and important aspects of the behavior of complex branches of the frequency equation were established by Onoe [11]. Further studies of the complex wave numbers corresponding to real frequencies were reported later, and it can now be said that the understanding of the complex roots of the frequency equation is fairly complete.

Figure 8 is an example of the frequency spectrum illustrating complex wave numbers, $\bar{\xi} = 2b\xi/\pi = \bar{\gamma} + i\bar{\sigma}$ in an isotropic, elastic plate; Fig. 8a representing the symmetric modes and Fig. 8b the antisymmetric modes. Mathematically, the frequency equation admits both $\pm\bar{\gamma}$ and $\pm i\bar{\sigma}$ as solutions and therefore the spectrum is symmetric about the $(\Omega; \bar{\gamma})$- and $(\Omega; \bar{\sigma})$-planes. For waves of the type $\exp i(\xi x_1 - \omega t)$, we consider only the branches with $\pm\text{Re}\,(\xi)$ and positive $\text{Im}\,(\xi)$. In general, each branch is made of real, imaginary, and complex segments, and complex segments in Fig. 8 are represented as dashed thick lines.

Each branch which is identified by a number, extends from zero to infinite frequency. A number with an overhead bar identifies the image of the corresponding branch about the imaginary plane. In Fig. 8a the first branch originates from $\Omega = 0$, $\bar{\xi} = 0$ and extends to infinity, lying entirely in the real $(\Omega; \bar{\gamma})$-plane. The complex segment of the second branch starts from a point which is labeled $(2, \bar{3})$ in the $\Omega = 0$-plane, and then ascends to a point which is the minimum point of the branch 2 in the real plane. As $\bar{\xi}$ increases, the real segment of branch 2 extends to infinity. The origin of the third branch is the image point of $(2, \bar{3})$ in the $(-\bar{\gamma} + i\bar{\sigma})$-plane (not shown in Fig. 8). It ascends to a minimum point in the negative $(\Omega; \bar{\gamma})$-plane as a complex segment, the mirror image of which is shown as $2, \bar{3}$ in the figure. It then changes to an ascending real segment, the mirror image of which is shown as $\bar{3}$. On reaching the Ω-axis it turns into a loop in the positive imaginary plane and finally ascends to infinity as a real segment. Thus the real segment $\bar{3}$ which has negative group velocity in $(\Omega, \bar{\gamma})$-plane should be construed as portion of third positive real branch with positive group velocity and negative phase velocity. Such an identification of the branches plays an important role in the analysis of transient motion in plates, using the saddle point method.

The rest of the higher branches can be traced in an analogous manner.

General properties of complex branches When α and β are eliminated from Eqs. (2.25) and (2.26), the frequency equation has an implicit representation of the type

$$F(\Omega; \bar{\xi}) = 0 \tag{2.49}$$

When $\Omega = 0$, we find from Eq. (2.26) that $\alpha = \beta = i\xi$ and the frequency Eq. (2.25) reduces to an identity. Therefore, it is not easy to find the intersection of the complex branches with the $\Omega = 0$-plane if we first set $\Omega = 0$ in Eqs.

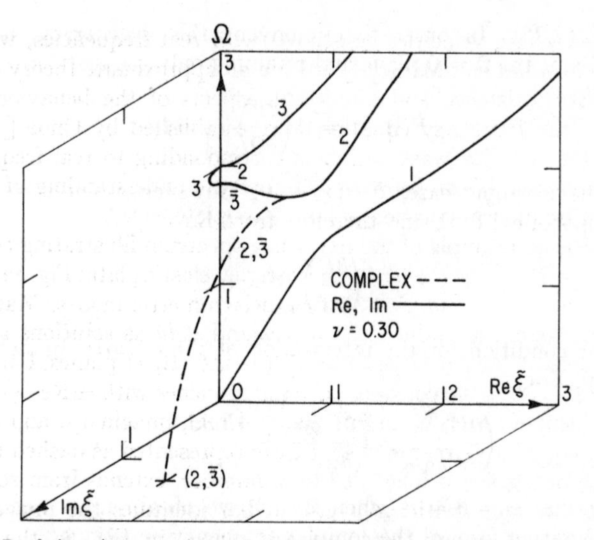

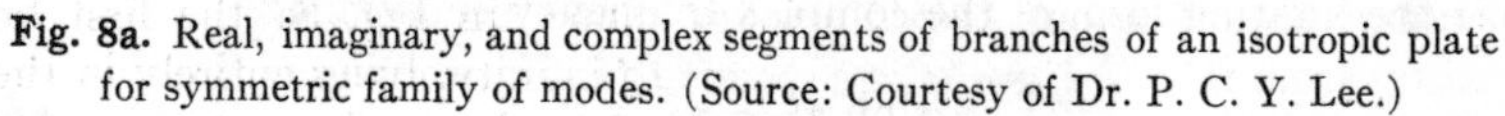

Fig. 8a. Real, imaginary, and complex segments of branches of an isotropic plate for symmetric family of modes. (Source: Courtesy of Dr. P. C. Y. Lee.)

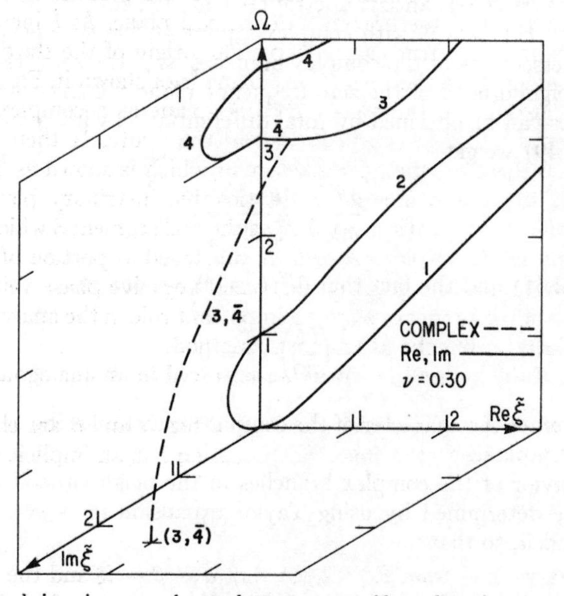

Fig. 8b. Real, imaginary, and complex segments of branches of an isotropic plate for antisymmetric family of modes. (Source: Courtesy of Dr. P. C. Y. Lee.)

(2.25) and (2.26). In order to circumvent this degeneracy, using Taylor expansion about the $\Omega = 0$-plane and retaining only the first term in Ω, we get

$$F(\Omega; \xi) = \Omega \left(\frac{\partial F}{\partial \Omega}\right)_{\Omega=0} = 0 \tag{2.50}$$

since for the isotropic case, $F(0; \xi) \equiv 0$. However, when $\Omega = 0$, $\partial F/\partial \Omega$ is linearly proportional to Ω, and therefore from Eq. (2.50)

$$\left(\frac{\partial F}{\partial \Omega^2}\right)_{\Omega=0} = 0 \tag{2.51}$$

which is the condition for the intersection of the complex branches with the plane $\Omega = 0$. But

$$\left(\frac{\partial F}{\partial \Omega^2}\right)_{\Omega=0} = \left(\frac{\partial F}{\partial \alpha}\frac{\partial \alpha}{\partial \Omega^2} + \frac{\partial F}{\partial \beta}\frac{\partial \beta}{\partial \Omega^2}\right)_{\Omega=0} = 0 \tag{2.52}$$

and noting that $\alpha = \beta = i\xi$ when $\Omega = 0$, we obtain from Eqs. (2.25) and (2.52) the limiting form of the complex frequency equations

$$\sinh \pi\xi \pm \pi\xi = 0 \tag{2.53}$$

for the symmetric (+), antisymmetric (−) modes, respectively. The roots of the frequency Eq. (2.49) with complex argument $\xi = \tilde{\gamma} + i\tilde{\sigma}$, determine the points of intersections of the complex branches with the zero-frequency plane.

In the neighborhood of the zero-frequency plane the character of the complex branches can be obtained by total differentiation with respect to Ω. Thus from Eq. (2.49) we get

$$\frac{\partial F}{\partial \Omega} + \frac{\partial F}{\partial \xi}\frac{\partial \xi}{\partial \Omega} = 0 \tag{2.54}$$

Using Eq. (2.51) and the fact that $\partial F/\partial \xi \neq 0$, we obtain

$$\frac{1}{\Omega}\left(\frac{\partial \xi}{\partial \Omega}\right)_{\Omega=0} = 0 \tag{2.55}$$

which determines the character of the complex branches in the neighborhood of zero-frequency plane.

The behavior of the complex branches in the neighborhood of real (Ω, ξ)-plane can be determined by using Taylor expansion in powers of imaginary wave numbers $\tilde{\sigma}$, so that

$$F(\Omega; \xi) = F(\Omega; \tilde{\gamma}) + i\tilde{\sigma}\left(\frac{\partial F}{\partial \xi}\right)_{\xi=\tilde{\gamma}} + \cdots = 0 \tag{2.56}$$

But for analytic functions of a complex variable,

$$\frac{\partial F}{\partial \xi} = \frac{\partial F}{\partial \tilde{\gamma}} = \frac{\partial F}{i\partial \tilde{\sigma}} \tag{2.57}$$

and therefore

$$F(\Omega; \xi) = F(\Omega; \tilde{\gamma}) + i\tilde{\sigma}\frac{\partial F}{\partial \tilde{\gamma}} = 0 \tag{2.58}$$

which implies

$$F(\Omega, \tilde{\gamma}) = 0; \qquad \tilde{\sigma}\frac{\partial F}{\partial \tilde{\gamma}} = 0 \tag{2.59}$$

From these two equations we have the real roots $F(\Omega; \tilde{\gamma}) = 0$ when $\tilde{\sigma} = 0$, and for $\tilde{\sigma} \neq 0$

$$F(\Omega; \tilde{\gamma}) = 0; \qquad \frac{\partial F}{\partial \tilde{\gamma}} = 0 \tag{2.60}$$

which govern the roots of the complex branches in the neighborhood of the real plane.

Again, from the frequency equation $F(\Omega; \tilde{\gamma}) = 0$, differentiating totally with respect to $\tilde{\gamma}$, we get

$$\frac{\partial F}{\partial \tilde{\gamma}} + \frac{\partial F}{\partial \Omega}\frac{\partial \Omega}{\partial \tilde{\gamma}} = 0 \tag{2.61}$$

From Eqs. (2.60) and (2.61) it follows that when $\partial F/\partial\Omega \neq 0$,

$$\frac{\partial \Omega}{\partial \tilde{\gamma}} = 0 \tag{2.62}$$

which implies that the maximum and minimum of the segments of the branches in the real $(\Omega; \tilde{\gamma})$-plane are a point of intersection, at normal incidence, with a complex segment. As pointed out by Kaul and Mindlin [M72], this is a general property of analytic functions of one real and one complex variable. Thus it can similarly be shown that the complex branches intersect, at maximum and minimum of the imaginary branches, at normal incidence. We have tentatively assumed in the above discussion that $\partial F/\partial\Omega \neq 0$. However, as shown by Mindlin [12] and Mindlin and Onoe [19], there are cases where $\partial F/\partial\Omega = 0$. These cases are exceptional and are not discussed here.

(E) *Bounded Plates*

The Rayleigh–Lamb frequency Eq. (2.25) is based on the plane-strain solutions of the system of Eqs. (2.5)–(2.8). This restriction can easily be removed by considering the x_1-coordinate in the solutions (2.19) and (2.22) to be the wave-normal which makes an angle θ with the x_1'-axis of the coordinate plane (x_1', x_3'). The axes x_1, x_3, x_1', and x_3' all lie in the middle plane of the plate, and $x_2 \equiv x_2'$. In terms of the primed axes, u_1 may be resolved into the two components u_1' and u_3', so that the phase of the plane wave is now given by $\exp i(\eta x_1' + \zeta x_3' - \omega t)$ where $\eta = \xi \cos\theta$, $\zeta = \xi \sin\theta$. Thus in the primed coordinate system the wave is "three-dimensional." Obviously, to obtain the frequency equation, it is sufficient to replace ξ^2 by $(\eta^2 + \zeta^2)$ in the original frequency Eqs. (2.25).

Such an extension of plane-strain solution to a three-dimensional wave was first noted by Rayleigh [4]. A similar result for cylindrical waves in a plate was recently shown by Sezawa [20] and Goodman [21]. However, such an extension should not be viewed only as a rotation of axes since the new solution in the primed coordinate system enables us to construct a variety of standing waves in the direction of x_1' or x_3', when additional boundary conditions are introduced on the end planes $x_1' = \pm a$, or $x_3' = \pm c$. However, except for a few special cases including one solved by Mindlin and Fox [M63], no general solution in closed form has been found for P- and SV-waves in a bounded plate with traction-free end planes.

The difficulty lies in the coupling of the P- and SV-waves at a traction-free surface. In the derivation of Eq. (2.18), we have seen that a combination of P-wave with angle of incidence and angle of emergence $\pm\theta_1$ and an SV-wave with angles $\pm\theta_2$, satisfies the traction-free boundary conditions at $x_2 = \pm b$. The waves reflected from these planes, each with a fixed angle of emergence, cannot be further recombined to satisfy the additional traction-free conditions on the planes $x_1 = \pm a$.

Further insight in this problem can be gained if we examine the effect of the end planes with the mixed boundary conditions

$$\tau_{11} = 0, \qquad u_2 = 0 \quad \text{on } x_1 = \pm a \tag{2.63}$$

Following Mindlin [M66] we consider only antisymmetric modes of motion. We first drop the time factor $\exp(-i\omega t)$ from the solution (2.22), and then take the real parts of Eqs. $(2.22)_2$ and $(2.23)_3$. We then find that the boundary conditions (2.63) are satisfied if

$$\xi a = \frac{s\pi}{2}, \qquad s = 0, 1, 2, \ldots \tag{2.64}$$

We thus see that a standing wave represented by Eq. (2.22) exists in a plate with traction-free boundary conditions on the two faces $x_2 = \pm b$, and the mixed boundary conditions on the two end planes $x_1 = \pm a$. Thus the frequency ω and the wave numbers α, β, and ξ are governed by Eqs. (2.24) and (2.64).

For convenience we redraw the two lowest branches of the antisymmetric part of the frequency spectrum, as shown in Fig. 9a. Instead of using Ω and ξ as coordinates as done in Fig. 4 we use now Ω and $1/\xi$ as the coordinates. Let the roots of the rth branch of the frequency Eq. (2.24) be represented by the relation

$$\frac{\pi}{2\xi b} = f_r(\Omega), \qquad r = 0, 1, 2, \ldots \tag{2.65}$$

Combining this with Eq. (2.64), we find that the frequency spectrum for the antisymmetric modes of vibration of a bounded plate are given by

$$\frac{a}{b} = s f_r(\Omega), \qquad r, s = 0, 1, 2, \ldots \tag{2.66}$$

This equation is a relation between the width to thickness ratio and the frequency of the bounded plate. This relation is shown graphically in Fig. 9b for the lowest two antisymmetric modes, $r = 0$ and 1.

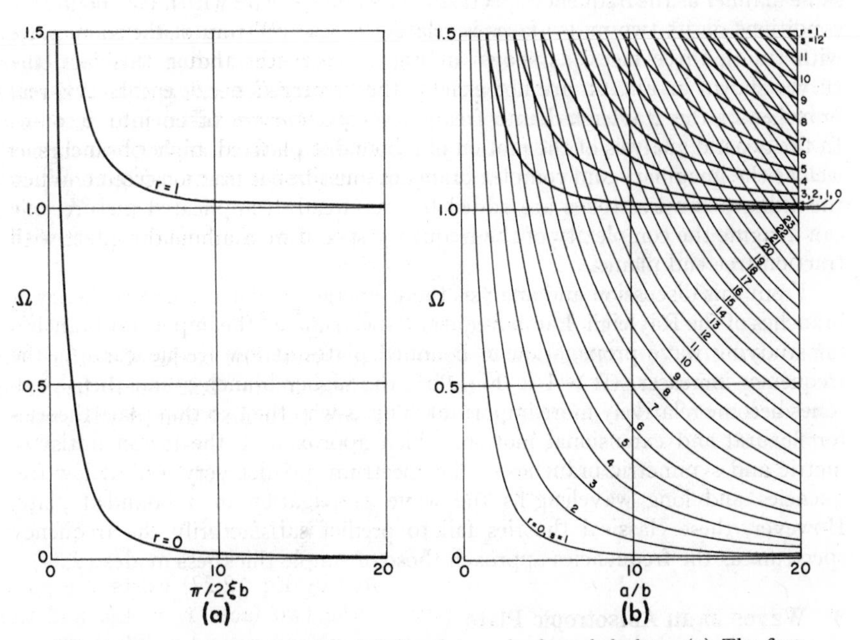

Fig. 9. Frequency spectra of an infinite plate and a bounded plate: (a) The first two antisymmetric branches of Rayleigh–Lamb equation. (b) Fundamental modes and anharmonic overtones of a bounded plate.

The two frequency branches in Fig. 9a are repeated again in Fig. 9b as $r = 0, 1$ and $s = 1$ $(a = \pi/2\xi)$. Each one of these is a fundamental mode of a family of modes for a bounded plate. The curves for $s = 2, 3, \ldots$ are obtained from the two branches $r = 0$ and $r = 1$ in Fig. 9a by multiplying the abscissa $\pi/2b\xi$, of each of these branches by the integers $s = 2, 3, \ldots$. They represent the anharmonic overtones of the fundamental mode of each branch.

Figure 9b presumably provides all necessary information about standing waves (free vibrations) in a bounded plate with traction-free faces and vanishing normal stress and vanishing tangential displacement on the two end planes. Given a width to thickness ratio (a/b) of the plate, the natural frequencies $\omega = \pi v_s(\Omega/2b)$, can be determined from these curves. The wave numbers ξ are given by Eq. (2.64) and the wave numbers α and β and the mode shape can be calculated from Eqs. (2.26) and (2.22), respectively.

The two end conditions (2.63) are not the practically encountered boundary conditions. A realistic condition would be that either all stresses, or all displacements vanish on $x = \pm a$. Exact solutions for a plate with traction-free faces and sides have not yet been found. However, the frequency spectrum shown in Fig. 9b is related to that of a plate with traction-free end planes in somewhat the same manner as the frequency spectrum of an infinite plate with mixed boundary conditions on its two major faces is related to the spectrum of the same plate with traction-free faces, as shown in Fig. 7. Notwithstanding this fact, the curves shown in Fig. 9b are not exactly the "bounds" because only two real branches of the Rayleigh–Lamb frequency equation are taken into account. In the general analysis of the motion of a bounded plate, all higher branches as well as the imaginary and complex branches must be taken into account. When the anharmonic overtones are added to the already complicated picture, one can imagine the complexity of the frequency spectrum of a bounded plate with traction-free end planes.

From this discussion and analysis there emerges a conclusion that the lower branches of the Rayleigh–Lamb frequency spectrum are the important branches for studying wave propagation in bounded plates at low frequencies. As the frequency increases, $(\Omega > 1$ in Fig. 9b), the higher branches and their overtones become relatively more important. This is why the two thin-plate theories for flexural and extensional motion, which approximate the lowest antisymmetric and symmetric branches of the spectrum, predict very well at low frequencies and long wavelengths the wave propagation in a bounded plate. However, these classical theories fail to predict satisfactorily the frequency spectrum as the frequencies approach those of simple thickness modes (2.39).

3 Waves in an Anisotropic Plate

In the case of anisotropy, the constitutive laws connecting the stresses τ_{ij} and the strain e_{ij} are given by

$$\tau_{ij} = c_{ijkl}e_{kl} = c_{ijkl}\partial_l u_k \tag{3.1}$$

where

$$c_{ijkl} = c_{klij} = c_{jikl} \tag{3.2}$$

are the constants of elastic stiffness. The infinitesimal strain tensor e_{ij} and the rotation tensor ω_{ij} are defined as

$$e_{ij} = \partial_{(i}u_{j)} = \tfrac{1}{2}(\partial_i u_j + \partial_j u_i)$$

$$\omega_{ij} = \partial_{[i}u_{j]} = \tfrac{1}{2}(\partial_i u_j - \partial_j u_i) \tag{3.3}$$

In the absence of body force, the three displacement equations of motion are given by

$$\partial_i \tau_{ij} = c_{ijkl}\partial_i \partial_k u_l = \rho \ddot{u}_j \tag{3.4}$$

For isotropic materials, Eq. (3.4) reduces to the familiar form (2.5).

As solutions of the equations of motion, we consider plane waves of the type

$$u_k = A_k \exp i(\xi_m x_m - \omega t) \tag{3.5}$$

where ξ_m are the components of propagation vector, ω is the angular frequency, and A_k are the components of unspecified amplitude vector. In this equation ω is real, but ξ_k and A_k are in general assumed to be complex, but the displacement u_k is understood to be real. The phase velocity is given by $v = \omega/|\xi|$, where $|\xi|^2 = \xi_1^2 + \xi_2^2 + \xi_3^2$. For the plane wave (3.5) to satisfy the equations of motion (3.4) in an infinitely extended anisotropic medium, the coefficients A_k, ω, and ξ_k must satisfy the system of homogeneous equations

$$A_k(\bar{c}_{ijkl}\bar{\xi}_j\bar{\xi}_l - \Omega^2\delta_{ik}) = 0 \tag{3.6}$$

which requires that Ω and $\bar{\xi}_m$ must satisfy the characteristic equation

$$|\bar{c}_{ijkl}\bar{\xi}_j\bar{\xi}_l - \Omega^2\delta_{ik}| = 0 \tag{3.7}$$

where

$$\bar{c}_{ijkl} = \frac{c_{ijkl}}{c_{1212}}; \qquad \bar{\xi}_i = \frac{2b\xi_i}{\pi}$$

$$\Omega = \frac{\omega}{\omega_s}; \qquad \omega_s = \frac{\pi}{2b}\left(\frac{c_{1212}}{\rho}\right)^{1/2}$$

and δ_{ik} is the Kronecker delta.

For an assigned direction of wave propagation $\xi_i/|\xi|$, three values for the phase velocity v may be determined from the roots of the characteristic equation. For each value of the phase velocity, the amplitudes A_k can be calculated up to an arbitrary factor from Eq. (3.6), which in general is neither parallel nor perpendicular to the direction of wave propagation. Thus in general, there are three types of waves, each with a distinct speed, which may propagate in an infinite anisotropic medium.

(A) *Waves in an Infinite Plate*

In a plate, bounded by two major surfaces $x_2 = \pm b$, the three waves mentioned above are coupled. Of major interest in this case are waves along the length (x_1-axis) or along the width (x_3-axis). We thus seek to determine the frequency equation, analogous to the Rayleigh–Lamb equation in the case of isotropic plates

The characteristic Eq. (3.7) suggests that for a given Ω and a given ratio $\xi_1:\xi_3$ there exist three wave numbers along the thickness direction, which are denoted by $\xi_2^{(n)}$, $(n = 1, 2, 3)$. For each of the wave numbers $\xi_2^{(n)}$, there exist three amplitude ratios $L_k^{(n)}$ governed by Eq. (3.6). Thus for standing waves in a plate, the displacement field can be written in the form (n is summed from 1 to 3)

$$\begin{aligned} u_1 &= \Sigma_n L_1^{(n)}[A_{(n)} f_1 \cos \xi_2^{(n)} x_2 + B_{(n)} f_2 \sin \xi_2^{(n)} x_2] \\ u_2 &= \Sigma_n L_2^{(n)}[B_{(n)} f_1 \cos \xi_2^{(n)} x_2 + A_{(n)} f_2 \sin \xi_2^{(n)} x_2] \\ u_3 &= \Sigma_n L_3^{(n)}[B_{(n)} f_1 \cos \xi_2^{(n)} x_2 + A_{(n)} f_2 \sin \xi_2^{(n)} x_2] \end{aligned} \tag{3.8}$$

where the time factor has been omitted and

$$f_1 = \sin(\xi_1 x_1 + \xi_3 x_3), \qquad f_2 = \cos(\xi_1 x_1 + \xi_3 x_3) \tag{3.9}$$

The unknown constants $A_{(n)}$ and $B_{(n)}$ are fixed by the six boundary conditions on the two major faces $x_2 = \pm b$. For the traction-free faces, the boundary conditions are

$$\tau_{2j} = c_{2jkl} \partial_l u_k = 0, \quad \text{on } x_2 = \pm b \tag{3.10}$$

Substituting Eq. (3.8) in Eq. (3.1) we get, for $i = 2$,

$$\begin{aligned} \tilde{\tau}_{2i} = {} & \Sigma_n A_{(n)}[M_i^{(n)} f_2 \cos \xi_2^{(n)} x_2 - N_i^{(n)} f_1 \sin \xi_2^{(n)} x_2] \\ & + \Sigma_n B_{(n)}[N_i^{(n)} f_2 \cos \xi_2^{(n)} x_2 - M_i^{(n)} f_1 \sin \xi_2^{(n)} x_2] \end{aligned} \tag{3.11}$$

where $\tilde{\tau}_{2i} = \tau_{2i}/c_{1212}$ and

$$\begin{aligned} M_i^{(n)} &= L_1^{(n)}(\xi_1 \tilde{c}_{2i11} + \xi_2 \tilde{c}_{2i13}) + \xi_2^{(n)}(L_2^{(n)} \tilde{c}_{2i22} + L_3^{(n)} \tilde{c}_{2i23}) \\ N_i^{(n)} &= \xi_1(L_2^{(n)} \tilde{c}_{2i12} + L_3^{(n)} \tilde{c}_{2i13}) + \xi_2^{(n)} L_1^{(n)} \tilde{c}_{2i12} \\ &\quad + \xi_3(L_2^{(n)} \tilde{c}_{2i23} + L_3^{(n)} \tilde{c}_{2i33}) \end{aligned} \tag{3.12}$$

From Eqs. (3.10) and (3.11) we obtain a set of six homogeneous equations in $A_{(n)}$ and $B_{(n)}$. For a nontrivial solution, the determinant of the coefficients must vanish, which when rearranged suitably, leads to the frequency equation

$$\begin{vmatrix} |M_i^{(n)} \cos \xi_2^{(n)} b| & |N_i^{(n)} \cos \xi_2^{(n)} b| \\ |N_i^{(n)} \sin \xi_2^{(n)} b| & |M_i^{(n)} \sin \xi_2^{(n)} b| \end{vmatrix} = 0, \qquad i, n = 1, 2, 3 \tag{3.13}$$

where every block matrix like

$$|M_i^{(n)} \cos \xi_2^{(n)} b|$$

represents an array of three rows for $i = 1, 2, 3$ and three columns for $n = 1, 2, 3$. This is a highly complicated transcendental frequency equation, the roots of which, in conjunction with the roots of Eq. (3.7), determine the dispersion spectrum of the crystal plate of infinite extent, finite thickness and with traction-free faces [M71].

(B) *Monoclinic Crystal Plate*

The frequency Eqs. (3.7) and (3.13) for the general anisotropy are so complicated that so far no detailed results are available. However, in the case of monoclinic crystals with twofold axis of symmetry (digonal), the frequency equations are considerably simplified.

Let the x_1-axis of the plate coordinates be parallel to the digonal axis of the crystal. Then the elastic constants c_{ij13} and c_{ij12} vanish for $ij = 11, 22, 33, 23$. We shall use in this subsection the contracted notation c_{pq} $(p, q = 1 \text{ to } 6)$ to replace c_{ijkl}. Thus c_{p5} and c_{p6} vanish when $p = 1, 2, 3, 4$ for monoclinic crystals. Further for convenience we define

$$\bar{c}_{pq} = \frac{c_{pq}}{c_{66}}, \qquad (p, q = 1 \text{ to } 6)$$

For straight crested waves propagating in the x_1-direction, $\xi_3 = 0$, and the characteristic Eq. (3.7) reduces to

$$\begin{vmatrix} (\bar{c}_{11}\xi_1^2 + \xi_2^2 - \Omega^2) & (1 + \bar{c}_{12})\xi_1\xi_2 & (\bar{c}_{14} + \bar{c}_{56})\xi_1\xi_2 \\ (1 + \bar{c}_{12})\xi_1\xi_2 & (\xi_1^2 + \bar{c}_{22}\xi_2 - \Omega^2) & \bar{c}_{56}\xi_1^2 + \bar{c}_{24}\xi_2^2 \\ (\bar{c}_{14} + \bar{c}_{56})\xi_1\xi_2 & \bar{c}_{56}\xi_1^2 + \bar{c}_{24}\xi_2^2 & (\bar{c}_{55}\xi_1^2 + \bar{c}_{44}\xi_2^2 - \Omega^2) \end{vmatrix} = 0 \qquad (3.14)$$

For each assigned pair of values for Ω and ξ_1, there are three roots for ξ_2 which are denoted as $\xi_2^{(n)}$ $(n = 1, 2, 3)$. When $\xi_1 = 0$, the three roots of Eq. (3.14) are $(\Omega/\xi_2)^{(n)} = v^{(n)}$ where

$$v^{(1)} = 1$$

$$v^{(2),(3)} = \{\tfrac{1}{2}(\bar{c}_{22} + \bar{c}_{44}) \pm [\tfrac{1}{4}(\bar{c}_{22} - \bar{c}_{44})^2 + \bar{c}_{24}^2]^{1/2}\}^{1/2} \qquad (3.15)$$

The $v^{(n)}$ are the ratios with respect to the reference wave velocity $(c_{66}/\rho)^{1/2}$, of the plane waves propagating along the x_2-axis.

When $\xi_2 = 0$ in Eq. (3.14), the three roots are $(\Omega/\xi_1)^{(n)} = \bar{v}^{(n)}$ $(n = 1, 2, 3)$, where

$$\bar{v}^{(2)} = (\bar{c}_{11})^{1/2}$$

$$\bar{v}^{(1),(3)} = \{\tfrac{1}{2}(1 + \bar{c}_{55}) \mp [\tfrac{1}{4}(1 - \bar{c}_{55})^2 + \bar{c}_{56}^2]^{1/2}\}^{1/2} \qquad (3.16)$$

Analogous to $v^{(n)}$, the $\bar{v}^{(n)}$ are the velocity ratios, with respect to $(c_{66}/\rho)^{1/2}$, of the plane waves propagating along the digonal axis (x_1-axis) of an unbounded monoclinic crystal.

Once $\xi_2^{(n)}$ are found for the general case ($\xi_1 \neq 0$), the amplitude ratios $L_i^{(n)}$ are the eigenvectors of the matrix in Eq. (3.14) and the displacements are obtained from Eq. (3.8) with $\xi_3 = 0$. Furthermore, the coefficients $M_1^{(n)}$, $N_2^{(n)}$, and $N_3^{(n)}$ in Eq. (3.12) vanish. The determinantal frequency Eq. (3.13) then takes the diagonal two-block form; the equations obtained by setting each block to zero are the Ekstein frequency equations [8]:

$$\begin{vmatrix} \{\xi_2^{(n)} L_1^{(n)} + \xi_1(L_2^{(n)} + L_3^{(n)}\bar{c}_{65})\} \sin \xi_2^{(n)} b \\ \{\bar{c}_{21}\xi_1 L_1^{(n)} + \xi_2^{(n)}(\bar{c}_{22}L_2^{(n)} + \bar{c}_{24}L_3^{(n)})\} \cos \xi_2^{(n)} b \\ \{\bar{c}_{41}\xi_1 L_1^{(n)} + \xi_2^{(n)}(\bar{c}_{24}L_2^{(n)} + \bar{c}_{44}L_3^{(n)})\} \cos \xi_2^{(n)} b \end{vmatrix} = 0 \qquad (3.17a)$$

$$\begin{vmatrix} \{\xi_2^{(n)} L_1^{(n)} + \xi_1(L_2^{(n)} + L_3^{(n)}\bar{c}_{65})\} \cos \xi_2^{(n)} b \\ \{\bar{c}_{21}\xi_1 L_1^{(n)} + \xi_2^{(n)}(\bar{c}_{22}L_2^{(n)} + \bar{c}_{24}L_3^{(n)})\} \sin \xi_2^{(n)} b \\ \{\bar{c}_{41}\xi_1 L_1^{(n)} + \xi_2^{(n)}(\bar{c}_{24}L_2^{(n)} + \bar{c}_{44}L_3^{(n)})\} \sin \xi_2^{(n)} b \end{vmatrix} = 0 \qquad (3.17b)$$

In the above determinants the columns are obtained by setting $n = 1, 2, 3$.

Corresponding to the frequency Eq. (3.17a), the displacement field is given by Eq. (3.8) when $B_{(n)} \equiv 0$ and, corresponding to the frequency Eq. (3.17b), $A_{(n)} \equiv 0$. In the first case the displacements u_1 and u_2 are symmetric about the middle plane and u_3 is antisymmetric about this plane and we call this an essentially symmetric family. In the second case when $A_{(n)} \equiv 0$, the displacement components u_1 and u_3 are antisymmetric about the middle plane, u_3 is symmetric about this plane, and we call this an essentially antisymmetric family.

As in the case of isotropic plates, all information pertaining to the wave propagation and vibration of an infinitely extended crystal plate are contained in the frequency Eqs. (3.7) and (3.13), or in the case of straight-crested waves in a monoclinic plate, Eqs. (3.14) and (3.17). The task of solving these equations is far from simple. The case of infinite wavelength ($\xi_1 = 0$) was treated by Koga [22]. For the case of thickness-shear mode at long wavelengths ($\xi_1 \leq 1$), Ekstein analyzed the roots of Eqs. (3.14) and (3.17) by using a perturbation method [8]. The general case of arbitrary wavelengths was first analyzed by Newman and Mindlin in 1957 [M59]. By constructing bounds and asymptotes, they mapped the frequency spectrum for real wave number ξ_1 for an AT-cut quartz crystal plate.

The imaginary and complex branches as well as the detailed behavior of the roots of Eqs. (3.14) and (3.17) at long wavelengths were investigated by Kaul and Mindlin and reported in two papers [M71, M72]. For the case of long wavelengths ($\xi_1 \ll 1$), they showed at first that the three roots of Eq. (3.14) are given approximately by [M71, Eq. (24)]

$$[\Omega^{(n)}]^2 = [v^{(n)}\xi_2^{(n)}]^2 + (\mathcal{K}^{(n)}\xi_1)^2 + \cdots \tag{3.18}$$

where $v^{(n)}$ are defined in Eq. (3.15) and $\mathcal{K}^{(n)}$ are constants the values of which depend on the elastic moduli c_{pq}, and $v^{(n)}$.

Next they showed the amplitude coefficients $L_i^{(n)}$ can be expressed in power series in the form of (Eq. (26) of [M71])

$$L_i^{(n)} = P_i^{(n)} + Q_i^{(n)}\beta_{(n)} + R_i^{(n)}\beta_{(n)}^2 + \cdots \tag{3.19}$$

where $\beta_{(n)} = \xi/\xi_2^{(n)}$. Finally, the frequency Eqs. (3.17) can be reduced to the following form [M71, Eq. (34)]

$$\begin{aligned}&[1 + (Q_2^{(1)} + \bar{c}_{56}Q_3^{(1)})\beta_{(1)}^2][B + C\beta_{(2)}^2 + D\beta_{(3)}^2 + \cdots](\tan \xi_2^{(1)})^{\pm 1} \\ &\quad + \beta_{(1)}\beta_{(2)}E(\tan \xi_2^{(2)}b)^{\pm 1} + \beta_{(1)}\beta_{(3)}F(\tan \xi_2^{(3)}b)^{\pm 1} + \cdots = 0\end{aligned} \tag{3.20}$$

where terms with $\beta_{(n)}^m$ for $m > 2$ are neglected. In Eqs. (3.19) and (3.20), the coefficients $P_i^{(n)}$, $Q_i^{(n)}$, $R_i^{(n)}$, B, C, D, E, and F all are constants with their values being dependent on $\bar{c}_{pq}$ and $v^{(n)}$. The positive exponent (in Eq. (3.20)) is associated with the essentially symmetric modes (3.17a) and the negative exponent with Eq. (3.17b).

Equation (3.20) resembles the frequency equation for the Rayleigh–Lamb wave (2.25). In fact it is shown in Ref. [M71] that Eq. (3.20) reduces to Eq. (2.25) for the case of isotropy. From Eq. (3.20), analytical expressions for the frequencies of the simple thickness modes ($\xi_1 = 0$, $\Omega \neq 0$) and the slopes and curvatures of the branches at these frequencies can be derived. The details of these analytical expressions are essential in the development of approximate plate theories.

Without going further into the details of the analysis, we summarize the final results below. In addition, a sample frequency spectrum (real, imaginary, and complex branches) is shown in Fig. 10 for an AT-cut quartz crystal plate. Quartz is a trigonal crystal (six elastic constants), and the AT-cut is a plate which contains a digonal axis and whose normal makes an angle of 35°15′ with the trigonal axis. When referred to rectangular axes in and normal to the plane of such a plate, the stress-strain relation has monoclinic symmetry.

Long wavelengths, low frequency When both ξ_1 and ξ_2 approach zero, so do the frequency, and the frequency equations become highly indeterminate. In the limit the behavior of the three real branches of the frequency spectrum passing through $\Omega = 0$ are given by

$$\begin{aligned}\text{Flexure Mode:}\quad & \Omega = \tfrac{1}{2}\pi(\tfrac{1}{3}\bar{\gamma}_{11})^{1/2}\xi_1^2 \\ \text{Face Shear Mode:}\quad & \Omega = (\bar{\gamma}_{55})^{1/2}\xi_1 \\ \text{Extension Mode:}\quad & \Omega = (\bar{\gamma}_{11})^{1/2}\xi_1\end{aligned} \tag{3.21}$$

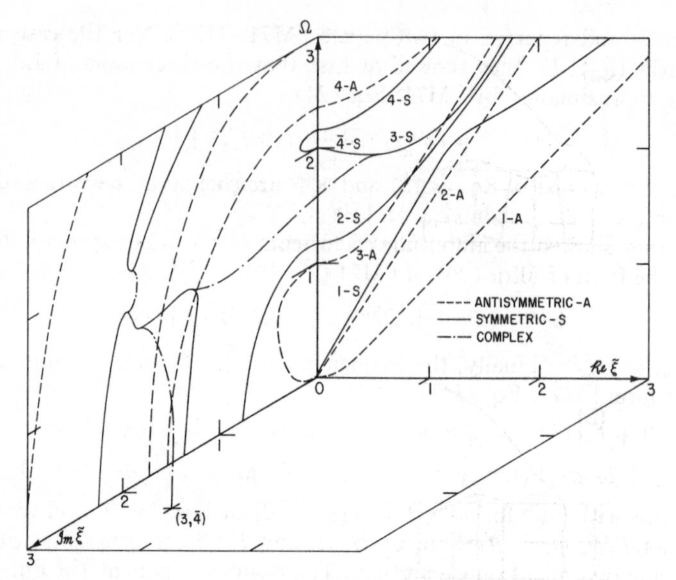

Fig. 10. Frequency spectrum for the first four antisymmetric modes and the first four symmetric modes of an infinite, AT-cut, quartz plate.

where $\Omega = \omega/\omega_s$, $\omega_s = (\pi/2b)(c_{66}/\rho)^{1/2}$, $\bar{\xi}_1 = 2b\xi_1/\pi$, $\tilde{\gamma}_{11} = \gamma_{11}/c_{66}$, $\tilde{\gamma}_{55} = \gamma_{55}/c_{66}$, and

$$\gamma_{11} = c_{11} - \frac{c_{12}{}^2 c_{44} + c_{14}{}^2 c_{22} - 2c_{12}c_{14}c_{24}}{c_{22}c_{44} - c_{24}{}^2}$$

$$\gamma_{55} = c_{55} - \frac{c_{56}{}^2}{c_{66}} \tag{3.22}$$

As shown in the next section, these modes are predicted by the Cauchy–Voigt equations for low-frequency extensional and flexural vibrations of thin crystal plates. The equations are two-dimensional in $x_1 - x_3$-coordinates, and when the motion is independent of x_3-coordinate (cylindrical bending in the case of flexural motion), they are

$$\text{Flexure:} \quad b^2\gamma_{11}\partial^4 u_2/\partial x_1{}^4 + 3\rho\ddot{u}_2 = 0$$

$$\text{Face shear:} \quad \gamma_{55}\partial^2 u_3/\partial x_1{}^2 - \rho\ddot{u}_3 = 0$$

$$\text{Extension:} \quad \gamma_{11}\partial^2 u_1/\partial x_1{}^2 - \rho\ddot{u}_1 = 0 \tag{3.23}$$

The dispersion relations of Eq. (3.21) can be derived also from the above equations by assuming $u_j = A_j \exp i(\xi_1 x_1 - \omega t)$. These three modes together with other thickness modes at higher frequencies are depicted in Fig. 11.

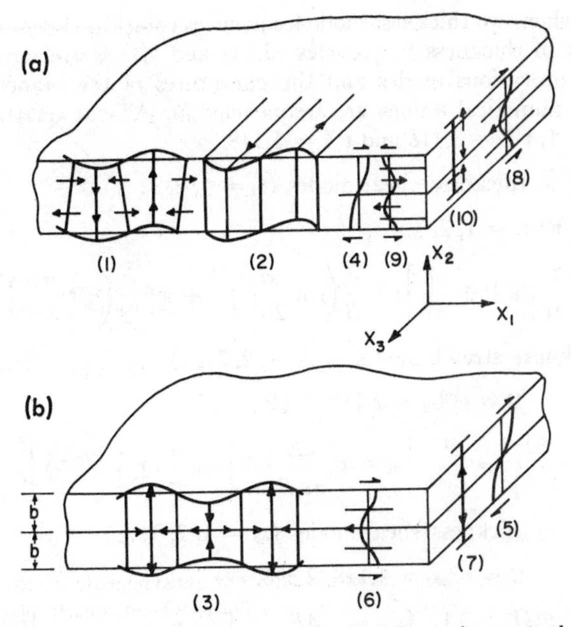

Fig. 11. Approximate shapes of 10 lower modes for AT-cut of quartz at long wavelengths: (a) Antisymmetric modes: (1) Flexure, $\Omega = 0.52\xi^2 b^2$; (2) Face-shear, $\Omega = 0.98\xi b$; (4) and (9) $x_1 - x_2$ thickness-shear, $\Omega = 1, 3$; (8) $x_2 - x_3$ thickness-shear, $\Omega = 2.29$; (10) Thickness-stretch, $\Omega = 4.21$. (b) Symmetric modes: (3) Extension, $\Omega = 1.08\xi b$; (5) $x_1 - x_3$ thickness-shear, $\Omega = 1.15$; (6) $x_1 - x_2$ thickness-shear, $\Omega = 2$; (7) Thickness-stretch, $\Omega = 2.12$.

For the AT-cut quartz, $(\bar{\gamma}_{11})^{1/2} = 1.7210$ and $(\bar{\gamma}_{55})^{1/2} = 1.5375$. Thus the lowest branch extending from $\Omega = 0$ is the flexure. The face-shear branch and extension branch are the next two in ascending order as shown in Fig. 10.

Simple thickness modes The frequencies for the simple thickness modes ($\xi_1 = 0$) are derivable either by assuming plane waves in x_2-direction with speed $v^{(n)}$ as given in Eq. (3.15) which satisfy the traction-free boundary condition (3.10), or simply by setting $\beta_{(n)} = \xi_1/\xi_2^{(n)} = 0$ in Eq. (3.20). The results are

$$\Omega = v^{(n)} s_n \qquad n = 1, 2, 3; \quad s = 1, 2, 3, 4, \ldots \tag{3.24}$$

The s_n are integers with the subscript n being attached to identify the direction of the predominant displacement. Since $v^{(2)}$ and $v^{(3)}$ depend strongly on the values of c_{pq}, the frequency of one thickness mode may coincide with that of another for certain plate materials, causing indeterminancy in the evaluation of the slopes and curvatures of the frequency branches [M71].

Except when two thickness mode frequencies coincide, slopes of all branches at the point of thickness frequencies (3.24) and $\xi_1 = 0$ are zero. The cut-off frequencies of various modes and the curvatures of the branches are listed below. The numerical values are frequencies for AT-cut quartz crystals for which $v^{(1)} = 1$, $v^{(2)} = 2.118$ and $v^{(3)} = 1.148$.

(1) $x_1 - x_2$ thickness–shear modes ($s_1 = 1, 2, 3\ldots$)

$$\Omega = v^{(1)}s_1 = 1, 2, 3, \ldots$$

$$\left.\frac{d^2\Omega}{d\xi_1{}^2}\right]_0 = \frac{1}{s_1}\left\{\mathcal{K}^{(1)} \mp \frac{4}{\pi s_1}\left[v^{(2)}\frac{E}{B}\left(\cot\frac{\pi s_1}{2v^{(2)}}\right)^{\mp 1} + v^{(3)}\frac{F}{B}\left(\cot\frac{\pi s_1}{2v^{(3)}}\right)^{\mp}\right]\right\} \tag{3.25a}$$

(2) thickness–stretch modes ($s_2 = 1, 2, 3\ldots$)

$$\Omega = v^{(2)}s_2 = 2.118, 4.236, \ldots$$

$$\left.\frac{d^2\Omega}{d\xi_1{}^2}\right]_0 = \frac{1}{s_2 v^{(2)}}\left\{\mathcal{K}^{(2)} \mp \frac{4E}{\pi s_2 B}v^{(2)}\left[\cot\frac{\pi}{2}(1 + s_2 v^{(2)})\right]^{\mp 1}\right\} \tag{3.25b}$$

(3) $x_3 - x_2$ thickness–shear modes ($s_3 = 1, 2, 3\ldots$)

$$\Omega = v^{(3)}s_3 = 1.148, 2.296, \cdots$$

$$\left.\frac{d^2\Omega}{d\xi_1{}^2}\right]_0 = \frac{1}{s_3 v^{(3)}}\left\{\mathcal{K}^{(3)} \mp \frac{4F}{\pi s_3 B}v^{(3)}\left[\cot\frac{\pi}{2}(1 + s_3 v^{(3)})\right]^{\mp 1}\right\} \tag{3.25c}$$

In the expressions for curvatures in Eq. (3.25), the negative exponent and negative sign refer to essentially symmetric modes, and positive exponent and positive sign refer to essentially antisymmetric modes. The coefficients $\mathcal{K}^{(n)}$, E, F, and B are the same as those which appear in Eqs. (3.18) and (3.20).

These thickness modes are depicted also in Fig. 11.

Short wavelength As discussed in Refs. [M59, M72], all real segments of branches except the lowest one for each family of symmetry approach asymptotically, as $\xi \to \infty$, to the line

$$\Omega = \bar{v}^{(1)}\xi_1 \tag{3.26}$$

where $\bar{v}^{(1)}$ is given by Eq. (3.16). For the AT-cut quartz, $\bar{v}^{(1)} = 0.9972$, $\bar{v}^{(2)} = 1.7291$ and $\bar{v}^{(3)} = 1.5418$. These include the face-shear branch which originates from $\Omega = 0$ and all higher branches which originate from the thickness mode cut-off frequencies. The flexure and extension branches approach as $\xi \to \infty$, the line

$$\Omega = v_s\xi_1 \tag{3.27}$$

where v_s is the velocity ratio, to $(c_{66}/\rho)^{1/2}$, of the surface wave in a semi-infinite crystal [M56].

Imaginary and complex segments of branches Waves in monoclinic crystals with imaginary and complex wave numbers (ξ_1) are discussed in details in Refs. [M71, M72]. The calculation of the complex segments of the branches is extremely complicated. Both imaginary and complex segments are shown in Fig. 10 for an AT-cut quartz plate.

(C) *Thickness-Twist Modes*

The straight-crested waves assumed in Eqs. (3.14) and (3.17) exclude the thickness-twist modes for waves in monoclinic plates. Thickness-twist modes are standing waves with displacement and wave-normal mutually orthogonal and parallel to the faces of the plate. In monoclinic crystal plate, such waves are possible if the displacement is in the direction of x_1-axis (digonal axis of the crystal) and wave-normal in the x_3-direction, or vice versa. Following Mindlin [M83], we consider solutions of Eq. (3.4) with boundary conditions (3.10) of the form

$$u_1 = U(x_2, x_3) \exp(i\omega t)$$
$$u_2 = u_3 = 0 \tag{3.28}$$

Then, with the exponential factor omitted, Eqs. (3.4) take the form

$$c_{66}\frac{\partial^2 U}{\partial x_2^2} + 2c_{56}\frac{\partial^2 U}{\partial x_2 \partial x_3} + c_{55}\frac{\partial^2 U}{\partial x_3^2} + \rho\omega^2 U = 0 \tag{3.29}$$

and the boundary conditions (3.10) reduce to

$$c_{56}\frac{\partial U}{\partial x_3} + c_{66}\frac{\partial U}{\partial x_2} = 0 \quad \text{on } x_2 = \pm b \tag{3.30}$$

since the other two boundary conditions are satisfied identically.

The differential Eq. (3.29) admits solutions of the type $\sin(\xi_2 x_2 + \xi_3 x_3)$ and $\sin(\xi_2 x_2 - \xi_3 x_3)$. In order to be able to satisfy the boundary conditions, we take a linear combination

$$U = C_1 \sin(\xi_2^{(1)} x_2 + \xi_3 x_3) + C_2 \sin(\xi_2^{(2)} x_2 - \xi_3 x_3) \tag{3.31}$$

Then the differential Eq. (3.29) requires that

$$\rho\omega^2 = c_{66}(\xi_2^{(1)})^2 + 2c_{56}\xi_2^{(1)}\xi_3 + c_{55}(\xi_3)^2$$
$$\rho\omega^2 = c_{66}(\xi_2^{(2)})^2 - 2c_{56}\xi_2^{(2)}\xi_3 + c_{55}(\xi_3)^2 \tag{3.32}$$

and therefore it follows that, either

$$\text{(i)} \quad \xi_2^{(1)} + \xi_2^{(2)} = 0,$$

or

$$\text{(ii)} \quad c_{66}(\xi_2^{(1)} - \xi_2^{(2)}) + 2c_{56}\xi_3 = 0 \tag{3.33}$$

Now from the two boundary conditions (3.30), displacement function (3.31) and the condition (ii) of (3.33), we get the characteristic equation

$$\sin (\xi_2^{(1)} + \xi_2^{(2)}) = 0 \tag{3.34}$$

or

$$\xi_2^{(1)} + \xi_2^{(2)} = \frac{m\pi}{b}, \qquad m = 0, 1, 2, \ldots \tag{3.35}$$

Thus from Eqs. (3.35) and $(3.32)_2$

$$\xi_2^{(1)} = \frac{m\pi}{2b} - \frac{c_{56}}{c_{66}} \xi_3$$

$$\xi_2^{(2)} = \frac{m\pi}{2b} + \frac{c_{56}}{c_{66}} \xi_3 \tag{3.36}$$

and therefore from Eq. (3.32) we get

$$\Omega^2 = (m^2 + \tilde{\gamma}_{55}\bar{\xi}_3{}^2) \tag{3.37}$$

where $\Omega = \omega/\omega_s$, $\bar{\xi}_3 = 2b\xi_3/\pi$ and $\tilde{\gamma}_{55} = \tilde{c}_{55} - \tilde{c}_{56}\tilde{c}_{56}$. The branches of the dispersion spectrum for real, positive $\bar{\xi}_3$ are quarter-hyperbolas and for imaginary, positive $\bar{\xi}_3$ are quarter-circles for the dimensionless ordinate Ω and abscissa $\bar{\xi}_3$. The spectrum is similar to that of the *SH*-wave in isotropic plates.

Although the thickness-twist mode solutions is far less complicated than the solutions for the Ekstein solution, it was not found until 1965, but its existence had been identified from an approximate theory of crystal plates developed by Mindlin (see next section). This solution, as applied to the AT-cut quartz is of interest in resonator and filter technology. The solution involves only the strains e_{12} and e_{13} which are directly linked, through the piezoelectric relations, to the electric field E_2. Thus the thickness modes can easily be excited by applying E_2 via electrodes on the surfaces $x_2 = \pm b$ of the crystal.

4 Approximate Equations of Plates

(A) *Various Order of Approximate Equations*

The lowest order of approximate theory for a plate in flexural motion is governed by the Germain–Lagrange Eq. (2.1). In the low-frequency long wavelength region, it has the asymptotic behavior

$$\Omega^2 = \pi\bar{\xi}^2 \left[\tfrac{1}{3} \left(1 - \frac{1}{k^2} \right) \right]^{1/2} \tag{2.31}$$

which agrees with the low-frequency portion of the first antisymmetric branch and the low-frequency portion of the imaginary part of the second antisymmetric branch of the Rayleigh–Lamb spectrum, both originating from $\Omega = 0$ in Fig. 3. This excellent agreement accounts for the success of the Germain–Lagrange equations in the low-frequency, long wavelength region ($0 \leq \Omega \leq 0.1$).

The next higher-order equations of flexure, extending the range to $0 \leq \Omega \leq 1.5$ are due to Mindlin [M35]. By taking into account the effect of rotatory inertia and shear, Mindlin derived the following equations of motion from the three-dimensional theory of elasticity,

$$D[(1-\mu)\nabla_1^2 u_1^{(1)} + (1+\mu)\partial_1\theta] - 2b\mu'(u_1^{(1)} + \partial_1 u_2^{(0)}) = \tfrac{2}{3}\rho b^3 \ddot{u}_1^{(1)}$$

$$D[(1-\mu)\nabla_1^2 u_3^{(1)} + (1+\mu)\partial_3\theta] - 2b\mu'(u_3^{(1)} + \partial_3 u_2^{(0)}) = \tfrac{2}{3}\rho b^3 \ddot{u}_3^{(1)}$$

$$b\mu'(\nabla_1^2 u_2^{(0)} + \theta) + q = \rho b \ddot{u}_2^{(0)} \qquad (4.1)$$

where $\mu' = \kappa^2\mu$, $D = 2\mu b^3/3(1-\mu)$, $\kappa^2 = \pi^2/12$, and $\theta = \partial_1 u_1^{(1)} + \partial_3 u_3^{(1)}$. The introduction of shear-correction factor, κ^2, in this equation, reproduces exactly the cut-off frequency of the second antisymmetric mode ($\Omega = 1$ in Fig. 3) and also reproduces exactly the lowest antisymmetric *SH*-mode. In the case of plane strain, these equations reduce to the same form as Timoshenko's beam equations [23]. At each point on the edge of the plate, these equations involve three boundary conditions which reduce to the two boundary conditions of Kirchhoff, if the shear deformation is neglected.

These higher-order equations of flexure have had numerous applications. Solutions have been obtained for reflection of waves at a free edge [24, 25] and for vibrations of rectangular plates [M53] and circular disks [26, M48]. Experimental results of Sharma [7], for the axially symmetric vibrations of a circular disk with free edges, verify the frequencies predicted by Eq. (4.1).

As discussed in Section 2, the lowest symmetric branch of the transcendental frequency Eq. (2.25) has the asymptotic behavior,

$$\Omega = 2\left(1 - \frac{1}{k^2}\right)^{1/2} \xi \qquad (2.32)$$

This is exactly the result given by the Poisson Eqs. (2.3). This excellent agreement accounts for the success of these equations in the low-frequency region $0 \leq \Omega \leq 0.5$. The Poisson extensional equations are the counterpart of the generalized plane stress in elastostatics, and they also reproduce exactly the lowest *SH*-branch in which the wave-normal is at right angles with the displacement and in the plane of the plate. The Poisson equations can be directly obtained from the equations of plane strain by replacing Lamé's constant λ by $\bar{\lambda}$ in the two-dimensional Eqs. (2.5). This means that any solution of the problem of plane strain can be converted into one of generalized plane stress, and vice versa.

The next higher approximation, for the symmetric modes of isotropic, elastic plates was obtained by Mindlin and Medick [M61]. These equations were derived from the three-dimensional equations of elasticity by using series expansion method. Instead of power series, the expansion used in this study is in terms of Legendre polynomials, which because of its orthogonality properties, avoids coupling of different orders of strains. The equations take into account the coupling between the extensional, symmetric thickness-stretch and symmetric thickness-shear modes and also include two face-shear modes (corresponding to the modes (3), (7), and (6) of Fig. 11b, and (2) of Fig. 11a). By taking into account these modes, Mindlin and Medick were able to reproduce the five lowest symmetric branches of the frequency spectrum as contained in the exact theory. They are, in Fig. 3, the branches originating from $\Omega = 0, k, 2$ and the two bounds from $\Omega = 0, 2$.

The theory also reproduces the imaginary loop discovered by Aggarwal and Shaw [17]; the frequency minimum of the second branch where group velocity is zero at a nonzero wave number [27]; anomalous behavior of the second and third branches; and finally, a part of complex branches. In addition, these equations reproduce exactly the lowest *SH*-branch and the next higher symmetric *SH*-branch approximately. Solutions of these equations, for a circular disc [M65], verify the experimental results previously obtained by Shaw [15], and in particular, the experimentally observed edge vibrations can be explained by the presence of the pair of complex branches in the Mindlin–Medick theory.

In the case of crystal plates, the zero-order equations are the Cauchy–Voigt equations of low-frequency extensional motion [6, 28]. The first-order equations of high-frequency coupled flexural and extensional motion of thin crystal plates are due to Mindlin [M69] (see Eq. (4.26) of this section). In these equations account is taken of the coupling with the two lowest thickness-shear modes and extensional modes. These equations are obtained from the three-dimensional equations of linear elasticity by a procedure based on series expansion method of Cauchy [6] and Poisson [3] and the variational method of Kirchhoff [2]. Comparison of the frequency spectrum of these equations with Ekstein's spectrum of the three-dimensional equations, shows close agreement between the two frequency spectra, of all five branches of the spectrum of the two-dimensional equations in the range $0 \leq \Omega \leq 1.5$.

(B) *Higher-Order Theory of Crystal Plates*

Because of the numerous applications of higher-order plate equations, it may be pertinent to bring out the salient points of the theory and the manner of derivation. Using series expansion in terms of thickness coordinate, the displacement field can be written as

$$u_i(x_1, x_2, x_3; t) = \Sigma_n x_2^n u_i^{(n)} \tag{4.2}$$

where $u_i^{(n)} \equiv u_i^{(n)}(x_1, x_3, t)$ are the independent nth-order displacement components. The strain tensor, which is the symmetric part of the displacement gradient takes the form

$$e_{ij} = \Sigma_n x_2^n e_{ij}^{(n)} \tag{4.3}$$

where

$$e_{ij}^{(n)} = \tfrac{1}{2}[\delta_{i\alpha}\partial_\alpha u_j^{(n)} + \delta_{j\alpha}\partial_\alpha u_i^{(n)} + (n+1)(\delta_{i2}u_j^{(n+1)} + \delta_{j2}u_i^{(n+1)})] \tag{4.4}$$

and Greek indices take the values 1 and 3 only. The zeroth- and first-order of strains are depicted in Fig. 12. [12].

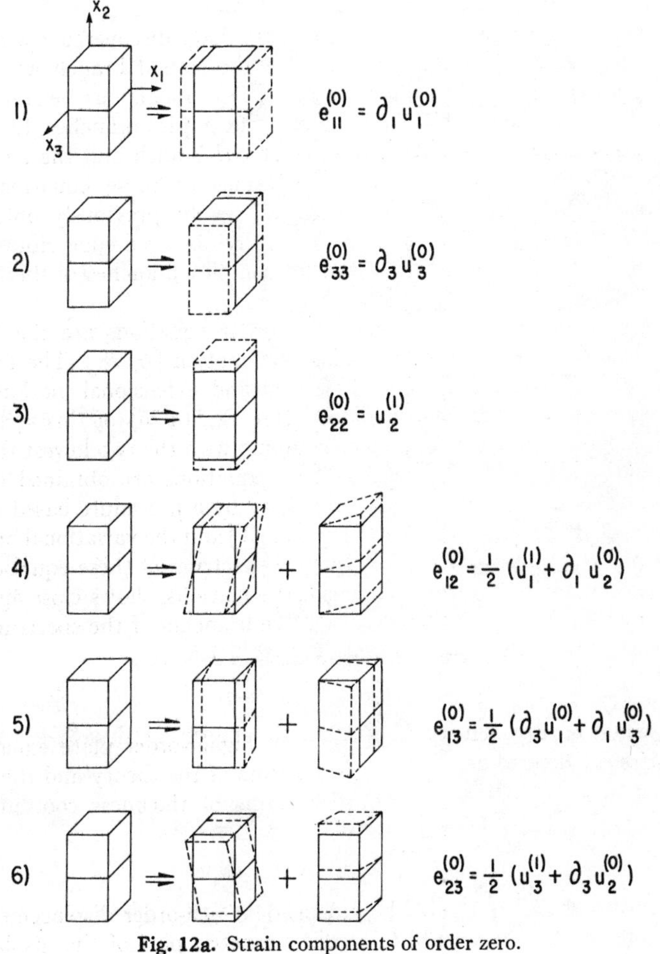

Fig. 12a. Strain components of order zero.

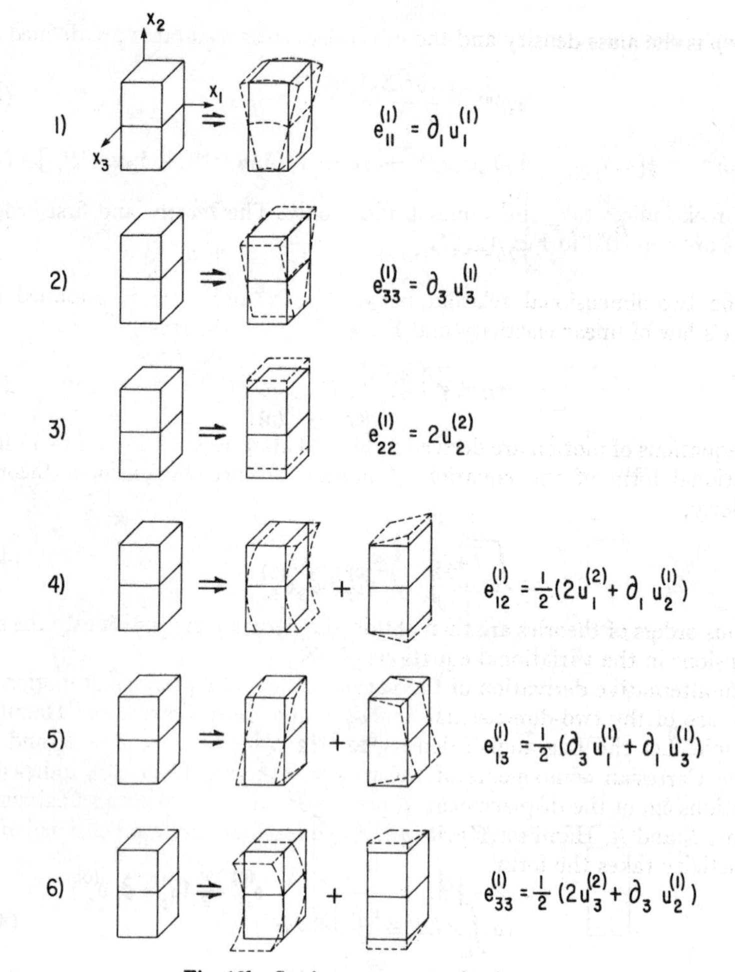

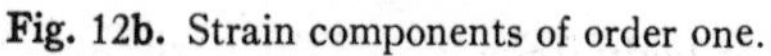

Fig. 12b. Strain components of order one.

The strain energy density $\bar{U}$ and kinetic energy density $\bar{K}$, per unit area of the plate are defined as

$$\bar{U} = \tfrac{1}{2} \int_{-b}^{b} \tau_{ij} e_{ij} dx_2 = \tfrac{1}{2} \sum_{r}^{\infty} \tau_{ij}{}^{(r)} e_{ij}{}^{(r)} \tag{4.5}$$

$$\bar{K} = \frac{1}{2} \int_{-b}^{b} \rho \left(\frac{\partial u_i}{\partial t}\right)^2 dx_2 = \tfrac{1}{2}\rho \sum_{m=r=0}^{\infty} \sum^{\infty} B_{(m,r)} \dot{u}_i{}^{(m)} \dot{u}_i{}^{(r)} \tag{4.6}$$

where ρ is the mass density and the nth-order stress resultants are defined as

$$\tau_{ij}^{(n)} = \frac{\partial \bar{U}}{\partial e_{ij}^{(n)}} = \int_{-b}^{b} x_2^n \tau_{ij} dx_2 \tag{4.7}$$

and

$$B_{(m,n)} = \begin{cases} 0, & m+n \quad \text{odd} \\ 2b^{m+n+1}/(m+n+1), & m+n \quad \text{even} \end{cases} \tag{4.8}$$

The two-dimensional relation between $\tau_{ij}^{(n)}$ and $e_{ij}^{(n)}$ is obtained from Hooke's law of linear elasticity and Eq. (4.7), with the result

$$\tau_{ij}^{(n)} = c_{ijkl} \sum_{m=0}^{\infty} B_{(m,n)} e_{kl}^{(m)} \tag{4.9}$$

The equations of motion are derived by substituting Eqs. (4.2) and (4.7) in the variational form of the equation of motion of three-dimensional theory of elasticity,

$$\int_V (\partial_i \tau_{ij} - \rho \ddot{u}_j) \delta u_j dV = 0 \tag{4.10}$$

Various orders of theories are then obtained by truncating judiciously the series expansions in the variational equations [M69].

An alternative derivation of the various orders of equations of motion is to make use of the two-dimensional analog of the three-dimensional Hamilton's principle. Let the Lagrangian density for the plate be $L \equiv T - U$ and let t_i be the Cartesian components of the traction vector. Then, for independent variations δu_i of the displacement u_i prescribed at the initial and final instants of time t_0 and t_1, Hamilton's principle for linearized three-dimensional theory of elasticity takes the form

$$\delta \int_{t_0}^{t_1} L dt + \int_{t_0}^{t_1} \delta \mathcal{W} dt = 0 \tag{4.11}$$

where in the absence of body force, $\delta \mathcal{W}$ is the variation of the work done by the external forces. For matter occupying some convex region V bounded by surface S consisting of a finite number of sectionally smooth parts

$$L = \int_V (T - U) dv; \qquad \delta \mathcal{W} = \int_S t_i \delta u_i da \tag{4.12}$$

The surfaces of the plate consist of two parallel faces $x_2 = \pm b$, and the edge with generators parallel to the x_2-axis. Therefore, the surface integral $(4.12)_2$ can be written as the sum of two parts consisting of the integral over the two faces of area A, and the boundary curve B. Thus in the case of a plate, the variation of the work done by the external forces can be written as

$$\delta W = \sum_r H_{(r)} \oint_B T_i^{(r)} \delta u_i^{(r)} ds + \sum_r H_{(r)} \int_A f_i^{(r)} \delta u_i^{(r)} dA \tag{4.13}$$

where s is the parameter of arc-length, dA the area of the face, and

$$T_i^{(r)} = \left(\frac{2r+1}{2b^{2r+1}}\right) \int_{-b}^{b} x_2^n [n_\alpha \tau_{i\alpha}]_B dx_2 \tag{4.14}$$

$$f_i^{(r)} = \left(\frac{2r+1}{2b^{2r+1}}\right) [x_2^r \tau_{2i}]_{-b}^{b} \tag{4.15}$$

$$H^{(r)} = \frac{2b^{2r+1}}{2r+1} \tag{4.16}$$

The $T_i^{(r)}$, $f_i^{(r)}$ are the edge-tractions and face-fractions, respectively. Two-dimensional analog of three-dimensional Hamilton's principle, for nth-order plate theory, can now be written as

$$\sum_r^n \left\{ \int_{t_0}^{t_1} \mathfrak{L}^{(r)} dt + H_{(r)} \int_{t_0}^{t_1} dt \left(\oint_B T_i^{(r)} \delta u_i^{(r)} ds + \int_A f_i^{(r)} \delta u_i^{(r)} dA \right) \right\} = 0 \tag{4.17}$$

where

$$\mathfrak{L}^{(r)} = \int_A \left\{ \sum_m^n \rho B_{(m,r)} \dot{u}_i^{(m)} \frac{\partial}{\partial t} \delta u_i^{(r)} - \tau_{ij}^{(r)} \delta e_{ij}^{(r)} \right\} dA \tag{4.18}$$

Substituting Eq. (4.4) in Eq. (4.18) and carrying out the variation, the Euler's equations of motion and natural boundary conditions can readily be determined. However, careful analysis reveals that the Euler's equations are accompanied by other natural conditions which are its variational concomitants [29].

For an nth-order plate theory, this formulation of the variational problem, allows amongst other things, free development of thickness-stretch $e_{22}^{(n)}$, $e_{22}^{(n-1)}$ and thickness-shear $e_{\alpha 2}^{(n)}$. In addition, the remaining independent variations are the components of displacements $u^{(k)}$. By using the procedure mentioned here briefly, the differential equations of motion, the constraint conditions and the natural boundary conditions, for nth-order theory of plate are

(a) Differential equations:

$$\partial_\alpha \tau_{\alpha\beta}^{(n)} - n\tau_{2\beta}^{(n-1)} + H_{(n)} f_\beta^{(n)} = \rho \sum_m B_{(m,n)} \ddot{u}_\beta^{(m)}$$

$$\partial_\alpha \tau_{\alpha i}^{(k)} - k\tau_{2i}^{(k-1)} + H_{(k)} f_i^{(k)} = \rho \sum_m B_{(m,k)} \ddot{u}_i^{(k)}, \qquad k = 0, 1, \ldots, (n-1) \tag{4.19}$$

(b) Variational concomitants:

$$\text{(i)}\ \tau_{2i}^{(n)} = 0, \qquad \tau_{2i}^{(n-1)} = 0$$

$$\text{(ii)}\ \dot{u}_2^{(n)} = 0,$$

$$\text{(iii)}\ f_2^{(n)} = 0 \qquad \text{in } A, \text{ and}$$

$$\text{(iv)}\ t_2^{(n)} = 0 \qquad \text{on } B \text{ of } A \tag{4.20}$$

(c) *Natural* boundary conditions on B:

$$n_\alpha \tau_{\alpha\beta}^{(n)} = H_{(n)} t_\beta^{(n)}$$

$$n_\alpha \tau_{\alpha i}^{(k)} = H_{(k)} t_i^{(k)}, \qquad k = 0, 1, \ldots, (n-1) \tag{4.21}$$

In the above formulation, we see that in addition to the equations of motion, we have simultaneously obtained certain additional conditions—the variational concomitants of the problem. In particular, for $n = 0$, the component of velocity $\dot{u}_2^{(0)} = 0$ and $\tau_{2i}^{(0)} = 0$. In the case of symmetric motion of isotropic plates, this implies that $\dot{u}_2^{(0)} = 0$ and $\tau_{22}^{(0)} = 0$. These are exactly the conditions used for the development of Poisson equations. For $n = 1$, we have the conditions $\dot{u}_2^{(1)} = 0$, $\tau_{22}^{(0)} = 0$, and $\tau_{2i}^{(1)} = 0$. For isotropic plates, we have the conditions $\dot{u}_2^{(1)} = 0$, $\tau_{22}^{(0)} = 0$, and $\tau_{2\alpha}^{(1)} = 0$. These are the conditions imposed by Mindlin in his development of the flexural equations [12, M35], to allow free development of thickness-stretch and thickness-shear. From the variational point of view, these are in fact a set of necessary conditions, for the problem to have a stationary value and arise conjointly with the system of equations of motion.

To construct first-order theory of crystal plates, it is necessary to retain zero- and first-order components of strain. Further, the stress resultants $\tau_{22}^{(0)}$ and $\tau_{2i}^{(1)}$ are both zero. Taking account of these two facts, the constitutive equations, for first-order theory of plate takes the form

$$\tau_{\alpha i}^{(0)} = 2b g_{\alpha i\gamma j} e_{\gamma j}^{(0)}$$

$$\tau_{\alpha\beta}^{(1)} = \tfrac{2}{3} b^3 \gamma_{\alpha\beta\mu\nu} e_{\mu\nu}^{(1)} \tag{4.22}$$

where

$$g_{\alpha i\gamma j} = \bar{c}_{\alpha i\gamma\sigma}\delta_{\sigma j} + 2\bar{c}_{\alpha i\gamma 2}\delta_{2j}$$

$$\bar{c}_{\alpha i\gamma j} = c_{\alpha i\gamma j} - \frac{c_{\alpha i22} c_{22\gamma j}}{c_{2222}} \tag{4.23}$$

and $\gamma_{\alpha\beta\mu\nu}$ is the reduced cofactor of $s_{\alpha\beta\mu\nu}$ in the determinant of $s_{\alpha\beta\mu\nu}$, where $s_{\alpha\beta\mu\nu}$ is the inverse of $\gamma_{\alpha\beta\mu\nu}$.

To account for expansion of strain tensor in the thickness direction and neglection of higher order of strains, it is necessary to adjust the strain energy density, so that the frequencies of the thickness modes be the cut-off frequencies of the exact spectrum. This necessary modification of strain energy density can be explained from the variational approach as follows: associated with the strains $e_{ij}^{(0)}$ and $e_{ij}^{(1)}$, we may postulate the existence of strain energy function $\bar{W}(e_{ij}^{(0)}, e_{ij}^{(1)})$. Since the stress resultants are derivable from the strain energy density $\bar{W}$, and since the strain energy is a homogeneous quadratic form of degree two in $e_{ij}^{(0)}$ and $e_{ij}^{(1)}$, it follows that

$$\tau_{\alpha i}^{(0)} = 2bg^*_{\alpha i\gamma j}e_{\gamma j}^{(0)}$$

$$\tau_{\alpha\beta}^{(1)} = \tfrac{2}{3}b^3\gamma^*_{\alpha\beta\mu\nu}e_{\mu\nu}^{(1)} \tag{4.24}$$

where $g^*_{\alpha i\gamma j} = \partial^2\bar{W}/\partial e_{\alpha i}^{(0)}\partial e_{\gamma j}^{(0)}$, $\gamma^*_{\alpha\beta\mu\nu} = \partial^2\bar{W}/\partial e_{\mu\nu}^{(1)}\partial e_{\alpha\beta}^{(1)}$. It therefore follows that existence of the function $\bar{W}$ implies the *existence* of the *plate constants* $g^*_{\alpha i\gamma j}$ and $\gamma^*_{\alpha\beta\mu\nu}$. These plate constants are different from the constants $g_{\alpha i\gamma j}$ and $\gamma_{\alpha\beta\mu\nu}$, but can be related to each other, if we require that the cut-off frequencies of the two-dimensional approximate theory may be the frequencies of the simple thickness modes of the exact spectrum. This gives us a procedure of determining the unknown plate constants in terms of the elastic constants. Alternatively, one can foresee in advance that the only new constants which will occur in Eq. (4.24) are those associated with $e_{21}^{(0)}$ and $e_{23}^{(0)}$, since these are the only two strains which effect the thickness frequencies. Therefore, following Mindlin we may postulate in advance that

$$g^*_{ijkl} = k_{i+j-2}{}^m k_{k+l-2}{}^n g_{ijkl} \quad \text{(no sum)}$$

$$\gamma^*_{\alpha\beta\mu\nu} = \gamma_{\alpha\beta\mu\nu} \tag{4.25}$$

where $m = \cos^2(ij\pi/2)$, $n = \cos^2(kl\pi/2)$, and $k_{i+j-2}{}^m$ (or $k_{k+l-2}{}^n$) is equal to k_1, k_3, or l according as $i+j$(or $k+l$) in g_{ijkl} is 3, 5 or neither, respectively.

Substituting Eq. (4.24) in the stress equations of motion (4.19), for $n = 1$, which implies that only zero- and first-order velocities are to be retained in the kinetic energy density, we obtain

$$\gamma_{\alpha\beta\gamma\sigma}\partial_{\alpha\gamma}u_\sigma^{(1)} - \frac{3}{b^2}[g^*_{2\beta\gamma\sigma}\partial_{\alpha\gamma}u_\sigma^{(0)} + g^*_{2\beta\gamma 2}(\partial_\gamma u_2^{(0)} + u_\gamma^{(1)})] + f_\beta^{(1)} = \rho\ddot{u}_\beta^{(1)}$$

$$g^*_{\alpha i\gamma\sigma}\partial_{\alpha\gamma}u_\sigma^{(0)} + g^*_{\alpha i\gamma 2}\partial_\alpha(\partial_\gamma u_2^{(0)} + u_\gamma^{(1)}) + f_i^{(0)} = \rho\ddot{u}_i^{(0)} \tag{4.26}$$

Equation (4.26) is the Mindlin equation for high-frequency coupled extensional flexural vibrations of thin crystal plates. In the equations account is taken of the coupling with the lowest two thickness-shear modes and extensional modes. If in these equations we suppress the extensional displacement $u_\alpha^{(0)}$, the equations of coupled thickness-shear and flexural motions are obtained [M36]. A derivation of these equations, along with many details of the dispersion spectrum are given in the fundamental monograph [12] and symposium report [M66].

It now remains to determine the plate constants $g^*_{\alpha i\gamma\sigma}$ and $g^*_{\alpha i\gamma 2}$, or what is the same thing as determining the two constants k_1 and k_3. The values of k_1 and k_3 are found by comparing the thickness-shear frequencies obtained from Eq. (4.26) with the corresponding ones obtained from the exact three-dimensional equations of elasticity. In the case of monoclinic symmetry with x_1 the digonal axis, $c_{1222} = 0$, $c_{1223} = 0$, and by equating corresponding frequencies of the thickness-shear modes, the two constants are given by

$$k_1^2 = \frac{\pi^2}{12}; \qquad k_3^2 = \frac{\pi^2 c_3}{12 g_{2323}} \tag{4.27}$$

where

$$c_3 = \tfrac{1}{2}\{c_{2222} + c_{2323} - [(c_{2222} - c_{2323})^2 + 4c_{2223}^2]^{1/2}\}$$

For predicting the frequencies of vibration of a bounded crystal plate, the accuracy of these equations of motion can be judged by a comparison of the five branches of the frequency spectrum of an infinite plate, as obtained from Eq. (4.26), with the first five branches of the spectrum obtained from the three-dimensional equations of elasticity. Detailed computations show correspondence to at least three significant figures over most of the range $\Omega < 1.5$, $\xi < 1.5$, where $\Omega = \omega/\omega_s$, $\xi = 2\xi b/\pi$, and $\omega_s = (\pi/2b)(c_{1212}/\rho)^{1/2}$. Furthermore, the limiting behavior of the four branches which intersect at $\Omega = 0$ are

1. Flexure: $\quad \Omega = \dfrac{\pi}{2}\left(\dfrac{\gamma_{1111}}{3c_{1212}}\right)^{1/2} \xi^2$

2. Face-shear: $\quad \Omega = \left(\dfrac{\gamma_{1313}}{c_{1212}}\right)^{1/2} \xi$

3. Extension: $\quad \Omega = \left(\dfrac{\gamma_{1111}}{c_{1212}}\right)^{1/2} \xi$

4. Thickness-shear: $\quad \Omega = -\dfrac{\pi}{2}\left(\dfrac{\gamma_{1111}}{3c_{1212}}\right)^{1/2} \xi^2 \qquad (4.28)$

These are in exact correspondence with the results obtained from the Ekstein's [8] solution of the three-dimensional equations. In addition, at zero frequency the thickness-twist mode, from the approximate equations has the imaginary wave number

$$i\xi = \left(\frac{c_3\gamma_{1111}}{\gamma_{1313} g_{1111}}\right)^{1/2}$$

and $i\xi = 0.7467$ in the exact equations. In the high-frequency region the important frequencies are near $\Omega = 1$ and the important branches are the thickness-shear and flexural branches. At $\Omega = 1$, the thickness-shear branch is exact, due to the choice $k_1^2 = \pi^2/12$; the flexural branch in the approximate equations has the value 1.2483, whereas the exact value is $\xi = 1.2417$.

(C) *Discussion*

In this section, we have presented a concise description of Mindlin's higher-order plate theory and in particular, the first-order equations of motion and their relation to the well-known classical theories. The higher-order theory of thin plate has a rational mathematical foundation and is derivable from variational principles. Furthermore, it is strongly supported by the fact that the frequency spectrum and mode shapes of a bounded crystal plate, as calculated from this theory, agree closely with the experimental observations [7, M69, M75, M88].

The first-order equations do have some limitations because they do not contain the segments of the complex branches. In those applications where contributions of the *edge modes* from the complex segment of the branches is important [15], or for frequencies higher than the range of frequencies for which the first-order theory is appropriate, one should make use of the next higher-order system of Eqs. (4.19). Such a system of higher-order equations is given by setting $n = 0, 1, 2$ in Eqs. (4.19)–(4.21).

In the last few years new attempts have been made to rederive plate equations by using asymptotic and perturbations methods. However, as demonstrated by Mindlin in his monograph [12] and in other research papers, the success of a plate theory in predicting in a certain frequency range the vibration characteristics of a thin plate, depends on properly taking into account the appropriate modes of displacements and strains (Fig. 12). This requires an understanding of Rayleigh's [4] and Ekstein's [8] exact solutions of an infinite plate, and this understanding is an aid in deciding what to retain and what to discard in various orders of approximations. It is this understanding and physical insight which Mindlin uses in the successful development of higher-order theories.

References

1. S. Germain, *Reserches sur la théorie des surfaces elastiques*, Courcier, Paris, 1821.
2. G. Kirchhoff, Über das Gleichgewicht und die Bewegung einer Elastichen Scheibe, *Crelles J.* **40,** 51–88 (1850). (= *Gesammelte Abhandlungen*, Leipzig, 1882, pp. 237–279.)
3. S.-D. Poisson, Mémoire sur l'équilibre et la mouvement des corps élastiques, *Mém. Acad. Sci.*, Série 2, **8,** 357–470, Paris (1829).
4. Lord Rayleigh, On the free vibrations of an infinite plate of homogeneous isotropic elastic matter, *Proc. London Math. Soc.* **20,** 225–234 (1889).
5. H. Lamb, On waves in an elastic plate, *Proc. Roy. Soc.* **A93,** 114–128 (1917).
6. A.-L. Cauchy, Sur L'équilibre et le mouvement d'une plaque élastique dont l'élasticité n'est pas la même dans tout les sens, *Exercices de Mathématique* **4,** 1–14, Bure Frères, Paris, (1829). (*Oeuvres*, Séries 2, Vol. 19, pp. 9–22, Gauthier-Villars, Paris, 1891.)

7. R. L. Sharma, Dependence of frequency spectrum of a circular disk on Poisson's ratio, *J. appl. Mech.* **24,** 641–642 (1957).
8. H. Ekstein, High frequency vibrations of thin crystal plates, *Phys. Rev.* **68,** 11–23 (1945).
9. M. G. Lamé, *Leçons sur la théorie mathématique de l'élasticite des corps solides*, 2nd edition, Gauthier-Villars, Paris, 1866.
10. A. N. Holden, Longitudinal modes of elastic waves in isotropic cylinders and bars, *Bell System Tech. J.* **30**(4), Part 1, 956–969 (1951).
11. M. Onoe, A study of the branches of the velocity dispersion equations of elastic plates and rods, *Report Joint Committee on Ultrasonics of the Institute of Electrical Engineers and Acoustical Society of Japan*, 1955.
12. R. D. Mindlin, *An Introduction to the Mathematical Theory of Vibrations of Elastic Plates*, U.S. Army Signal Corps Engineering Laboratories, Fort Monmouth, N.J., 1955.
13. R. D. Mindlin, Investigations in the mathematical theory of vibrations of anisotropic bodies, *Columbia University Report No. CU-4-56-SC-64687-CE*, 1956. (= Signal Corps Engineering Laboratories; Fort Monmouth, N.J.)
14. J. N. Goodier and R. E. D. Bishop, A note on critical reflections of elastic waves at free surfaces, *J. appl. Phys.* **23,** 124–126 (1952).
15. E. A. G. Shaw, On the resonant vibrations of thick barium titanate disks, *J. Acoust. Soc. Amer.* **28,** 38–50 (1956).
16. M. Onoe, Contour Vibrations of Thin Rectangular Plates, *J. Acoust. Soc. Amer.* **30,** 1159–1162 (1958).
17. R. R. Aggarwal and E. A. G. Shaw, Axially symmetric vibrations of a finite isotropic disk. IV, *J. Acoust. Soc. Amer.* **26,** 341–342 (1954).
18. R. H. Lyon, Response of an elastic plate to localized driving forces, *J. Acoust. Soc. Amer.* **27,** 259–265 (1955).
19. R. D. Mindlin and M. Onoe, Mathematical theory of vibrations of elastic plates, *Proceedings of the Eleventh Annual Symposium on Frequency Control*, U.S. Army Signal Corps Engineering Laboratories, Fort Monmouth, N.J., 1957, pp. 17–40.
20. K. Sezawa, On the accumulation of energy of high-frequency vibrations of an elastic plate on its surfaces, *Proc. Third Int. Cong. Appl. Mech.* **3,** 167–172, Stockholm, 1930.
21. L. E. Goodman, Circular crested vibrations of an elastic solid bounded by two parallel surfaces, *Proc. First U.S. Nat'l Cong. Appl. Mech.* 65–73 (1951).
22. I. Koga, Thickness vibrations of piezoelectric oscillating crystals, *Physics*, **3,** 70–80 (1932).
23. S. P. Timoshenko, On the transverse vibrations of bars of uniform cross-section, *Philosophical Magazine*, Ser. 6, **43,** 125–131 (1922).
24. T. R. Kane, Reflection of Flexural Waves at the Edge of a Plate, *J. appl. Mech.* **21,** 213–220 (1954).
25. C. C. Chao and Y. H. Pao, On the flexural motion of plates at the cut-off frequency, *J. appl. Mech.* **31,** 22–24 (1964).
26. H. Deresiewicz, Symmetric flexural vibrations of a clamped circular disk, *J. appl. Mech.* **23,** 319 (1956).
27. I. Tolstoy and E. Usdin, Wave propagation in elastic plates: Low and high mode dispersion, *J. Acoust. Soc. Amer.* **29,** 37–42 (1957).
28. W. Voigt, *Lehrbuch der Kristallphysik*, 2nd edition, B. G. Teubner, Leipzig, 1928, pp. 675–698.
29. R. Courant and D. Hilbert, *Methods of Mathematical Physics*, Interscience, New York, 1953, Vol. 1, pp. 231–242.

Vibrations and Wave Propagation in Rods

HUGH D. McNIVEN

University of California, Berkeley, California 94720

and

JOHN J. McCOY

The Catholic University of America, Washington, D.C. 20017

Abstract—This article devoted to vibrations and wave propagation in rods does not attempt to cover all aspects of the subject. The enormous amount of research on the dynamic behavior of rods and the extensive literature that has resulted has forced a selection of the topics that are described. In deciding what to omit and what to include the guide has been Professor Mindlin's own research. Included are those topics on which he himself worked or which he directly influenced.

The article describes theoretical studies directed to isotropic, linearly elastic rods. Space has been assigned to the various topics roughly in proportion to the attention each has received in the literature. The first section, which is devoted to extensional waves in rods is, accordingly, the longest. Torsional vibrations of rods is treated briefly and flexural vibrations in detail. The section on other vibrations in circular rods is short, after which vibrations in rods of non-circular section is discussed at greater length. In the final section, research is described in which Professor Mindlin was not directly involved but in which his theories were instrumental in the solution of the problem.

Introduction

Vibrations and wave propagation in rods have long been attractive subjects for scientific and engineering research. Interest began over two hundred years ago with Daniel Bernoulli, and since this beginning the tempo of research has increased, reaching its peak in the years since the Second World War.

It is possible to speculate on the reasons why the dynamics of rods has aroused so much attention. Of all bounded bodies of arbitrary length, the rod has the simplest enclosing boundary. It is a possible mechanism for delay lines and wave guides. Furthermore, wave propagation in rods has been studied very recently to identify material properties.

The enormous amount of research on the dynamic behavior of rods and the extensive literature that has resulted, however, makes it difficult to summarize the research in a single article. Thus it was not possible to include all aspects of vibrations and waves in rods and the decision was made to discuss certain parts of the research in some detail at the expense of omitting other parts altogether. In deciding what to omit and what to include we were guided by Professor Mindlin's own research and have included those topics on which he himself worked or which he directly influenced. As his involvement was extensive and his influence profound, some of the surviving topics have had to be rather briefly described.

The article includes only theoretical studies devoted to isotropic, linearly elastic rods. Much important theoretical research has had to be omitted. We have had to prune discussion of work on anisotropically elastic, visco-elastic, and plastic rods. Nor have we been able to include important research on hollow or composite rods. From the numerous papers describing transient wave propagation in rods, we have been able to include only those that use an approximate theory derived by Mindlin and his associates.

All of these theoretical studies have their counterpart in experiments. However, this equally important part of the work on vibrations and wave propagation in rods has reluctantly had to be excluded.

For those topics that are included, we have tried to assign space in the article roughly in proportion to the attention each topic has received in the literature. The first section, which is devoted to extensional waves in rods is, accordingly, the longest. We next treat torsional vibrations of rods briefly and flexural vibrations in detail. The section on other vibrations in circular rods is short, after which we discuss at greater length vibrations in rods of noncircular section. In the final section we describe research in which Professor Mindlin was not directly involved but in which his theories were instrumental in the solution of the problem.

Extensional Waves in Circular Rods

The history of the study of axially symmetric or "extensional" motions of isotropically elastic rods is characterized, as are most such studies, by short periods of intense activity interspersed with long periods of neglect. The study can be divided, for purposes of organization, into two distinct activities, activities which unfortunately cannot be separated historically, as attention has alternated between the two over the various periods of research.

The first activity is centered around the "exact" theory governing the axially symmetric motions of elastic rods of infinite extent, the cylindrical surface of which is free of traction. The term "exact" is used because the displacements, strains, and stresses describing such motions satisfy the field equations of isotropically elastic materials. The exact theory, established almost a hundred years ago, is in the form of a frequency equation dictating the relationship between frequency and wavelength that axially symmetric wave trains must satisfy for the cylindrical surface of the rod to be free of traction.

The frequency equation, though it looks innocent enough, contains a wealth of information, much of it requiring subtle techniques to extract. This section will describe both the establishment of the equation and the gradual unfolding of its secrets by several investigators.

Because the exact equation is complicated and because it contains an infinite number of "modes" of axisymmetric motion, it cannot in general be used when end boundaries are introduced or when the propagation of transient waves is studied. These problems require a simpler theory. The second activity, then, has been to create approximate theories in which the motions do not satisfy the field equations of elasticity. They do, however, produce a good relationship between frequency and wavelength for the lower frequencies and longer wavelengths, and they are considerably easier to use.

It seems appropriate in this review to point out that it was Professor Mindlin, working with his associates, who brought the two activities together. They probed the exact equation in depth and the frequency spectra that resulted from the study became the guide and model after which they constructed approximate theories.

It is tempting to review the development of knowledge of axially symmetric waves in rods by studying the exact and the approximate theories separately, but the choice here is to present the findings as they occurred in chronological order.

The mathematical theory of extensional or longitudinal motions in elastic bars has its beginnings in studies made by Daniel Bernoulli in 1741. Bernoulli's investigation was not on elastic rods but on the vibration of a column of air. However, the theory governing the motions of the column of air has the same classical form as the bar theory, namely the one-dimensional wave equation,

$$C^2 \frac{\partial^2 u_z}{\partial z^2} = \frac{\partial^2 u_z}{\partial t^2} \tag{1}$$

The equation governs the displacement u_z parallel to the axis of the bar, and for the elastic bar, the constant phase velocity is

$$C = \left(\frac{E}{\rho}\right)^{1/2} \tag{2}$$

where E is the modulus of elasticity and ρ is the mass density. The theory is based on the assumptions that u_z is the only displacement component existing in the motion and that it is independent of the radius, i.e., plane sections translate during motion. It will be shown that the theory, though very simple, predicts the correct relationship between frequency and wavelength only for very low frequencies and very long wavelengths.

In 1876 L. Pochhammer [1] developed what has become a milestone in our understanding of waves in rods, the exact frequency equation. Pochhammer's equation was not restricted to axially symmetric motions, but for this special case it is now usually written

$$(2\zeta^2 - \Omega^2)^2 \mathcal{J}_1(\delta\alpha) + 4\alpha^2\zeta^2 \mathcal{J}_1(\delta\beta) - 2\alpha^2\Omega^2 = 0 \tag{3}$$

In Eq. (3),

$$\alpha^2 = \Omega^2 k^{-2} - \zeta^2, \qquad \beta^2 = \Omega^2 - \zeta^2$$

$\Omega = \omega/\omega_s$, a normalized frequency
$\zeta = \gamma a/\delta$, a dimensionless propagation constant
$\omega =$ circular frequency
$\omega_s = (\delta/a) V_2$
$\gamma =$ wave number along axis of rod
$a =$ radius of rod
$\delta =$ lowest nonzero root of $J_1(\delta_n) = 0$
$k =$ ratio of dilatational (V_1) to equivoluminal (V_2) wave velocities
$\mathcal{J}_1(x) = xJ_0(x)/J_1(x)$ is the modified quotient of Bessel functions of order one.

The theory leading to Eq. (3) is formulated so that the displacement, stress, and strain fields resulting from axisymmetric motions satisfy the field equations of linear, isotropic elasticity. The equation itself is a statement that the cylindrical boundary of the rod, which is of infinite length, is free of traction. The equation has the form

$$F(\Omega, \zeta; \nu) = 0 \tag{4}$$

It implicitly relates the normalized frequency Ω and the dimensionless propagation constant ζ, and in the equation, Poisson's ratio ν acts as a parameter. It is this equation that was to prove such a challenge to investigators in the twentieth century.

Ten years later, in 1886, C. Chree, obviously unaware of Pochhammer's work, produced a paper [2] that was essentially a duplicate of Pochhammer's. Many investigators seem unaware that Chree first presented the exact frequency equation in Ref. [2] and not in his very long paper [3] of 1889.

It is interesting that Pochhammer and Chree not only produced the same exact frequency equation, but also, in Refs. [1, 2], established the same approximate theory that was the first improvement on the classical. Both wanted to study finite bars and thus developed the approximate theory for bars in which the radius is small. The approximate theory was based purely on mathematical considerations. They simply replaced the Bessel functions in Eq. (3) by the first two terms of their series expansions which are power series involving the radius of the rod. When the Bessel functions are so replaced, Eq. (3)

degenerates to

$$\frac{\omega}{\gamma} = \left(\frac{E}{\rho}\right)^{1/2} \left(1 - \frac{\nu^2 a^2}{4} \gamma^2\right)^2 \tag{5}$$

or, in the notation of Eq. (3), to

$$\Omega^2 = 2(1 + \nu)\zeta^2 \left[1 - \left(\frac{\nu\delta}{2}\right)^2 \zeta^2\right]^2 \tag{6}$$

This is the first approximate theory that allows for the dispersion of waves traveling in a rod. Rayleigh was apparently the first to point out that if the Bessel functions in the exact equation are replaced by only one term of their series expansions, the classical equation results.

Lord Rayleigh derived an equation almost identical to Eq. (6) which apparently appeared first in the second edition of his book, *Theory of Sound* [4] in 1894. However, unlike Pochhammer and Chree, his theory was based entirely on physical reasoning. The theory follows from the assumption that the axial stress exists alone and that the stress-strain relations are only satisfied if the radial displacement exists as well as the axial. These relations give

$$\begin{aligned} \epsilon_{\theta\theta} &= -\nu\epsilon_{zz} \\ \epsilon_{rr} &= -\nu\epsilon_{zz} \end{aligned} \tag{7}$$

each of which gives

$$u_r = -\nu r \frac{\partial u_z}{\partial z} \tag{8}$$

Radial displacement gives rise to radial inertia, and it is by accommodating the inertia that Rayleigh's theory extends the classical. Rayleigh established the frequency equation

$$\frac{\omega}{\gamma} = \left(\frac{E}{\rho}\right)^{1/2} \frac{1}{[1 + (2\pi^2\nu^2a^2/\Lambda^2)]^{1/2}} \tag{9}$$

In the notation of Eq. (3) this has the form

$$\Omega^2 = 2(1 + \nu)\zeta^2 \left[1 + 2\left(\frac{\nu\delta}{2}\right)^2 \zeta^2\right]^{-1} \tag{10}$$

Love [5], following closely the work of Rayleigh, used a variational equation involving the kinetic and potential energy densities to derive the approximate equation of motion

$$\left(\frac{E}{\rho}\right)\frac{\partial^2 u_z}{\partial z^2} = \frac{\partial^2 u_z}{\partial t^2} - \frac{\nu^2 a^2}{2}\frac{\partial^4 u_z}{\partial z^2 \partial t^2} \tag{11}$$

in which the radial inertia is reflected in the last term. When the trial solution

$$u_z = Ae^{i\gamma(z-Vt)} \tag{12}$$

is substituted in Eq. (11), Eq. (9) results as it should.

Even though the Pochhammer–Chree and the Rayleigh–Love approximate theories were developed quite differently, one on mathematical approximations, and the other on the physics of the problem, the relationship between frequency and wave number contained in each is almost identical (at least for $\nu = 0.25$). The closeness of the two spectral lines is displayed in Fig. 1. With the establishment of the approximate equation by Love, interest in approximate theories was not to be renewed for almost fifty years. Attention turned to an exploration of the exact equation.

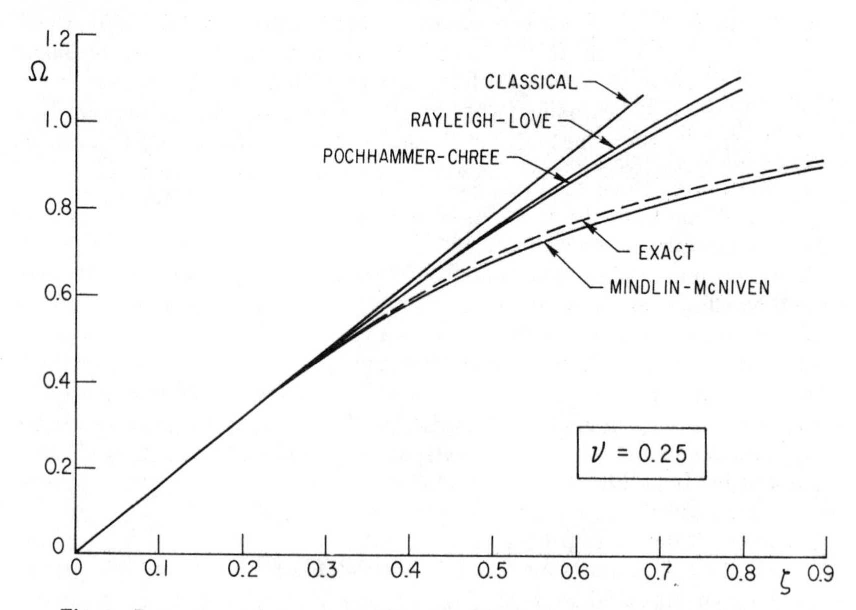

Fig. 1. Comparison of the spectral lines from several approximate theories with the fundamental line from the exact theory for extensional vibrations.

Even though the exact equation was established by Pochhammer in 1876 and received prominence by being included and discussed at length in Love [5], no systematic exploration of its roots was undertaken until 1941. In that year Bancroft [6] studied the relation between phase velocity and wavelength contained in the lowest spectral line formed from an array of roots of the exact equation. He found that the shape of this lowest branch varied with changes in Poisson's ratio, and he presented a spectrum that showed its influence. He found that as the wavelength of vibration becomes smaller, the phase velocity approaches the velocity of Rayleigh surface waves. Although Bancroft focused primarily on the fundamental branch, he also found that for a given wavelength, the transcendental exact equation supplies an infinite number of phase velocities. He also made the important discovery, that, again as the wavelength becomes small, the phase velocity of all of the higher branches approaches the velocity of equivoluminal waves in an infinite medium. Both Bancroft and, a little later, Hudson [7], because they lacked today's high-speed computers, were forced to make ingenious choices of the physical parameters and to recognize asymptotic properties of Bessel functions to extract the information they did. Perhaps Hudson's most important contribution was to show that even though the fundamental branch changes with changes in Poisson's ratio, there is one combination of phase velocity and wavelength that remains invariant to such a change.

In 1948 R. M. Davies [8], studying the Hopkinson Pressure Bar, extended the knowledge contained in the exact equation by finding the lowest three roots of the equation, representing wave numbers, for each normalized phase velocity for steel rods whose Poisson's ratio was taken to be 0.29. From these roots, the lowest three spectral lines were formed and displayed along with the spectral line from the Rayleigh–Love approximate theory. The choice of the phase velocity as the ordinate axis for displaying the spectral lines proved to be unfortunate. With this choice, both the second and third spectral lines extend to infinity as the wave number becomes very small so that the strong coupling between these two modes, particularly for small wave numbers, remained undiscovered. Davies also showed that when an arbitrary pulse travels along a rod, its shape will change owing to dispersion and that it is the waves with long wavelengths and low frequencies that will be predominant in the eventual wave shape.

It was in about 1950 that Mindlin first became interested in the vibration of rods. He was led to a more refined approximate theory, having had considerable success in developing higher-order approximate theories for elastic plates [M36,† 9, 10]. This first study resulted in a paper by Mindlin and Herrmann [M39] which appeared in 1951. Mindlin and Herrmann were prompted to develop their more extended approximate theory because they were interested in studying the vibration of finite rods and transients in long rods with steep wave fronts. On the one hand, the Pochhammer theory was too complex, and on the other, the Rayleigh–Love theory, the best approximate theory then available, was suitable only for waves with low frequencies and long wavelengths. Their theory had its origin in the physics of the problem, as did Rayleigh's. They reasoned that radial shear was as important to axisymmetric

† Numbers preceded by the letter M in brackets refer to the Publications of R. D. Mindlin listed after the Preface to this volume.

motions as radial inertia, and if one was to be accounted for, both should. Their theory began by assuming displacements of the form:

$$u_r = ra^{-1}u(z, t), \quad u_\theta = 0, \quad u_z = w(z, t) \tag{13}$$

In separating out the radial dependence, in the form of polynomials, from the length and time dependence as he had done with the thickness dependence of plates, Mindlin acknowledged his indebtedness to Poisson [11] and Cauchy [12].

The displacements of Eqs. (13) do not satisfy the field equations of linear elasticity, so other governing equations had to be produced. To this end a variational equation was used, following the lead of Kirchhoff [13], which, after considerable development, resulted in the displacement equations of motion

$$\begin{aligned} a^2K^2\mu u'' - 8K_1^2(\lambda + \mu)u - 4aK_1\lambda w' + 4aR &= \rho a^2\ddot{u} \\ 2K_1a\lambda u' + a^2(\lambda + 2\mu)w'' + 2aZ &= \rho a^2\ddot{w} \end{aligned} \tag{14}$$

In Eqs. (14),

$$\lambda, \mu = \text{Lamé constants}$$

R and Z are the radial and tangential components of stress at the surface.

K and K_1 are adjustment factors described below.

In Eqs. (14) prime and dot denote differentiation with respect to z/a and t, respectively.

The term $\rho a^2\ddot{u}$ is the radial inertia correction and $a^2K^2\mu u''$ is the radial shear correction.

When u and w are taken in the form of waves traveling in the positive z-direction, the frequency equation, or in this case the velocity equation, results. The authors display the roots of the frequency equation in the same way Davies did.

It is worth noting that the two generalized displacements introduced in Eqs. (13) lead to a set of two governing second-order differential equations and two spectral lines representing the roots of the frequency equation.

Adjustment factors K and K_1 are introduced into the equations so that the spectral lines produced from them can be adjusted to match more closely the comparable lines from the exact. Both factors are used to improve the shape of the fundamental spectral line. The factor K is used to insure that as the wavelength becomes very small the phase velocity will approach the Rayleigh velocity. The factor K_1 is used to make the line pass through the coordinates of the point that Hudson found to be invariant to changes in Poisson's ratio. It follows that K will depend on Poisson's ratio, K_1 will not.

Even though the Mindlin–Herrmann theory proved to be a marked improvement over previous approximate theories, it was later found that a two-mode theory is inappropriate except perhaps when Poisson's ratio is small. For realistic values of ν, the second mode is strongly coupled to a third mode, and indeed for higher values the radial mode, which is the second mode of the Mindlin–Herrmann theory, is not the second mode but the third. The shortcoming of the Mindlin–Herrmann theory, which resulted in its being replaced by another theory, arose because in 1951 the Pochhammer equation had not been sufficiently explored.

In the same year that the Mindlin–Herrmann approximate theory appeared, a paper by A. N. Holden [14] presented some important findings about the Pochhammer equation. Although most of the paper is devoted to the Rayleigh–Lamb exact equation for plates, the analogy between this equation and the Pochhammer equation for rods was also noted. The important properties of the spectral lines presented for the first time in this paper are common to both rods and plates. Holden showed that the frequency equation for each is satisfied by a combination of a real frequency and a pure imaginary wave number. He also introduced the important idea that the spectral lines can be traced quite accurately over extended domains by identifying lines that act as bounds to the spectral lines and points through which the spectral lines must pass. In one of those odd coincidences that occur in science, Mindlin also adopted, in the same year, the idea of using bounding curves and intersections in a paper on the flexural vibrations of crystal plates [M36]. Each was unaware, at that time, of the other's work in using bounds to help trace frequency spectra.

The discovery that the Pochhammer equation accommodates imaginary wave numbers was extended by J. Adem in 1954 [15] to complex wave numbers. Adem's study was of a semi-infinite rod, and when he found that complex wave numbers satisfied the exact equation, he correctly deduced that such wave numbers give rise to waves that are harmonically distributed along the axis of the rod but whose amplitudes decay exponentially. Such wave forms, though not admissible for a rod of infinite length, can properly describe motions near a rod end, making them pertinent for the problem Adem was studying.

Two papers by Mindlin and his associates, which followed a few years after Adem's study, form the major part of Mindlin's work in the study of axisymmetric vibrations of rods. A large part of the discoveries and techniques revealed in both papers represents the culmination of research which Mindlin had devoted to the study of the vibration of plates during the years immediately preceding the study of rods [M36, 9, 10].

The first paper, by Mindlin and McNiven [M62], was devoted to the derivation of a three-mode approximate theory, and the second, by Onoe, McNiven, and Mindlin [M74], presented the results of a study of the Pochhammer equation. Though the papers appeared more than two years apart, research for the two was carried out simultaneously. The study of the exact equation appeared in the latter paper but will be discussed here first, as it is necessary in the development of the approximate theory to borrow so much from the findings which emerged from the exact.

Perhaps the major contribution of the Pochhammer study is a presentation of the roots of the exact equation in the form of spectral lines on the plane where the propagation constant is real, the plane where it is imaginary, and the domain between the two planes in which the propagation constant is complex. Some eighteen spectral lines are presented in detail, showing the pattern of the lines to be extremely intricate, and one might be surprised that such a complex array could emerge from such an innocent-looking equation. What is also interesting is that the complete set of spectral lines are drawn, not by finding actual roots of the equation, but by finding the roots of fairly simple algebraic equations.

There are four families of algebraic equations and, as in the theory of plates, one family represents purely equivoluminal motions and one purely dilatational. Unlike plates, the other two families cannot be identified with any physical phenomena. The roots of the algebraic equations are used to form two sets of grid lines on both the $(\Omega, \mathrm{Re}\,\zeta)$- and $(\Omega, \mathrm{Im}\,\zeta)$-planes, and the coordinates of the points of intersection of grid lines are established. The slopes of the spectral lines at certain of these intersections are found, and where spectral lines and grid lines coincide at cut-off ($\zeta = 0$), the curvatures of each are found and compared.

It is shown that the grid lines in pairs form bounds of the spectral lines and that the spectral lines can cross grid lines, but only at pair intersections. By knowing the intersection through which a spectral line enters a small bounded domain, its slope on entering, and its slope on leaving the same domain, a spectral line can be traced in each bounded area with little possibility for error. Extended arrays of spectral lines, constructed section by section as described, are shown in Fig. 2.

Both Bancroft and Hudson had pointed out that the roots of the exact equation change with changes in Poisson's ratio, but the extent and importance of its influence is only understood when the extended spectrum is examined. The change in the spectrum can be appreciated when one realizes that when the value of Poisson's ratio is raised, the cut-offs of the spectral lines representing the axial shear modes remain fixed, and the intersections of the radial lines move up the spectrum. During the change, a "loop" on the imaginary plane can sever; one part will then unite with a neighbor, and the other extend in a tortuous way to infinity. The circumstances under which this phenomenon will take place are established.

In addition to the real and imaginary spectral lines, the nature of the complex branches is studied in detail. Study of the general properties of complex branches shows that they form loops in complex space between real or imaginary branches and do so by emanating from and reentering the real or imaginary plane at stationary points of the branches on those planes. It is also established that all branches, except the fundamental, progress downward from high frequencies and eventually intersect the $\Omega = 0$-plane as com-

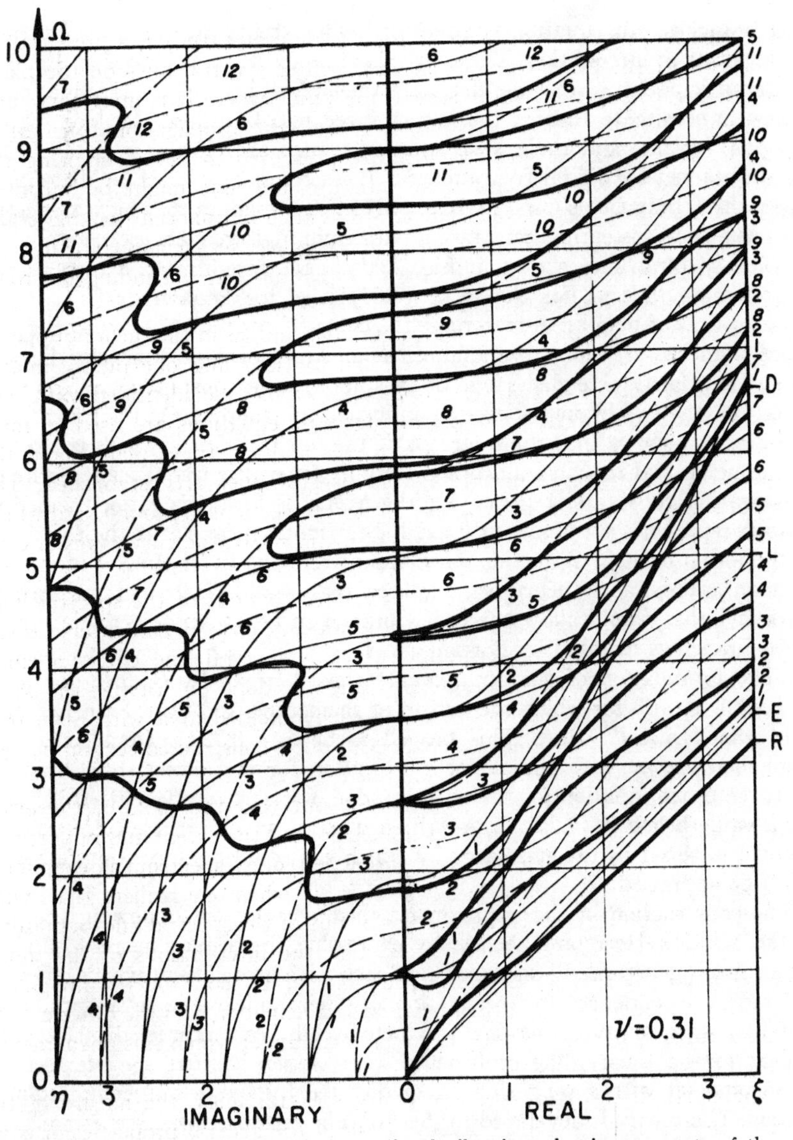

Fig. 2. Frequency spectrum for extensional vibrations showing segments of the lowest eighteen branches for real and imaginary propagation constants.

plex branches. An equation is developed whose roots are the coordinates of these points of intersection. These complex branches are displayed in Fig. 3.

In anticipation of the role the exact theory will play in the formulation of a new approximate theory, the lowest three real branches, the loop on the imaginary plane, and the lowest complex branch are plotted in detail, using roots obtained by digital computer, for Poisson's ratio of 0.31. A diagram of these three branches shows, in Ref. [M74], the manner in which a spectral line, as a separate entity, can trace a path from high frequencies to zero.

The spectral lines shown in Fig. 2 also gave insight into a higher-order approximate theory. For the value (0.31) used for Poisson's ratio, a value close to that of many real materials, the cut-off frequencies of the second and third branches are close to one another, and it is obvious from the diagram that, for a fairly extended range of the propagation constant away from the intercept, the two branches strongly influence one another. On the other hand, the diagram shows that the intercept of the fourth branch, which is fixed, is much higher and that its influence is felt only for higher frequencies and larger propagation constants. Further, for this value of Poisson's ratio, the second branch represents predominantly axial shear motions, and it is the third that is the radial branch. With this information from the exact theory, it was felt that an approximate theory was required that would result in three spectral lines and that would allow the radial branch to be above or below the axial shear branch as dictated by Poisson's ratio.

The formulation of the approximate theory began by finding the value of Poisson's ratio for which the cut-off of the second and third branches coincide, the "critical" value. This proved to be $\nu_c = 0.2833$. Detailed spectra were then drawn, with the aid of a computer, for the lowest three branches from the exact equation for the critical value, for a value above it (0.31), and for a value below it (0.25). These three spectra, which included the complex branch, were set as the standards by which the resulting approximate theory could be appraised.

The construction of the approximate theory continued, following the lead of the Mindlin–Herrmann theory, by expressing the displacements u_r and u_z in terms of "generalized" displacements each with a coefficient which reflected the radial dependence. In anticipation of their appearance in a variational equation and to avoid having products of the generalized displacements appear in the theory, the coefficients were chosen so that they formed an orthogonal set with a weighting factor dictated by the cylindrical coordinate system. The theory is developed to this point in a general way, accommodating an arbitrary number of modes, and the coefficients are Jacobi polynomials of the radial coordinate.

For a theory containing only three modes, the displacements are expressed as

$$u_r = \alpha u(z, t), \qquad u_z = w(z, t) + (1 - 2\alpha^2)\psi(z, t) \tag{15}$$

where $\alpha = r/a$.

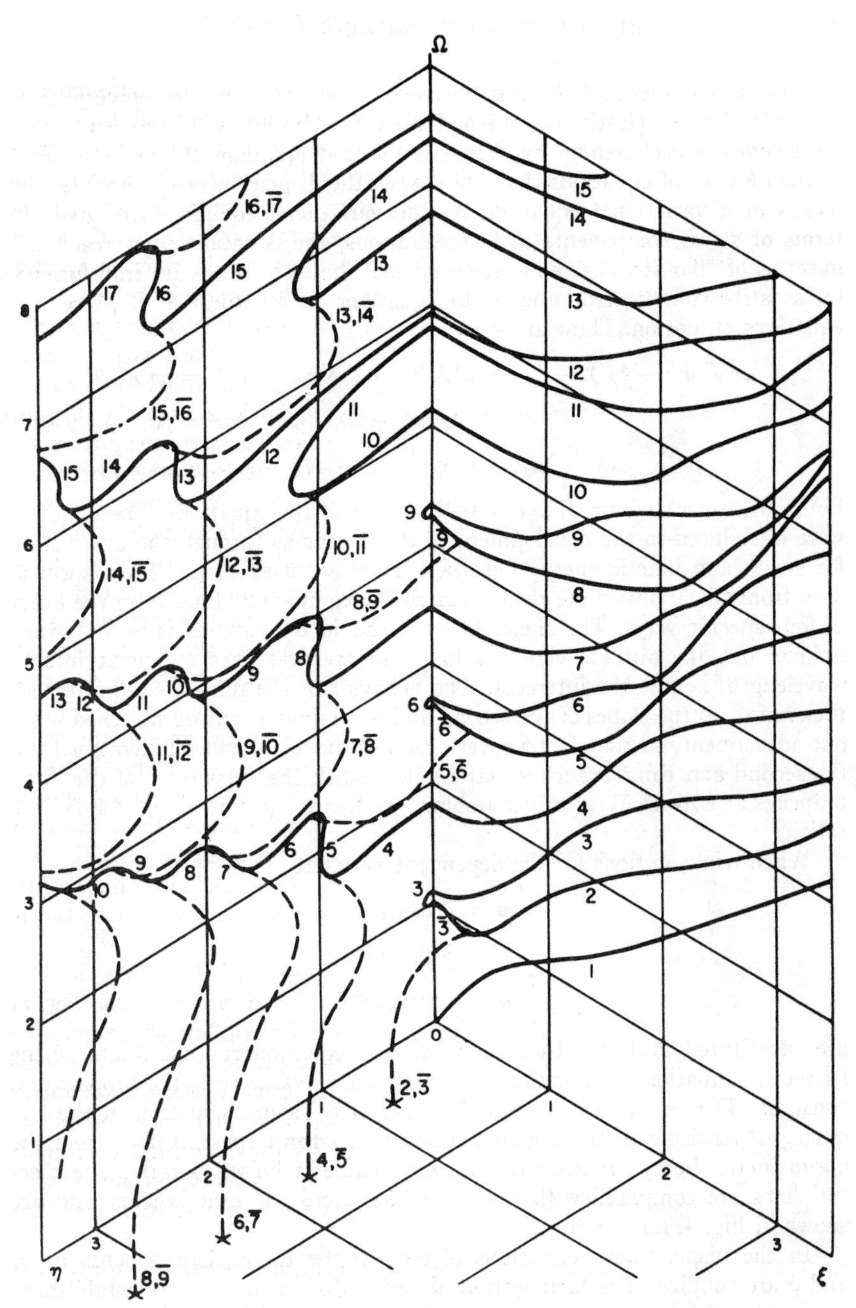

Fig. 3. Illustration of the linkage of real, imaginary, and complex segments to form branches for extensional vibrations.

The development of the displacement equations of motion governing u, w, and ψ follows closely the method used by Mindlin and Herrmann. The displacements as chosen no longer satisfy the field equations of linear elasticity, so the best set of equations that can govern the displacements is produced by means of a variational equation. As the variational equations are given in terms of stress components and accelerations, the equations that result are in terms of "bar stresses" and accelerations. These equations are transformed, by a fairly complicated process, to equations in the form of displacement equations of motion. These are

$$\mu K_2^2(u'' - 4\psi') - 8(\lambda + \mu)K_1^2 u - 4\lambda K_1 w' + 4aR = \rho a^2 K_3^2 \ddot{u}$$
$$(\lambda + 2\mu)w'' + 2\lambda K_1 u' + 2aZ = \rho a \ddot{w}$$
$$(\lambda + 2\mu)\psi'' + 6\mu K_2^2(u' - 4\psi) - 6aZ = \rho a^2 K_4^2 \ddot{\psi} \qquad (16)$$

Four adjustment factors K_i ($i = 1$–4) appear in the equations. These factors were introduced in the development of the equations through the expressions for strain and kinetic energy densities. They are used to make the spectral lines from the approximate theory match the comparable lines from the exact in four specific ways. The choice of what should be matched is by no means unique, but the authors chose to have the spectral lines match at infinite wavelength, i.e., at the intercept. The behavior of the fundamental line near the origin and the slopes of the second and third lines at cut-off matched without adjustment, so that factors were used to make the cut-off frequencies of the second and third branches match, along with the curvatures of the same branches at cut-off. With this matching, the factors are functions of Poisson's ratio.

When trial solutions for the dependent variables

$$u = A \cos \gamma z e^{i\omega t}$$
$$w = B \sin \gamma z e^{i\omega t}$$
$$\psi = C \sin \gamma z e^{i\omega t} \qquad (17)$$

are substituted in Eqs. (16), the frequency equation results. Roots of the frequency equation are found by assigning first a value to ζ, the propagation constant. The equation then has the form of a bicubic equation in the frequency Ω resulting in three values. In this way three spectral lines from the approximate theory are generated for each value of Poisson's ratio. The spectral lines are compared with the same lines from the exact theory and are shown in Fig. 4, for $\nu = 0.31$.

In the displacement equations of motion the three displacements u, w, and ψ are coupled. The final section of Ref. [M62] is devoted to establishing potentials of these displacements in such a way that the differential equations

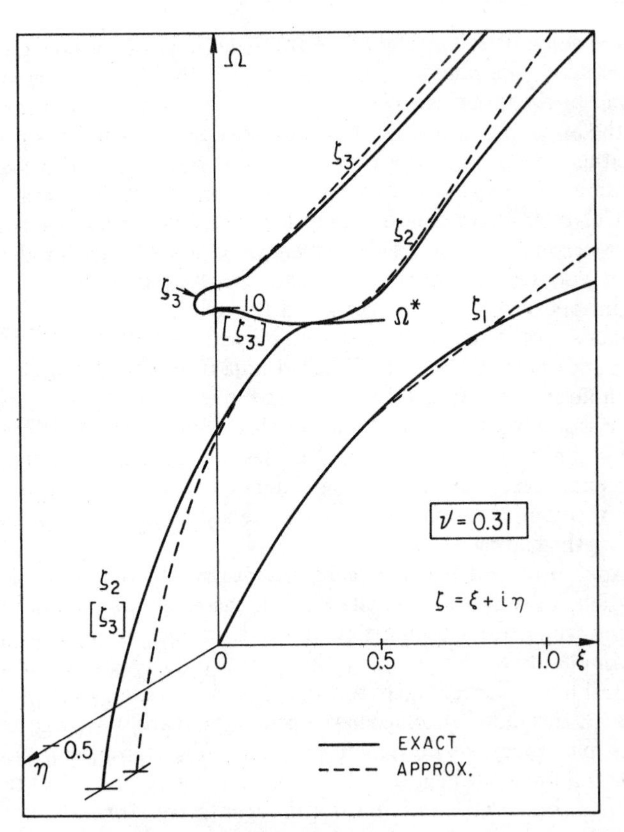

Fig. 4. Comparison between the branches from the Mindlin–McNiven theory and the lowest three from the exact.

governing the potentials are uncoupled. These potentials were extremely useful in later work, when end boundaries were introduced.

To gain insight into the geometric dispersion of wave pulses in rods and the velocity with which such pulses transmit energy, it is useful to recognize certain properties of the spectra. The ratios of the phase velocity (V) and the group velocity (Vg) to the velocity of equivoluminal waves (V_2) can be easily visualized from examining the spectra since

$$\frac{V}{V_2} = \frac{\Omega}{\zeta}$$

$$\frac{Vg}{V_2} = \frac{d\Omega}{d\zeta} \tag{18}$$

Thus the slope of the straight line from the origin to a point on a branch is proportional to the phase velocity, and the slope of a branch at a point is proportional to the group velocity.

Since the appearance of these last two papers there has been little activity directed either toward an improved approximate theory or toward a further understanding of the exact. Indeed, as far as the authors are aware, the Mindlin–McNiven approximate theory has not been replaced and is still the theory chosen when an extended frequency range and short wavelengths are to be accommodated. Concerning the exact theory, a paper by J. Zemanek Jr. [16] which appeared in 1972, may serve the role of completing the study of Pochhammer's equation. Zemanek established quantitatively what Onoe, McNiven, and Mindlin had established qualitatively. Using a high-speed digital computer, Zemanek traced out the spectral lines from roots of the exact theory for real, imaginary, and complex wave numbers. It is remarkable how closely the spectrum traced from these roots and shown as Fig. 1 in Zemanek's paper resembles Fig. 6 (Fig. 3 here) of the paper by Onoe, McNiven, and Mindlin which was constructed with the aid of grid lines and the general properties of the complex roots.

The paper by Zemanek also confirms other findings about the axisymmetric motions of solid rods by establishing the radial and axial displacement distributions which exist for particular combinations of frequency and wave number, combinations which form the coordinates of points on the lowest three spectral lines. For instance, the distributions confirm that near the origin of the spectra the radial displacements are very small and that, for axial displacements, plane sections do remain plane, and that for a point on the fundamental spectral line for which the phase velocity is close to that of the Rayleigh velocity, both the radial and axial displacements are largest at the surface of the rod, diminishing rapidly away from the surface.

Torsional Vibrations of Circular Cylindrical Rods

The elementary, or strength-of-materials, theory for the axisymmetric torsional vibrations of cylindrical rods dates back to the works of Coulomb [17], Poisson [11], and Saint Venant [18]. The fundamental postulates of the theory are twofold. First, there is no in-plane distortion of the cross section formed by a plane cut orthogonal to the central line. Second, whereas some out-of-plane distortion is allowed, inertial forces associated with the accompanying motions are neglected. The theory leads to the one-dimensional wave equation,

$$c^2 \frac{\partial^2 u_\theta}{\partial z^2} = \frac{\partial^2 u_\theta}{\partial t^2} \tag{19}$$

where

$$c = \left(\frac{\mu}{\rho}\right)^{1/2}$$

and u_θ is the circumferential component of displacement. The elementary theory is a nondispersive, single-mode theory.

Pochhammer [1] studied the problem of the torsional wave train solutions that can exist in a circular, cylindrical rod of infinite extent. He showed that the frequency equation governing such waves is

$$\beta[\mathcal{J}_1(\beta) - 2] = 0 \tag{20}$$

where

$$\beta^2 = \Omega^2 - \zeta^2$$

and

$$\Omega = \frac{\omega a}{V_2}; \qquad \zeta = \gamma a \tag{21}$$

Note that in Eq. (21) Ω is defined differently than it was for extensional vibrations.

The root

$$\beta = 0 \tag{22}$$

leads to the mode

$$\zeta = \Omega \tag{23}$$

This lowest mode is nondispersive and is exactly reproduced by the elementary theory.

The higher spectral lines have their shapes described by

$$\zeta_n^2 = \Omega^2 - \beta_n^2 \tag{24}$$

where the β_n are the roots of

$$\mathcal{J}_1(\beta_n) - 2 = 0 \tag{25}$$

The form of Eq. (24) is the same as the form of the bounds for the exact theory for extensional waves so that the nature of the spectral lines representing torsional waves can be seen by observing the bounds in Fig. 2.

The complex branches associated with torsional waves have not been explored. They no doubt exist, but because the stationary points of the spectral lines all lie on the ordinate axis, it is difficult to predict the paths they will follow.

The remarkable thing about torsional waves is that, of all possible waves in a circular rod, they are the only ones in which the modes, represented by spectral lines, are uncoupled.

Flexural Vibrations of Circular Cylindrical Rods

The problem of flexural vibrations of rods has received a great deal less attention during its history of study than its counterpart, the problem of extensional vibrations. It is not easy to identify the reasons, but perhaps it is because, for flexural waves, the approximate theories were easier to formulate and the exact theory a great deal more difficult to explore than the comparable theories for extensional waves. Attention focused alternately on approximate and exact theories.

The elementary theory of flexural vibrations dates back to L. Euler and Daniel Bernoulli [19]. The fundamental postulates are three: (1) there is no distortion of the cross section formed by a plane cut orthogonal to the central line; (2) the section orthogonal to the central line before deformation remains orthogonal during the deformation; and (3) inertial effects associated with the rotatory motion of a section about the transverse axis are ignored. The theory leads to the following partial differential equation governing the transverse displacement w of the central line,

$$EI\,\frac{\partial^4 w}{\partial z^4} = \rho A\,\frac{\partial^2 w}{\partial t^2} \tag{26}$$

It is important to note that even for low frequencies and long wavelengths, flexural waves display dispersion. The Bernoulli–Euler theory, even though it is a single-mode theory, reflects this dispersion. Experimental evidence shows it to be a valid low-frequency theory. A physical weakness of the theory is its prediction that the velocity of a signal goes to infinity along with its frequency content.

At the same time that he developed the exact theory for extensional waves, Pochhammer [1] developed the exact theory for flexural waves for rods of circular section and infinite length. The equation, as before, relates the normalized frequency Ω and the wave number, or propagation constant ζ, and states that the cylindrical surface is free of traction.

The equation is

$$J_1(\alpha)\,J_1^2(\beta)[f_1\mathcal{J}_1^2(\beta) + f_2\mathcal{J}_1(\alpha)\mathcal{J}_1(\beta) + f_3\mathcal{J}_1(\beta) + f_4\mathcal{J}_1(\alpha) + f_5] = 0 \tag{27}$$

where

$$\alpha^2 = k^{-2}\Omega^2 - \zeta^2; \qquad \beta^2 = \Omega^2 - \zeta^2$$

$$\Omega = \frac{\omega a}{v_2}; \qquad \zeta = \gamma a$$

and

$$f_1 = 2(\beta^2 - \zeta^2)^2, \qquad f_2 = 2\beta^2(5\zeta^2 + \beta^2)$$

$$f_3 = \beta^6 - 10\beta^4 - 2\beta^4\zeta^2 + 2\beta^2\zeta^2 + \beta^2\zeta^4 - 4\zeta^4$$

$$f_4 = 2\beta^2(2\beta^2\zeta^2 - \beta^2 - 9\zeta^2)$$

$$f_5 = \beta^2(-\beta^4 + 8\beta^2 - 2\beta^2\zeta^2 + 8\zeta^2 - \zeta^4) \tag{28}$$

The equation is so complicated that it is no wonder it remained unexplored for so many years.

In 1921 Timoshenko [20] introduced an approximate theory, the first improvement over the classical. Even though the paper was very brief and was presented over fifty years ago, it is still the pertinent approximate theory when any but the lowest frequencies are to be included. The theory is presented as an improvement on the classical by relaxing some of the constraints imposed by that theory.

The Timoshenko beam theory relaxes the requirement that the rotated cross section remain orthogonal to the central line while still retaining the no-distortion postulate. It also incorporates the inertial effects that are associated with this rotatory motion. The theory leads to the pair of coupled partial differential equations

$$EI\frac{\partial^2\psi}{\partial z^2} + k\mu A\left(\frac{\partial w}{\partial z} - \psi\right) = \rho I\frac{\partial^2\psi}{\partial t^2}$$

$$k\mu A\left(\frac{\partial^2 w}{\partial z^2} - \frac{\partial\psi}{\partial z}\right) = \rho A\frac{\partial^2 w}{\partial t^2} \tag{29}$$

where

I = second moment of area of the cross section
A = area of the cross section
w = deflection
ψ = angle of rotation of an element of the cross section.

The factor k is a correction factor similar to those introduced in discussing extensional vibrations. It is sometimes termed a shape factor. The Timoshenko theory is a dispersive two-mode theory with the lower-order mode a modification of the single-mode predicted by the Bernoulli–Euler theory. In the low-frequency, long wavelength limit the two theories predict identical solutions. Further, the Timoshenko theory predicts a finite velocity of energy propagation for all signals. The higher-order mode of the Timoshenko beam theory has no counterpart in the Bernoulli–Euler theory. This second mode is an approximation of the second lowest mode predicted by the exact three-dimensional formulation for flexural waves in a circular cylindrical rod. Thus the added solutions of the Timoshenko beam theory are physically meaningful. Since the Timoshenko theory does not give rise to mathematical difficulties that are significantly greater than those encountered using the elementary theory, it is quite common to find it being used even in situations for which the elementary theory would suffice.

The next major activity in the study of flexural waves in circular rods was a study in depth of the Pochhammer equation by Mindlin and Pao. They were anxious to know whether the technique of constructing bounds of the spectral lines for flexural waves would be as powerful as it was for extensional waves.

The description of the study is presented in two papers, the first by Pao and Mindlin [M64], and the second by Pao [21]. They found that the bounding method used for extensional waves was pertinent to the study of flexural waves but was much more complicated for the latter.

Since we have already discussed the idea of first drawing a grid of bounds through which the desired spectrum can then be sketched, we need only consider the aspects that are unique to the case of flexural waves. The main difference is that one cannot find two families of simple algebraic curves that are capable of bounding the spectrum predicted by Eq. (27). Only one such family can be found. Lacking the second family of simple bounds, Pao and Mindlin [M64] introduced a two-step procedure. First, they identified a family of curves that can serve as the missing family of bounds, which unfortunately are as complicated as the branches themselves. The advantage of this new family of bounds is that they may be bounded in turn by simpler bounds in a manner analogous to that followed in analyzing the longitudinal wave dispersion or the plate wave spectra.

To illustrate the difficulty encountered with flexural waves and the manner in which it was surmounted, we consider the following. Equation (27) will be satisfied by solutions of

$$\mathcal{J}_1(\beta) - 1 = 0 \tag{30}$$

if and only if the solutions simultaneously satisfy

$$\mathcal{J}_1(\alpha) - 1 - \frac{\beta^2 + \zeta^2}{2\beta^2(\beta^2 - 2)} = 0 \tag{31}$$

To verify this the reader can substitute Eq. (30) in Eq. (27). Equation (30) defines a family of curves in the frequency-wave number space. Restricting attention to the real plane, i.e., to nondecaying waves, the family of curves is a family of hyperbolas that are asymptotic to $\Omega = \zeta$ as ζ becomes unboundedly large. They are illustrated by the notation B_1 in Figs. 5 and 6, which are reprinted from Ref. [M64]. These hyperbolas are termed bounds because a branch of the dispersion spectrum cannot cross them except at points of intersection of the hyperbolas and the family of curves defined by Eq. (31). Pao and Mindlin actually differentiate between a barrier and a bound, a bound having the additional requirement that it coincide with a branch at $\zeta = 0$. The hyperbolas given by Eq. (30) do satisfy this additional requirement.

For α real, i.e., the region of the (Ω, ζ)-plane that lies to the left of the line OD in Figs. 5 and 6, Eq. (31) may be approximated by

$$\mathcal{J}_1(\alpha) - 1 = 0 \tag{32}$$

Hence, the family of curves that define the crossover points for the bounds given by Eq. (30) are approximated by hyperbolas. These hyperbolas are illustrated by the notation I_1 in Figs. 5 and 6. For α imaginary, i.e., the region of the (Ω, ζ)-plane to the right of the line OD, Eq. (31) has only one root for each value of β. Thus for each bound B_1, there is one point in this region of the (Ω, ζ)-plane at which a branch crosses the bound. The point denoted by d_2 is such a point. Analogous points for the higher bounds fall outside the region of the (Ω, ζ)-plane that is shown in the figures.

In the construction of bounds for the longitudinal wave dispersion spectra, curves analogous to I_1, in addition to giving the crossover points of the bounds B_1, were themselves bounds that could only be crossed at these same intersection points. Here this is not so. There are crossover points on I_1 in addition to the intersection points with the bounds B_1. One could substitute Eq. (31) in Eq. (27) to obtain an equation to be satisfied by all of the crossover points. But to extract the roots of this equation, a large amount of computation would be required. Further, it is noted that the I_1 curves do not coincide with branches of the dispersion spectrum at $\zeta = 0$. Thus I_1 could be, at best, barriers.

Unlike the extensional wave spectrum, therefore, only a single family of bounds are provided by simple hyperbolas. The second family of bounds must be a family of curves of a more complicated geometry. Pao and Mindlin consider the family of curves given by the equation

$$J_1(\alpha)\,J_1(\beta)[6\mathcal{J}_1(\alpha) + 6\mathcal{J}_1(\beta) - 24 + 3\beta^2 - \zeta^2] = 0 \tag{33}$$

This set does coincide with those branches of the dispersion spectrum at $\zeta = 0$ that are not given by the solutions of Eq. (30). Unfortunately, the presence of both $\mathcal{J}_1(\alpha)$ and $\mathcal{J}_1(\beta)$ immediately leads to the suspicion that a direct determination of this set of curves represents a numerical analysis of a difficulty comparable to that of finding the branches. The value of this last set of curves is that they may be bounded by sets of much simpler curves. The details are tedious, as is the verification that the curves do provide bounds. The interested reader should refer to the original paper. Here we shall simply indicate that the pairs of families denoted by (B_{21}, I_{21}), by (B_{22}, I_{22}), and by (B_{23}, I_{23}) all define barriers to which the curves given by Eq. (33) must conform. Once these barriers are drawn on the (Ω, ζ)-plane, the curves given by Eq. (33) may be readily traced. They are illustrated by the notation B_2 in Fig. 5. The curves are only shown in the region to the left of the line OD, since it is only in this region that the curves are required. To the right of OD, a sufficient number of restrictions are placed directly on the branches by the auxiliary barriers, and by a bound given by the straight line $\zeta = \beta$, that the branches can be traced without making recourse to B_2. Traces of the branches themselves are shown in Fig. 6.

In a subsequent paper, Pao [21] extended the analysis to cover a larger region of the $(\Omega, \mathrm{Re}\,\zeta)$-plane. He also investigated the intersection of the

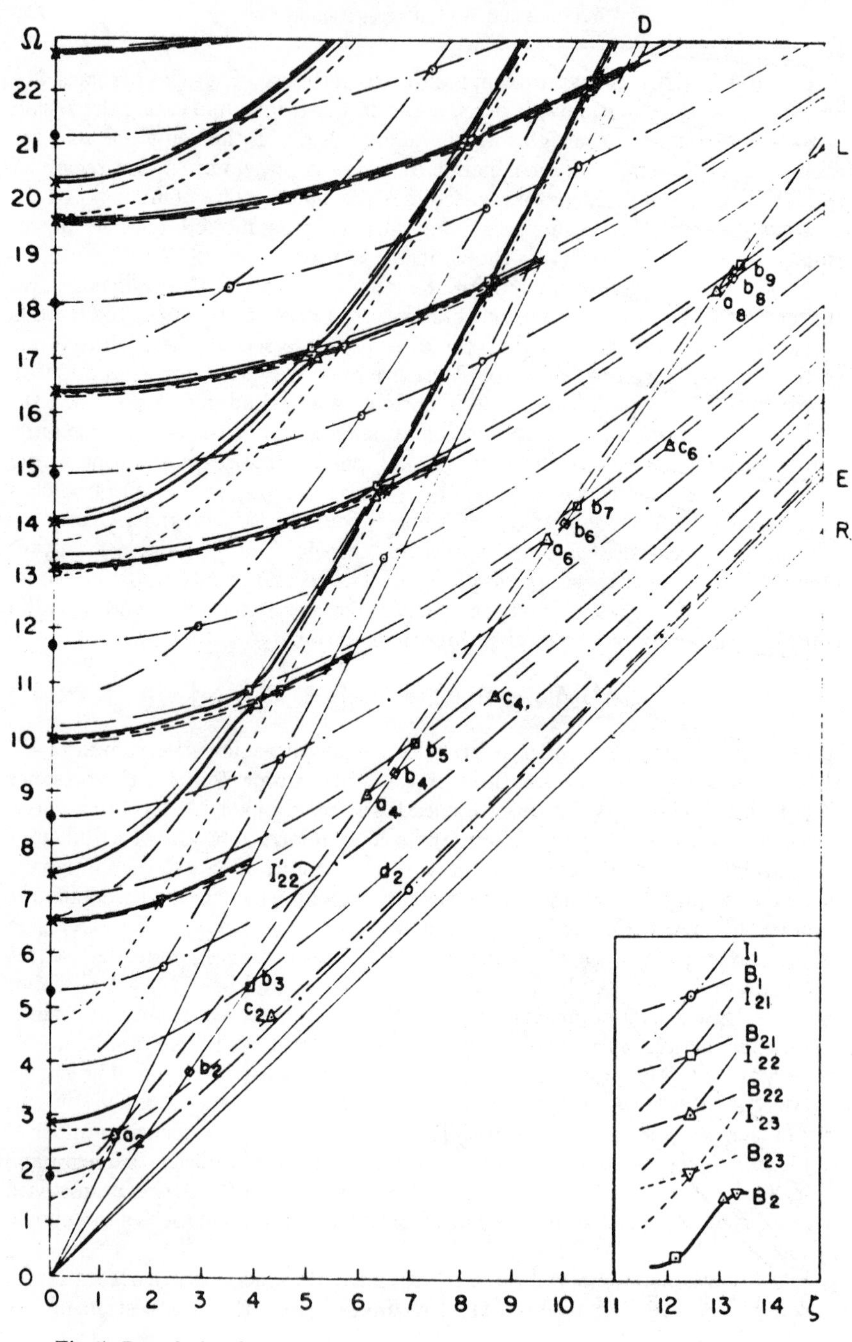

Fig. 5. Bounds, barriers, auxiliary barriers, and intersectors for flexural vibrations.

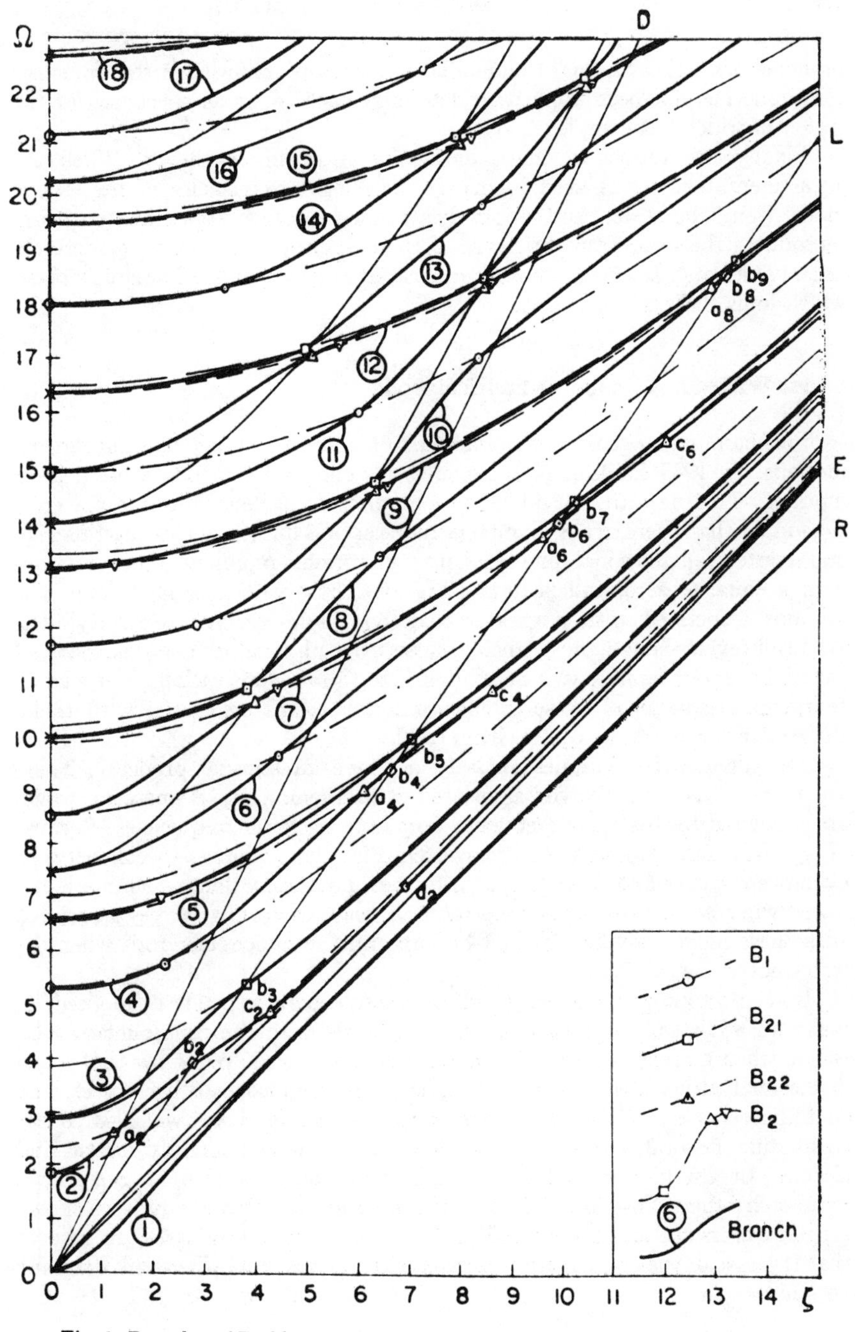

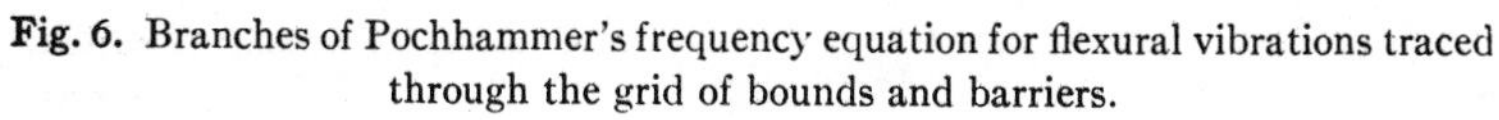

Fig. 6. Branches of Pochhammer's frequency equation for flexural vibrations traced through the grid of bounds and barriers.

branches with the (Ω, Im ζ)-plane. To the authors' knowledge the bounding technique has not been applied to determine the dispersion spectrum for the region in which ζ is complex.

Finally, we return to the Timoshenko approximate theory. When the adjustment factor k is used to match the cut-off frequencies of the second mode from the exact and approximate theories, the approximate theory reproduces the lowest two branches of the exact spectrum out to moderately large values of ζ. It also reproduces accurately the loop on the imaginary plane which joins these two branches.

Other Waves in Circular Cylindrical Rods

Pochhammer's equation was limited to extensional, torsional, and flexural motions. In 1943 Hudson [7] extended the theory to include more general motions. This new theory allows the motions to be distributed on a cross section so that there are n diameters representing lines of nodes. In this general context, extensional and torsional vibrations represent those motions with n equal to zero, and flexural vibrations those with n equal to one. The motions associated with n greater than one have received relatively little attention even though they can of course exist physically. The discussion of the higher-order families will be limited here to two observations. For a more thorough discussion of these families and the waves associated with them, the reader is referred to W. A. Green [22].

The higher-order families ($n > 1$) do have an unusual property. There are no low-frequency waves associated with them. Only evanescent waves are predicted by the exact frequency equations at lower frequencies. Because of this fact and because of the knowledge that the coupling is weak between evanescent waves and those propagating with no attenuation, which are introduced when an end boundary is introduced, one expects that waves associated with these higher families will not be important in noncircular rods when the frequency is low.

It seems appropriate here, where the more general attitude is adopted, to make an additional observation. This has to do with the consequences that follow when a traction-free end boundary is introduced. We know that when extensional waves strike such an end, they are coupled to other waves, and that these waves are also extensional waves. Flexural waves will also couple with other flexural waves. That is: when a rod is isotropically elastic and circular, the "families of motion" (identified by the value of n) do not couple with each other. Now, one learns in the next section that for rods with rectangular cross sections these families do in fact couple. One is led to suspect that the case of the circular cross section is the special case where this coupling does not occur.

Waves in Rods of Noncircular Cross Section

Up to this point, our article has been devoted to the vibration of rods of circular cross section. The lengthy attention paid to circular rods, compared to the brief description below of waves in rods of noncircular cross section, reflects how differently the two types of sections have been able to attract the attention of scientists. The reasons for this are very real and will become apparent in the discussion of waves in rods of rectangular section.

There is one shape for a rod which is noncircular and yet does not generate the difficulties inherent to the rectangular rod, and that is the ellipse. An ellipse represents the only other simple shape for which the cylindrical surface is given by assigning a constant value to one variable of the coordinate system. If one follows the development of Pochhammer's equations for circular rods, one realizes that no substantial difficulty is encountered when the cylindrical coordinate system for circular rods is replaced by the elliptical coordinate system for elliptical rods. Equations comparable to Pochhammer's could be derived in which the Bessel functions would be replaced by Mathieu functions, but to develop such equations probably would be academic.

The most important noncircular cross section is the rectangle. This cross section introduces two major difficulties not previously encountered. The first, mentioned in the previous section, is the coupling introduced between "families" of motions. Torsional motions are coupled to extensional motions, and each is coupled to flexural motions, etc. To understand this coupling one need only refer to the static torsion of a rod. If the rod is of circular section, the displacements of a plane cross section are all contained in the plane, so that the plane cross section remains plane during twist. When the cross section is rectangular, the displacements in the plane of the cross section are coupled to displacements normal to the section, causing the section to warp.

The simplest approximate theory for torsional vibration of rectangular rods is contained in the equation

$$c^2 \frac{\partial^2 u_\theta}{\partial z^2} = \frac{\partial^2 u_\theta}{\partial t^2} \tag{34}$$

In this equation the wave speed c is influenced, not only by the material properties of the rod, but also by the properties of its cross section. This empirical approach was modified by Love [5], who developed a theory which accounted, not only for the inertial effects of tangential displacements, but also for those due to axial displacements induced by coupling.

The second difficulty is perhaps more serious. We have learned from the work of Mindlin and his associates, concerning both plates and rods, a systematic method of establishing approximate theories of vibration. One first studies the exact equation in depth to learn as much as possible of the relationship between wavelength and frequency in the form of spectral lines and to develop a feeling for displacement distributions. After this, an approximate theory is constructed based on what one has learned from the exact equation, and finally, the approximate theory is judged by comparing its predictions with those of the exact.

This general approach is not possible with rods of rectangular section because it is not possible to establish exact frequency equations. It follows that the approximate equations require substantial intuition to construct and can be evaluated after they have been established only by comparison with experiments.

To understand why the exact equations cannot be established for a bar of rectangular cross section, one must first consider the plane-wave solutions for an infinite elastic plate (see article by Pao and Kaul in this volume). For a plate with mixed boundary conditions to be satisfied on the two parallel faces, i.e., when the normal displacement component and the tangential traction component are zero, the dispersion spectrum is readily given by the uncoupled equivoluminal and irrotational modes and their anharmonic overtones. For a plate with traction-free faces, however, the modes of an infinite medium are coupled by the plane cuts that form the faces. The resulting dispersion spectra—there are two, one corresponding to symmetry about the middle plane and one corresponding to skew symmetry—exhibit an intricacy comparable to the spectra of rods of circular section. Mindlin [M62, M64] has shown that suitable bounds for the branches of the traction-free spectra are provided by the much simpler spectra for mixed boundaries. He also illustrated the nature of the coupling in the traction-free problem by introducing an elastically restrained face. The mixed and traction-free boundary conditions define the two extremes of the elastically restrained face.

These same ideas can be applied to the bar of rectangular cross section. The dispersion spectra for the bar, for which two parallel faces are traction-free and the remaining two are mixed, are given by the modes for an infinite plate and their anharmonic overtones. Thus we begin with an infinite number of fundamental wave modes, in contrast to the two that must be considered in going from the infinite medium to the plate. For the rod with all four faces traction-free, the coupling which leads to the intricacies of the branches of the infinite plate spectra comes into play further complicating the already intricate traction-free, mixed spectra. Furthermore, the branches of the traction-free mixed spectra do not in this case provide exact bounds for the completely traction-free spectra. The reasons for this are, first, that there are more than two families of these spectra, and, second, they have complex parts in addition to real and imaginary parts.

Lacking, therefore, an ability either to exactly construct the desired spectra or to exactly construct bounds that would enable us to sketch the desired spectra, we are forced to look to special cases in which a wave train solution can be exactly determined. These special cases require that the relative dimensions of the sides of the rectangular section conform in some way with the reflection of a pressure wave and a shear wave to form a traction-free surface. For example, a well-known special case in which no mode conversion occurs is the reflection of a shear wave at an incident angle of 45° from a traction-free surface. For a rod of square section, therefore, a wave of this type can reflect successively to form a closed path. Two such waves, traveling in opposite directions, form the Lamé wave modes. Mindlin [M66] and Mindlin and Fox [M63], consider several special solutions that locate isolated points on the dispersion spectra for rods of rectangular cross section with certain ratios of width to depth.

Concerning approximate theories, the expansion and truncation procedures that were developed by Mindlin for plates and rods of circular cross section were extended by Medick [23] and by Bleustein and Stanley [24] to develop approximate theories for analyzing rods of noncircular cross section. Medick restricted his attention to rods of rectangular section and investigated the case of "essentially" extensional motions. Bleustein and Stanley treated rods of a more general sectional geometry and investigated the case of "essentially" torsional motions.

Both Medick and Bleustein–Stanley refer the rod to Cartesian coordinates with one of the axes directed along the central line of the bar. Medick expands the displacement, stress, and strain fields in terms of a product series of Legendre polynomials in the lateral coordinates. Bleustein and Stanley expand the displacement field as a double power series. In both studies an infinite set of coupled ordinary differential equations is then obtained by making use of a variational principle. In carrying out the required manipulations, Medick makes use of the rectangular shape of the section, whereas Bleustein and Stanley only restrict their development to sections having two planes of geometrical symmetry which are at right angles to one another and which intersect at the axis of the rod. In both treatments, the equations separate into four sets of uncoupled equations—one governing essentially extensional motions, two governing essentially flexural motions, and the last one governing essentially torsional motions. Medick studies the first set, Bleustein and Stanley consider the last.

The development of a useful one-dimensional approximate theory from the set of infinite equations requires one crucial additional step, the introduction of a truncation procedure. Here, too, in both studies the procedure used is a natural extension of that used by Mindlin. Medick considers both what he terms a symmetric truncation and an asymmetric truncation. Symmetry requires that the truncation be of the same order in each of the lateral coordinates. Bleustein and Stanley consider only the case of symmetric truncation. Of particular interest in the Bleustein–Stanley study is the light it sheds on the nature of the coupling between the torsional, the in-plane distortion, and the warping motions.

Recent Research on Circular Rods

The studies described briefly in this last section are devoted to work in which Professor Mindlin was not directly involved. The studies have been carried out since his attention was channeled to other problems of applied mechanics, but in each case, the research conducted earlier by Mindlin and his associates have been instrumental in solving the problems.

The influence of Professor Mindlin's work on later studies is mainly through the Mindlin–McNiven approximate theory for extensional waves in rods. The theory has been directly applied to two types of problems. The first is the study of the resonant, steady-state vibrations of semi-infinite and finite rods, and the second is the study of the response of a rod to a transient input on its end.

The Mindlin–McNiven theory has been particularly useful for the study of semi-infinite and finite rods because it not only reproduces a relationship between frequency and wavelength for a frequency range from zero to just below the cut-off frequency of the fourth spectral line, but also because it reproduces faithfully the lowest branch, which represents complex wave numbers. This complex branch led to an explanation of an "end mode" which appeared experimentally at a traction-free end.

The introduction of a traction-free plane end of a rod results in a coupling of the modes of propagation of the infinite rod spectrum. The evanescent waves that are excited by the interaction of the traction-free end and a forcing mechanism occur in conjugate pairs. The net result is the setting up of a vibratory motion, or standing wave, the amplitude of which decays with distance from the end. This vibratory motion is coupled to the propagating waves that can exist in the rod at the same frequency. Thus, for example, any signal that reflects from the end of a rod will lose some of its energy to the vibratory motion set up by the evanescent waves. The energy of the vibratory motion will then gradually be "leaked" back along the rod via the same propagating waves from whence it came.

McNiven [25] considered the consequence of an incident plane wave striking the end of a semi-infinite rod within the framework of the Mindlin–McNiven equations. He evaluated the amplitude of the end vibration that resulted as a function of the frequency. The resulting graph exhibits the typical behavior of the response of a slightly damped single-degree-of-freedom system to a harmonic excitation. Further, McNiven was able to identify the frequency of the end resonance with the frequency of an "end mode" that Oliver [26] had observed in a rod excited by a strain signal initiated by a pressure pulse applied to the free end of the rod. In a subsequent paper, McNiven and Perry [27] calculated the frequency spectrum for a finite, elastic rod. Of interest was the appearance of the familiar terrace-like structure that indicates a mode coupling at this same frequency.

More recently, McCoy [28] presented some further results on a solution previously obtained by Kaul and McCoy [29]. In these studies, the experimental problem of Oliver was solved using the Mindlin–McNiven equations. The solution was first obtained in transform (i.e., frequency) space. McCoy numerically investigated the transformed solution and located two pole type singularities near to, but immediately below, the real frequency axis. He was able to show that it is the presence of these poles that give rise to the end resonance observed by Oliver and explained by McNiven.

The Mindlin–McNiven equations have also been used effectively to study the response of a circular rod to a transient input on its end. The equations have been used with two separate mathematical methods, the first the method of integral transforms and the second the method of characteristics.

In 1964 Kaul and McCoy [29] used the theory to determine the radial strains in a semi-infinite, circular rod when a sudden pressure is applied to its end. They used integral transforms and asymptotic methods for evaluating the integrals. The solutions were valid for stations far from the end of the rod and revealed the response for the head of the pulse.

In 1970 Mengi and McNiven [30] used the Mindlin–McNiven equations with the method of characteristics to solve much the same problem. The response in a circular rod was found to a uniform pressure on the end of the rod which has a step dependence on time. The method of characteristics is appropriate for the problem because the approximate equations are hyperbolic, and the wave fronts coincide with characteristic lines. Further, the response can be found near the end of the rod, the area of major interest, and the response can be found for a considerable distance behind the head of the pulse. The method of characteristics is really useful only for problems in which there are only two independent variables, which is the reason an approximate theory is required.

References

1. L. Pochhammer, Über die Fortpflanzungsgeschwindigkeiten kleiner Schwingungen in einem unbegrenzten isotropen Kreiscylinder, *Journal fur reine und angewandte Mathematik* **81,** 324 (1876).
2. C. Chree, Longitudinal vibrations of a circular bar, *Quart. J. Pure and Appl. Math.* **21,** 287 (1886).
3. C. Chree, The equations of an isotropic elastic solid in polar and cylindrical coordinates their solution and application, *Trans. Cambr. Phil. Soc.* **14,** 250 (1889).
4. Lord Rayleigh, *Theory of Sound*, Dover Publications, New York, 1945, p. 251.
5. A. E. H. Love, *The Mathematical Theory of Elasticity*, 4th edition, Cambridge University Press, 1927.
6. D. Bancroft, The velocity of longitudinal waves in cylindrical bars, *Phys. Rev.* **59,** 588 (1941).
7. G. E. Hudson, Dispersion of elastic waves in solid circular cylinders, *Phys. Rev.* **63,** 46 (1943).
8. R. M. Davies, A critical study of the hopkinson pressure bar, *Phil. Trans. Roy. Soc.* **A240,** 375 (1948).

9. R. D. Mindlin, An introduction to the mathematical theory of vibrations of elastic plates, U.S. Army Signal Engin. Labs., Fort Monmouth, N.J., 1955.
10. R. D. Mindlin and M. Onoe, Mathematical theory of vibrations of elastic plates, *Proc. 11th Ann. Sym. on Freq. Control* **17** (1957), U.S. Army Signal Engin. Labs., Fort Monmouth, N.J.
11. S. D. Poisson, Mémoire sur l'équilibre et le mouvement des corps élastiques, *Mém. Acad. Sci.*, Ser. 2, **8,** 357 Paris, 1829.
12. A. L. Cauchy, Sur l'équilibre et le mouvement d'une plaque solide, *Exercise de Mathématique* **3,** 328 (1828).
13. G. Kirchhoff, Über das Gleichgewicht und die Bewegung einer elastischen Scheibe, *Crelles J.* **40,** 51 (1850).
14. A. N. Holden, Longitudinal modes of elastic waves in isotropic cylinders and slabs, *Bell Sys. Tech. J.* **30,** 956 (1951).
15. J. Adem, On the axially-symmetric steady wave propagation in elastic circular rods, *Quart. Appl. Math.* **12,** 261 (1954).
16. J. Zemanek, Jr., An experimental and theoretical investigation of elastic wave propagation in a cylinder, *J. Acoust. Soc. Amer.* **51**(1), 265 (1972).
17. C. A. Coulomb, *Histoire de l'Academie of* 1784, Paris, 1787.
18. B. de Saint Venant, Mémoire sur les vibrations tournantes des verges élastiques, *C.r. hebd. Seanc. Acad. Sci. Paris* **28,** 1180 (1843).
19. D. Bernoulli, letter to L. Euler, Correspondence mathématique et physique, St. Petersburg, 1843.
20. S. Timoshenko, On the correction for shear of the differential equation for transverse vibrations of prismatic bars, *Philosophical Magazine*, Ser. 6 **41,** 744 (1921).
21. Y. H. Pao, The dispersion of flexural waves in an elastic, circular cylinder: Part 2, *Trans. ASME* (*J. Appl. Mech.*) **84E,** 61 (1962).
22. W. A. Green, Dispersion Relations for Elastic Waves in Bars, *Progress in Solid Mechanics*, edited by I. N. Sneddon and R. Hill, North-Holland Publishing, Amsterdam, 1960.
23. M. A. Medick, One-dimensional theories of wave propagation and vibrations in elastic bars of rectangular cross-section, *J. appl. Mech.* **33,** 489 (1966).
24. J. L. Bleustein and R. M. Stanley, A dynamical theory of torsion, *Int. J. Solids Struct.* **6,** 569 (1970).
25. H. D. McNiven, Extensional waves in a semi-infinite elastic rod, *J. Acoust. Soc. Amer.* **33,** 23 (1961).
26. J. Oliver, Elastic wave dispersion in a cylindrical rod by a wide-band short-duration pulse technique, *J. Acoust. Soc. Amer.* **29,** 189 (1957).
27. H. D. McNiven and D. C. Perry, Axially symmetric waves in finite, elastic rods, *J. Acoust. Soc. Amer.* **34**(4), 433 (1962).
28. J. J. McCoy, The effect of edge vibrations on the propagation of a strain transient, *J. Acoust. Soc. Amer.*, **45,** 1104 (1969).
29. R. K. Kaul and J. J. McCoy, Propagation of axisymmetric waves in a circular semi-infinite elastic rod, *J. Acoust. Soc. Amer.* **36,** 653 (1964).
30. Y. Mengi and H. D. McNiven, Analysis of the transient excitation of an elastic rod by the method of characteristics, *Int. J. Solids Struct.* **6,** 871 (1970).

Piezoelectric Crystals and Electro-elasticity

P. C. Y. LEE

Department of Civil and Geological Engineering, Princeton University, Princeton, New Jersey 08540

and

D. W. HAINES

College of Engineering, University of South Carolina, Columbia, South Carolina 29208

Abstract—The classical, linear theory of piezoelectricity is briefly reviewed. Applications to piezoelectric vibrations are discussed for plates of infinite and finite extents, of uniform and variable thicknesses, with full and partial coatings, and under initial stresses.

Recently developed theories of elastic dielectrics, i.e., the polarization gradient theory and the theory of diatomic dielectrics, are presented. Their accommodations to many physical phenomena not accounted for by the classical theory and their convergences and relations to the long-wave limits of dynamical theories of crystal lattices are discussed.

Introduction

Since the development of the piezoelectric resonator by Cady in 1918, oscillators and filter circuits have relied heavily on piezoelectric crystals for ultra-precision requirements in frequency standards and for frequency control. Bars, cylinders, and especially plates, cut from the crystals and coated with thin metallic electrodes, are the elements most often used in such applications. They operate at mechanical resonance with an alternating voltage of high frequency impressed on the electrodes. The classical, linear theory of piezoelectricity is sufficient for the analysis in most instances. In this article, we will first briefly review this basic theory and consider its application to the study of plate vibrations.

Quartz is probably the most widely used crystal [1]. Since the electromechanical coupling, i.e., the piezoelectric effect, is weak in quartz, the analysis of quartz plate vibrations can be simplified for many applications by considering only the elastic field. Hence, frequent reference will be made to studies of quartz in which the piezoelectric effect is neglected.

In the remaining portion of the article, some recent developments in electro-elasticity are introduced. It is known that the differential equations of the classical piezoelectricity are not the long-wave, low-frequency limit of the difference equations of the dynamical theory of crystal lattices with polarizable atoms [M97].† Mindlin showed that as the gradient of electronic polarization is added to the energy density of deformation and polarization, the resulting equations of the augmented theory are the correct limit [M90]. Furthermore, this theory permits specification of both the electric potential and polarization on surface, provides an extra electromechanical interaction even for materials with centrosymmetric properties, and accommodates physical phenomena of surface energy of deformation and polarization, anomalous capacitance of thin dielectric films, acoustical activity, and, with the inclusion of the magnetic field, optical activity. All of these are not accounted for by the classical theory.

To extend further the range of the polarization gradient theory to frequencies of optical vibrational modes for diatomic dielectrics, the distinction and interaction of the positive and negative ions within the same cell of sodium chloride-type lattices are modeled by two interpenetrating and interacting continua. The continuum equations of diatomic crystals are the long-wave limit, not restricted to low frequencies, of the difference equations of lattices of both mass particles and polarizable atoms and they give both acoustical and optical branches with correct long-wave behavior [M103]. With the inclusion of magnetic field, these equations yield a dispersion relation exhibiting the long-wave portions of the coupled acoustic, optical and electromagnetic branches [M104].

Classical Piezoelectricity

It is possible to summarize the three-dimensional, classical theory of piezoelectricity with twenty-five equations [2]:

$$T_{ij,i} = \rho \ddot{u}_j \tag{1}$$

$$S_{ij} = \tfrac{1}{2}(u_{j,i} + u_{i,j}) \tag{2}$$

$$D_{i,i} = 0 \tag{3}$$

$$E_i = -\phi_{,i} \tag{4}$$

$$T_{ij} = c_{ijkl}{}^E S_{kl} - e_{kij} E_k \tag{5}$$

$$D_i = e_{ikl} S_{kl} + \epsilon_{ik}{}^S E_k \tag{6}$$

$$P_i = D_i - \epsilon_0 E_i \tag{7}$$

where $i, j, k, l = 1, 2, 3$, T_{ij}, S_{ij}, u_i, D_i, E_i, and P_i are the components of stress, strain, mechanical displacement, electric displacement, electric field,

† Numbers preceded by the letter M in brackets refer to the Publications of R. D. Mindlin listed after the Preface to this volume.

and polarization, respectively, and ϕ is the electric potential. Material properties in Eqs. (1)–(7) are the elastic stiffness constant c_{ijkl}^E, measured in a constant electric field, the dielectric constant ϵ_{ij}^S, measured at constant strain; the piezoelectric constant e_{ijk}, the mass density ρ, and ϵ_0, the permittivity of vacuum. Note the following conditions which reduce the number of independent constants:

$$c_{ijkl} = c_{ijlk} = c_{jikl} = c_{klij}$$

$$\epsilon_{ik} = \epsilon_{ki}, \quad e_{ijk} = e_{ikj}$$

These equations are valid for the most general anisotropic case, i.e., for triclinic crystals. The mechanical and the quasistationary electrical fields are coupled through the constitutive relations (5) and (6). Tiersten has shown why this quasistationary approximation, rather than the full generality of Maxwell's equations, is an adequate representation [3].

Equations (1)–(6) may be reduced to four equations in four unknowns:

$$c_{ijkl}^E u_{k,li} + e_{kij}\phi_{,ki} = \rho \ddot{u}_j$$

$$e_{ikl}u_{k,li} - \epsilon_{ik}^S \phi_{,ki} = 0 \tag{8}$$

A uniqueness theorem analogous to Neumann's [4] in the linear theory of elasticity can be extended to the piezoelectricity [2]. Accordingly, a unique solution of Eqs. (1)–(7) can be insured by specifying u_j, $\dot{u}_j$, and E_i at each point of the body at initial time t_0, and by specifying at each point of the surface at all times one member of each of the products

$$t_1u_1, \quad t_2u_2, \quad t_3u_3, \quad \phi(n_iD_i) \tag{9}$$

where n_i is the unit outward normal and t_j $(= n_iT_{ij})$ is the surface-traction. Mindlin has extended Eqs. (1)–(7) and boundary conditions to include coupling with thermal fields [M108].

Since an electric field can exist beyond the boundaries of the material, the last in Eq. (9) should be replaced by the conditions

$$\phi = \hat{\phi}, \quad n_iD_i = n_i\hat{D}_i$$

on the boundary and

$$\nabla^2\hat{\phi} = 0$$

in the surrounding space, where $\hat{\phi}$ and $\hat{D}_i$ are the electric potential and displacement, respectively, in the surrounding space.

The convenience of neglecting the surroundings by the use of relations in Eq. (9) is permissible if ϵ_{nn}^S in the material is significantly larger than the dielectric constant of the surrounding region. However, some solutions can be overlooked by use of this approximation, e.g., the surface wave found by Bleustein [5].

An alternative to the field equations of Eqs. (1)–(4) and the boundary conditions is the variational formulation which can be derived from Hamilton's principle for a classical linear piezoelectric medium [6]:

$$\delta \int_{t_0}^{t_1} dt \int_V (\tfrac{1}{2}\rho \dot{u}_j \dot{u}_j - H)\, dV + \int_{t_0}^{t_1} dt \int_S (\bar{t}_k \delta u_k - \bar{\sigma}\delta\phi)\, dS = 0 \tag{10}$$

where

$$H = \tfrac{1}{2} c_{ijkl}{}^E S_{ij} S_{kl} - e_{ijk} E_i S_{jk} - \tfrac{1}{2}\epsilon_{ij}{}^S E_i E_j \tag{11}$$

is the electric enthalpy density, $\bar{t}_k$ is the specified surface-traction, and $\bar{\sigma}$, the applied surface charge. A general form of the variational equation derived from Hamilton's principle, suitably modified, is given by Tiersten [7, 8] for a bounded region containing an internal surface of discontinuity $S^{(d)}$ which divides the region into two parts designated by the superscripts (1) and (2):

$$\begin{aligned}
\int_{t_0}^{t_1} dt \sum_{m=1}^{2} \Bigg\{ & \int_{V^{(m)}} [(T_{kl,k}{}^{(m)} - \rho^{(m)} \ddot{u}_l{}^{(m)})\, \delta u_l{}^{(m)} + D_{k,k}{}^{(m)} \delta\phi^{(m)}]\, dV \\
& + \int_{S_N{}^{(m)}} [(\bar{t}_l{}^{(m)} - n_k{}^{(m)} T_{kl}{}^{(m)})\, \delta u_l{}^{(m)} - (\bar{\sigma}^{(m)} + n_k{}^{(m)} D_k{}^{(m)})\, \delta\phi^{(m)}]\, dS \\
& + \int_{S_C{}^{(m)}} n_k{}^{(m)} [(u_l{}^{(m)} - \bar{u}_l{}^{(m)})\, \delta T_{kl}{}^{(m)} + (\phi^{(m)} - \bar{\phi}^{(m)})\, \delta D_k{}^{(m)}]\, dS \Bigg\} \\
& + \int_{t_0}^{t_1} dt \int_{S^{(d)}} n_k{}^{(d)} \tfrac{1}{2} [(T_{kl}{}^{(2)} - T_{kl}{}^{(1)})(\delta u_l{}^{(1)} + \delta u_l{}^{(2)}) \\
& + (u_l{}^{(1)} - u_l{}^{(2)})(\delta T_{lk}{}^{(1)} + \delta T_{lk}{}^{(2)}) + (D_k{}^{(2)} - D_k{}^{(1)})(\delta\phi^{(1)} + \delta\phi^{(2)}) \\
& + (\phi^{(2)} - \phi^{(1)})(\delta D_k{}^{(1)} + \delta D_k{}^{(2)})]\, dS = 0
\end{aligned} \tag{12}$$

where $S_N{}^{(m)}$ and $S_C{}^{(m)}$, respectively, stand for the portion of the mth surface on which the traction $\bar{t}_k{}^{(m)}$ and/or the charge $\bar{\sigma}^{(m)}$ are prescribed, and the portion of the mth surface on which mechanical displacement $\bar{u}_k{}^{(m)}$ and/or the electrical potential $\bar{\phi}^{(m)}$ are prescribed.

With appropriate forms of this variational equation, Eer Nisse [9] and Holland and Eer Nisse [10] have analyzed vibrating piezoelectric structures of complex geometries. These approximate techniques which incorporate Rayleigh–Ritz methods have produced accurate results. They require a minimum of physical insight and can be quite useful with the availability of modern computer facilities.

However, for basic understanding of the nature of plate vibrations, it is essential to pursue the subject in stages from simple cases to more complex ones, starting with infinite plates.

Waves and Vibrations in Infinite Plates

An investigation of waves and vibrations in infinite plates, where it is possible to obtain exact solutions which satisfy the boundary conditions on the faces, is a logical starting point. This knowledge not only contributes to the basic understanding of the subject, but also provides essential information for establishing frequency ranges in which each of the approximate theories are valid.

Consider an infinite plate bounded by planes $x_2 = \pm h$. For traction-free surfaces with infinitesimally thin, perfectly conducting electrodes which are shorted, the boundary conditions on $x_2 = \pm h$ are

$$T_{2j} = 0, \qquad \phi = 0 \tag{13}$$

If the electrodes are not shorted and an alternating potential difference is applied to them, the second of Eqs. (13) is replaced by

$$\phi|_{x_2=\pm h} = \pm\phi_0 \cos \omega t \tag{14}$$

where ω is the circular frequency and ϕ_0 is a constant. Note the perfectly conducting electrodes require $\phi_{,1} = \phi_{,3} = 0$ on $x_2 = \pm h$. With no electrodes present we have

$$D_2|_{x_2=\pm h} = 0 \tag{15}$$

instead of the second of Eqs. (13).

For a ferroelectric ceramic poled in the x_2-direction, Tiersten formulated the frequency equation obtained from solutions of Eqs. (8) and (13) of the form

$$\begin{aligned} u_1 &= A_1 \cos \eta x_2 \sin (\xi x_1 - \omega t) \\ u_2 &= A_2 \sin \eta x_2 \cos (\xi x_1 - \omega t) \\ u_3 &= 0 \\ \phi &= B \sin \eta x_2 \cos (\xi x_1 - \omega t) \end{aligned} \tag{16}$$

An independent set of solutions are obtained by interchanging $\cos \eta x_2$ and $\sin \eta x_2$ in Eqs. (16). Further, Tiersten also considered solutions for the case of the plate poled in its plane [11].

For the thickness vibrations (solutions independent of x_1 and x_3) driven by an alternating potential difference at $x_2 = \pm h$, Tiersten formulated and investigated aspects of the solution for certain ferroelectric ceramic and rotated Y-cut quartz plates. He found that the resonance frequencies are in each case less than their corresponding values in the purely elastic case which are integral multiples of the fundamental frequencies [12].

The boundary conditions at $x_2 = \pm h$ must be altered when actual electrodes coat these surfaces. For most applications the electrodes are so thin that only their mass need be considered. Consequently, if both faces of the plate are equally coated with perfectly conducting material of density ρ', thickness $2h' \ll h$, the traction-free boundary condition is replaced by

$$T_{2j} = \mp 2h'\rho'\ddot{u}_j \tag{17}$$

If the stiffness of the electrode is significant, then the boundary conditions must incorporate this stiffness in addition to the mass [M80].

Recalling that the electromechanical coupling in quartz is weak, for many cuts we note that the study of waves propagating along the digonal axis of an infinite AT-cut plate was covered by Kaul and Mindlin's solution of Ekstein's frequency equation [M72].

Piezoelectric surface waves are receiving increased attention for use in ultrasonic applications, particularly signal processing devices. The amplitudes of such waves decay exponentially from the surface similarly to the well-known Rayleigh waves in isotropic materials. Deresewicz and Mindlin described the surface wave that propagates along the digonal axis of an AT-cut quartz plate [M56]. Coquin and Tiersten extended that work to all rotated Y-cuts and also considered surface waves in any direction on X-cuts; in addition they discussed aspects of excitation and detection [13]. For isotropic materials, Tiersten examined the properties of surface waves guided by thin films [14]. Bluestein described a piezoelectric surface wave which, unlike the others mentioned, has no counterpart in the purely elastic case [5]. An extensive review of the applications of surface waves was given by White [15].

Vibrations of Finite Piezoelectric Plates

There are very few exact solutions for bounded piezoelectric plates. Most well-known solutions for finite piezoelectric plates are obtained from approximate two-dimensional equations which are extensions of those earlier developed by Mindlin for the purely elastic case [M69].

A system of five two-dimensional piezoelectric plate equations, applicable to frequencies as high as those of the fundamental thickness-shear modes, was derived by Tiersten and Mindlin [M70] based on an approximation involving the early terms of power series expansions of mechanical and electric displacements in a variational principle. Recently, based on a different variational principle and a return to an earlier expansion [M38] of the electric potential, instead of the electric displacement, Mindlin obtained two-dimensional equations which are in simpler forms, with fewer terms, without numerical coefficients and without the appearance of surface boundary values in the constitutive relations [M107]. The results of this work are summarized here.

By the substitution of following expansions

$$u_j = \sum_{n=0}^{g} x_2{}^n u_j{}^{(n)}, \qquad \phi = \sum_{n=0}^{g} x_2{}^n \phi^{(n)}$$

$$E_j = \sum_{n=0}^{g} x_2{}^n E_j{}^{(n)}, \qquad S_{ij} = \sum_{n=0}^{g} x_2{}^n S_{ij}{}^{(n)} \tag{18}$$

into a variational principle which is similar to Eq. (10) and by the employment of truncation and correction procedures analogous to those discussed in article by Pao and Kaul in this volume, the following equations are obtained.

Constitutive relations:

$$\begin{aligned} T_{ij}{}^{(0)} &= 2h(c_{ijkl}{}^{(0)} S_{kl}{}^{(0)} - e_{kij}{}^{(0)} E_k{}^{(0)}) \\ T_{ab}{}^{(1)} &= \tfrac{2}{3}h^3(c_{abcd}{}^{(1)} S_{cd}{}^{(1)} - e_{cab}{}^{(1)} E_c{}^{(1)}) \\ D_i{}^{(0)} &= 2h(e_{ijk}{}^{(0)} S_{jk}{}^{(0)} + \epsilon_{ij} E_j{}^{(0)}) \\ D_a{}^{(1)} &= \tfrac{2}{3}h^3(e_{abc}{}^{(1)} S_{bc}{}^{(1)} + \epsilon_{ab} E_b{}^{(1)}) \end{aligned} \tag{19}$$

Equations of motion:

$$\begin{aligned} T_{ij,i}{}^{(0)} + 2hT_j{}^{(0)} &= 2h\rho\ddot{u}_j{}^{(0)} \\ T_{ab,a}{}^{(1)} - T_{2b}{}^{(0)} + \tfrac{2}{3}h^3 T_b{}^{(1)} &= \tfrac{2}{3}h^3\rho\ddot{u}_b{}^{(1)} \end{aligned} \tag{20}$$

Equations of electrostatics:

$$\begin{aligned} D_{i,i}{}^{(0)} + 2hD^{(0)} &= 0 \\ D_{a,a}{}^{(1)} - D_2{}^{(0)} + \tfrac{2}{3}h^3 D^{(1)} &= 0 \end{aligned} \tag{21}$$

Strain-displacement relations:

$$\begin{aligned} S_{ij}{}^{(0)} &= \tfrac{1}{2}(u_{i,j}{}^{(0)} + u_{j,i}{}^{(0)} + \delta_{2j}u_i{}^{(1)} + \delta_{2i}u_j{}^{(1)}) \\ S_{ab}{}^{(1)} &= \tfrac{1}{2}(u_{a,b}{}^{(1)} + u_{b,a}{}^{(1)}) \end{aligned} \tag{22}$$

Electric field-potential relations:

$$\begin{aligned} E_i{}^{(0)} &= -\phi_{,i}{}^{(0)} - \delta_{i2}\phi^{(1)} \\ E_a{}^{(1)} &= -\phi_{,a}{}^{(1)} \end{aligned} \tag{23}$$

Note $i, j, k, l = 1, 2, 3$, and $a, b, c, d = 1, 3$. Terms in Eqs. (17)–(21) not previously defined are,

$$D_i{}^{(n)} = \int_{-h}^{h} x_2{}^n D_i dx_2, \qquad \text{the } n\text{th-order electric displacement}$$

$$D^{(n)} = B_n{}^{-1}[x_2{}^n D_2]_{-h}{}^{h}, \qquad \text{the } n\text{th-order face-charge}$$

$$T_i{}^{(n)} = B_n{}^{-1}[x_2{}^n T_{2i}]_{-h}{}^{h}, \qquad \text{the } n\text{th-order face-traction}$$

where $B_n = 2b^{2n+1}/(2n+1)$. The material properties are

$$c_{ijkl}^{(0)} = (\kappa_{i+j-2})^{\mu}(\kappa_{k+l-2})^{\nu}(c_{ijkl} - c_{ij22}c_{22kl}/c_{2222}), \quad \text{(not summed)}$$

$$e_{kij}^{(0)} = (\kappa_{i+j-2})^{\mu}(e_{kij} - e_{k22}c_{ij22}/c_{2222}), \quad \text{(not summed)}$$

$$c_{abcd}^{(1)} = A_{abcd}/|S_{abcd}|$$

$$e_{cab}^{(1)} = d_{cde}c_{abde}^{(1)}$$

where $(\kappa_{i+j-2})^{\mu}$ and $(\kappa_{k+l-2})^{\nu}$, ($\mu = \cos^2(ij\pi/2), \nu = \cos^2(kl\pi/2)$) are correction factors, introduced in order for the dispersion curves to match those of the three-dimensional theory at the thickness-shear frequencies. The elastic compliances s_{ijkl} and piezoelectric constants d_{ijk} are defined such that

$$s_{ijmn}c_{ijkl} = I_{mnkl}, \qquad d_{kmn} = s_{ijmn}e_{kij}$$

where I_{mnkl} is the fourth rank unit tensor. The determinant $|S_{abcd}|$ is given by

$$|S_{abcd}| = \begin{vmatrix} S_{1111} & S_{3311} & S_{1311} \\ S_{1133} & S_{3333} & S_{1333} \\ S_{1113} & S_{3313} & S_{1313} \end{vmatrix}$$

and A_{abcd} is the cofactor of S_{abcd} in $|S_{abcd}|$.

For a right cylindrical or prismatic boundary in a curve C, the edge conditions are

$$\begin{aligned} n_a T_{aj}^{(0)} &= 2ht_j^{(0)} \quad &&\text{or } u_j^{(0)} = \bar{u}_j^{(0)} \\ n_a D_a^{(0)} &= 2hd^{(0)} \quad &&\text{or } \phi^{(0)} = \bar{\phi}^{(0)} \\ n_a T_{ab}^{(1)} &= \tfrac{2}{3}h^3 t_b^{(1)} \quad &&\text{or } u_b^{(1)} = \bar{u}_b^{(1)} \\ n_a D_a^{(1)} &= \tfrac{2}{3}h^3 d^{(1)} \quad &&\text{or } \phi^{(1)} = \bar{\phi}^{(1)} \end{aligned} \tag{24}$$

where $\bar{u}_j^{(n)}$ and $\bar{\phi}^{(n)}$ are the prescribed edge displacements and potentials, respectively, and the edge-tractions $t_j^{(n)}$, and edge-charges $d^{(n)}$, are given by

$$t_j^{(n)} = B_n^{-1}\int_{-h}^{h} x_2^n (n_a T_{aj})_c dx_2, \qquad d^{(n)} = B_n^{-1}\int_{-h}^{h} x_2^n (n_a D_a)_c dx_2$$

Two-dimensional equations for thermoelastic, crystal plates were derived by Tasi and Herrmann [16] based on power series expansions of the mechanical displacement and the temperature change in a variational principle that requires an adjoint field to receive the energy dissipated in the thermoelastic medium. By employing a dissipation function [17] and expanding the mechanical displacement, temperature change and electric potential in power series in an integral form of the principle of conservation of elastic, thermal and electrical energy [M68], two-dimensional equations of thermopiezoelectric vibrations of plates were obtained by Mindlin [M108].

Two-dimensional equations of successively higher orders of approximation for elastic isotropic and crystal plates have been deduced from the three-dimensional theory of elasticity by Lee and Nikodem [18, 19] based on a series expansion in terms of simple thickness modes for infinite plates. This approach leads to a generalized nth-order matrix from which accurate frequency equations of any desired range can be generated directly.

The addition of electrode coating to the surfaces $x_2 = \pm h$ requires changes in the equations of motion. We again assume a symmetrically coated crystal where only the mass of the electrodes is significant. Equations (20) are thereby altered by replacing $2h\rho\ddot{u}_j^{(0)}$ by $2h\rho(1 + R)\ddot{u}_j^{(0)}$ and $\frac{2}{3}h^3\rho\ddot{u}_b^{(1)}$ by $\frac{2}{3}h^3\rho(1 + 3R)\ddot{u}_b^{(1)}$ where $R = 2\rho'h'/\rho h$, the plate-back ratio [M86]. Generalizations to cases of asymetrically coated plates where the stiffness of the electrodes is significant were given by Mindlin [M80].

It is possible to obtain closed-form solutions that satisfy Eqs. (19)–(23) and boundary conditions on one pair of parallel edges, free of traction. Gazis and Mindlin obtained the intricate frequency spectrum from the one-dimensional solutions of these equations, reduced to the purely elastic case, for AT-cut quartz plates [M75]. This spectrum compares extremely well with experimental results.

The study of thickness-shear resonances in the AT-cut is further simplified if the weak face-shear coupling is ignored. Mindlin made use of this approximation in earlier investigations [M36, M38] where excellent agreement with experiments was found. Mindlin and Forray examined thickness-shear modes in contoured plates, i.e., plates of varying thickness, by neglecting the coupling with flexural modes, thus obtaining a manageable approximation with closed-form solutions [M43]. Jerrard included the coupling of thickness-shear and flexure in his study of plates with exponentially varying thickness [20].

The one-dimensional solutions previously discussed apply to rectangular plates, although in each case one pair of boundary conditions was omitted from consideration. In the solutions cited, this is of no consequence as many modes exist for which reflections from such edges are not significant. However, it is possible to excite anharmonic overtones of thickness-shear and flexural resonances, so-called twist overtones. These overtones, which in general are electrically weak resonances because of charge cancellation, are caused by reflections from the free edges previously ignored. To describe them, Mindlin and Spencer [M88] made use of the special anisotropy of the AT-cut of quartz and reduced the equations of Ref. [M69] to obtain equations of motion for rectangular plates with one set of edges parallel to the diagonal axis:

$$k_1^2c_{66}(u_{2,11}^{(0)} + u_{1,1}^{(1)}) - \tfrac{1}{3}h^2\gamma_{55}(u_{2,1133}^{(0)} - u_{1,133}^{(1)}) = \rho\ddot{u}_2^{(0)}(1 + R)$$

$$\gamma_{11}u_{1,11}^{(1)} - \gamma_{55}(u_{2,133}^{(0)} - u_{1,33}^{(1)}) - \frac{3}{h^2}k_1^2c_{66}(u_{2,1}^{(0)} + u_1^{(1)}) = \rho\ddot{u}_1^{(1)}(1 + 3R) \quad (25)$$

where

$$k_1^2 = \frac{\pi^2(1+3R)}{12(1+R)}$$

$$\gamma_{11} = c_{11} - \frac{c_{12}^2}{c_{22}} - \frac{(c_{14} - c_{12}c_{24}/c_{22})^2}{c_{44} - c_{24}^2/c_{22}}$$

$$\gamma_{55} = c_{55} - \frac{c_{56}^2}{c_{66}}$$

In these equations, the coefficients of elastic stiffness are denoted by abbreviated notion. Traction-free boundary conditions for rectangular plates of dimensions $2l \times 2c$ were established as

$$[T_{12}^{(0)} + T_{13,3}^{(1)}]_{x_1=\pm l} = 0$$

$$[T_{11}^{(1)}]_{x_1=\pm l} = 0$$

$$[T_{13}^{(1)}]_{x_3=\pm c} = 0 \qquad (26)$$

and $T_{13}^{(1)}$ continuous at the four corners, where in this approximation:

$$T_{12}^{(0)} = 2hk_1^2c_{66}(u_{2,1}^{(0)} + u_1^{(1)})$$

$$T_{11}^{(1)} = \tfrac{2}{3}h^3\gamma_{11}u_{1,1}^{(1)}$$

$$T_{13}^{(0)} = -\tfrac{2}{3}h^2\gamma_{55}(u_{2,13}^{(0)} - u_{1,3}^{(1)}) \qquad (27)$$

The solutions of Eqs. (25)–(27) for $R = 0$, i.e., plates without electrodes, presented in Ref. [M88], represent the first closed-form solutions satisfying traction-free conditions on all four edges of a rectangular plate. Close agreement with experiment, both of frequencies and mode shapes was found.

Lee and Spencer [21] solved the Mindlin–Spencer equations with the addition of rectangular strip electrodes which span the full width $(-c \leq x_3 \leq c)$ but only part of the length $(-a \leq x_1 \leq a,\ l > a)$. Again comparison with experiments is excellent.

Lee and Chen [22] extended these solutions to rectangular, contoured plates with partial electrodes. Jerrard's technique of approximating the thickness variation was utilized.

Energy Trapping

As a mass loading, electrodes covering an infinite plate lower the frequencies of the plate, in particular, the cut-off frequencies. In a partially coated infinite plate, important resonances lie between the corresponding cut-off frequencies of uncoated and completed coated infinite plates. Under these circumstances, standing waves exist in the coated portion, but waves with amplitude exponentially decaying away from the electrode edges are in the uncoated portion of an infinite plate. Thus vibrational energy is essentially trapped between the electrodes where it contributes to the charge on the electrodes. A localized mode in lattice vibrations is an analogous phenomenon. Another analogy exists with electromagnetic waves in a cavity resonator, the example to which Mortley referred in the first description of the concept of energy trapping in crystal plates [23].

In practice, it is desirable that resonances be as pure as possible. One serious difficulty is the presence of closely spaced anharmonic overtones in the vicinity of the desired frequency. Bechmann observed that reductions of the electrode area in a partially coated plate eliminates such undesirable overtones one by one until only the desired resonance—the fundamental remains [24]. The determination of Bechmann's number (B), equal to the ratio of electrode length to plate thickness (a/h) below which only the fundamental exists, depends on the material properties of the plate and R, the plate-back ratio. As the plate area should be as large as possible to maximize the charge on the electrodes, Bechmann's number establishes an important upper limit on the electrode length.

For the fundamental thickness-shear resonance in the AT-cut quartz plate, Mindlin and Lee [M86] found the criterion for Bechmann's number to be that the electrode length be equal to the wavelength of the thickness-shear wave in the coated infinite plate at the cut-off frequency of the uncoated infinite plate. The same criterion was found to be true for the harmonic overtones of thickness-shear by Haines [25]. Mindlin employed this principle to calculate Bechmann's number for thickness-twist modes in AT-cut quartz plates [M87]. Bechmann's number for these cases can be summarized by the formula

$$B = \pi K_n (Rn)^{-1/2}, \qquad (Rn \ll 1) \tag{28}$$

where n is the order of the overtone ($n = 1$ for the fundamental thickness-shear mode) and $K_n/2$ is the curvature of the particular branch at its cut-off frequency in the usual case where the branch is sufficiently described by a parabola in that region. Under these conditions $K_n = 0(1/n)$. Exceptions to this are discussed in Ref. [25].

For the fundamental thickness-shear mode, Bleustein and Tiersten examined the influence of including the piezoelectric effect and also the benefits of using plates with discontinuous electrodes [26].

If the length of the plate is greater than twice the electrode length, i.e., $l > 2a$, the infinite plate model we have described is sufficient. However, for smaller ratios, anharmonic overtones caused by reflections require analysis of finite plates. Such studies were performed by Byrne, Lloyd, and Spencer [27] and Lee and Spencer [21].

One of the more significant applications of energy trapping theory is the monolithic filter in which a series of electrodes is installed on a single piezoelectric plate. If the plate is quartz, either thickness-shear or thickness-twist modes can be established. Beaver gave a full analysis of this device [28].

Crystal Plates Under Initial Stress

The effect of initial stress and deformation on the velocity of sound waves of small amplitude, propagating in infinite media has been studied by Toupin and Bernstein [29] and Thurston and Brugger [30]. From the measurements of sound wave velocity and piezoelectric current as functions of initial deformation, material coefficients of higher orders than those in Eqs. (5) and (6) can be determined [31, 32]. These coefficients, associated with nonlinear terms in constitutive relations, play an important role in studies of all anharmonic phenomena, such as the interaction of acoustic and thermal phonons and equations of state.

Two-dimensional theories for linear, elastic plates with initial stress have been derived by Green and Zerna [33] and Herrmann and Armenàkas [34]. A system of five, two-dimensional equations for crystal plates, governing small vibrations superposed on static, initial deformation and accommodating the coupling of the fundamental thickness-shear, thickness-twist, flexure, extension, and face-shear modes and all their anharmonic overtones, has been derived by Lee, Wang and Markenscoff [35]. These equations were applied to the rotated Y-cuts of quartz, and the changes in the resonance frequencies of simple thickness-shear mode were computed as a function of the azimuth angle (between the X crystallographic axis and the direction of an applied force for a given cut) and also as a function of angle of cut for a given direction of force. The predicted results compare closely with experimental data. It was found that in predicting frequency variation due to initial stress the contributions from third-order elastic coefficients (associated with nonlinear terms) are comparable to those from second-order coefficients in constitutive relations.

Elastic, Dielectric Continuum with Polarization Gradient

The differential equations of the classical, linear theory of piezoelectricity, Eqs. (1)–(7), can be derived from Toupin's variational principle [36] by assuming that the stored energy density is a function of strain and polarization [M90]. The electromechanical interactions are presented by couplings between the effective local electric force and strain or, alternatively, between the polarization and stress. The coupling is achieved by means of material coefficients of third rank tensors. Since no central symmetric third rank tensor exists, there can be no electromechanical coupling or piezoelectric effect in central symmetric materials. By including polarization gradient, in addition to strain and polarization, in the stored energy function and by means of an extension of a linear version of Toupin's variational principle for the classical theory, Mindlin obtained an augmented theory of elastic, dielectric continuum [M90].

In a body occupying a volume V bounded by a surface S, separating V from an outer vacuum V', the extension of Toupin's variational principle takes the form

$$\delta \int_{t_0}^{t_1} dt \int_{V''} (\tfrac{1}{2}\rho\dot{u}_i\dot{u}_i - H)dV + \int_{t_0}^{t_1} dt \left[\int_V (f_i\delta u_i + E_i^0\delta P_i)dV + \int_S t_i\delta u_i dS\right] = 0 \quad (29)$$

where $V'' = V + V'$, H is the electric enthalpy density, u_i is the displacement, P_i is the polarization and f_i, E_i^0, and t_i are the external body force, electric field, and surface-traction, respectively.

The electric enthalpy density is defined

$$H = W^L(S_{ij}, P_i, P_{j,i}) - \tfrac{1}{2}\epsilon_0\phi_{,i}\phi_{,i} + \phi_{,i}P_i \quad (30)$$

In Eq. (30) the functional dependence of the stored energy W^L is extended to include polarization gradient in additional to strain and polarization, S_{ij} is the strain,

$$S_{ij} = \tfrac{1}{2}(u_{j,i} + u_{i,j}) \quad (31)$$

ϕ is the potential of the Maxwell self-field,

$$E_i = -\phi_{,i} \quad (32)$$

and ϵ_0 is the permittivity of a vacuum.

For independent variation of u_i, P_i, and ϕ between fixed limits at times t_0 and t_1 and by the chain rule of differentiation,

$$\delta H = -T_{ij,i}\delta u_j - (E_{ij,i} + \bar{E}_j - \phi_{,j})\delta P_j - (-\epsilon_0\phi_{,ii} + P_{i,i})\delta\phi + (T_{ij}\delta u_j)_{,i} + (E_{ij}\delta P_j)_{,i} + [(P_i - \epsilon_0\phi_{,i})\delta\phi]_{,i} \quad (33)$$

where T_{ij}, $\bar{E}_i$ are the stress and effective local electric force and E_{ij} is derivable from W^L as defined by

$$T_{ij} \equiv \frac{\partial W^L}{\partial S_{ij}}, \quad \bar{E}_i \equiv -\frac{\partial W^L}{\partial P_i}, \quad E_{ij} \equiv \frac{\partial W^L}{\partial P_{i,j}} \quad (34)$$

Since δu_j vanishes at t_0 and t_1, it can be shown

$$\delta \int_{t_0}^{t_1} dt \int_V \tfrac{1}{2}\rho\dot{u}_i\dot{u}_i dV = -\int_{t_0}^{t_1} dt \int_V \rho\ddot{u}_j\delta u_j dV \quad (35)$$

Substituting Eqs. (33) and (35) into Eq. (29) and applying the divergence theorem, one obtains

$$\int_{V''} [(T_{ij,i} + f_j - \rho \ddot{u}_j)\delta u_j + (E_{ij,i} + \bar{E}_j - \phi_{,j} + E_j^0)\delta P_j$$

$$+ (-\epsilon_0 \phi_{,ii} + P_{i,i})\delta\phi]dV + \int_S [(t_j - n_i T_{ij})\delta u_j + n_i E_{ij} \delta P_j$$

$$+ n_i(P_i - \epsilon_0[[\phi_{,i}]])\delta\phi]dS = 0 \tag{36}$$

where n_i is the unit outward normal to S and $[[\phi_{,i}]]$ is the jump in $\phi_{,i}$ across S.

From Eq. (36), one obtains the Euler equations in V:

$$T_{ij,i} + f_j = \rho \ddot{u}_j$$

$$E_{ij,i} + \bar{E}_j - \phi_{,j} + E_j^0 = 0$$

$$-\epsilon_0 \phi_{,ii} + P_{i,i} = 0 \tag{37}$$

and, in V':

$$\phi_{,ii} = 0 \tag{38}$$

and the boundary conditions on S:

$$n_i T_{ij} = t_j \quad \text{or specification of } u_j$$

$$n_i E_{ij} = 0 \quad \text{or specification of } P_j$$

$$n_i(P_i - \epsilon_0[[\phi_{,i}]]) = 0 \quad \text{or specification of } \phi \tag{39}$$

If one takes

$$W^L = b_{ij}^0 P_{j,i} + \tfrac{1}{2} a_{ij} P_i P_j + \tfrac{1}{2} b_{ijkl} P_{j,i} P_{l,k} + \tfrac{1}{2} c_{ijkl} S_{ij} S_{kl}$$

$$+ d_{ijkl} P_{j,i} S_{kl} + f_{ijk} S_{ij} P_k + g_{ijk} P_i P_{k,j} \tag{40}$$

then from Eq. (34), one finds constitutive equations

$$-\bar{E}_j = a_{jk} P_k + g_{jkl} P_{l,k} + f_{klj} S_{kl}$$

$$E_{ij} = g_{kij} P_k + b_{ijkl} P_{l,k} + d_{ijkl} S_{kl} + b_{ij}^0$$

$$T_{ij} = f_{ijk} P_k + d_{klij} P_{l,k} + c_{ijkl} S_{kl} \tag{41}$$

Equations (31), (37), (38), and (41), with boundary conditions (39) form the equations of Mindlin's theory. These equations can be reduced to those of the classical theory of piezoelectricity in the form given by Toupin by discarding the $E_{ij,i}$ term from Eqs. (37), the second set of boundary conditions from Eqs. (39) and, for the constitutive equations, the second equation and the terms associated with g_{ijk} and d_{ijkl} in the first and third equations of Eqs. (41). The relations between the new constants a_{ij} (for fixed strain), f_{ijk} and c_{ijkl} (for fixed polarization) in Eqs. (41) and those in Eqs. (5)–(17) were found by Mindlin [M109].

The inclusion of polarization gradient, in addition to strain and polarization, in the stored energy function of deformation and polarization W^L can be justified from the viewpoint of modern dynamical theory of ionic crystal lattices. By considering a one-dimensional lattice theory of the Cochran-type [37] based on the Dick–Overhauser shell model [38] of the atom, Mindlin [M97] showed that the equations of this extended theory, rather than the classical equations of elastic dielectrics, are the correct, low-frequency, long-wave limit of the lattice theory. He also exhibited the relations between the force constants of the lattice and the pertinent material coefficients in the constitutive Eqs. (41).

A review of some recent developments in the area between the dynamical theory of crystal lattices, in the harmonic approximation, and the classical, linear theories of elasticity and piezoelectricity was given in a comprehensive paper by Mindlin [M109]. Detailed discussions of theories of crystal lattices and continua has been covered in the article by Gazis and Gong in this volume.

In a paper by Askar, Lee, and Cakmak [39], a three-dimensional dynamical theory was formulated for cubic ionic crystal lattices with shell model. By suppressing the difference in displacements of positive and negative ions within the same cell and taking the long-wave approximation, the potential energy density function was shown to include strain, polarization as well as polarization gradient in an identical form proposed by Mindlin. Expressions relating the material coefficients of continuum theory to the lattice properties were also established. For materials of central symmetric cubic symmetry, the material coefficients in Eqs. (41) are reduced to [1]

$$f_{ijk} = 0, \quad g_{ijk} = 0, \quad b_{ij}{}^0 = b_0\delta_{ij}, \quad a_{ij} = a\delta_{ij}$$

$$b_{ijkl} = b\delta_{ijkl} + b_{12}\delta_{ij}\delta_{kl} + b_{44}(\delta_{ik}\delta_{jl} + \delta_{il}\delta_{jk}) + b_{77}(\delta_{ik}\delta_{jl} - \delta_{il}\delta_{jk})$$

$$c_{ijkl} = c\delta_{ijkl} + c_{12}\delta_{ij}\delta_{kl} + c_{44}(\delta_{ik}\delta_{jl} + \delta_{il}\delta_{jk})$$

$$d_{ijkl} = d\delta_{ijkl} + d_{12}\delta_{ij}\delta_{kl} + d_{44}(\delta_{ik}\delta_{jl} + \delta_{il}\delta_{jk})$$

where

$$b = b_{11} - b_{12} - 2b_{44}, \quad c = c_{11} - c_{12} - 2c_{44}, \quad d = d_{11} - d_{12} - 2d_{44}$$

δ_{ij} is the Kronecker delta and δ_{ijkl} is unity if all indices are alike and zero otherwise. Among the above coefficients, c_{ijkl} is the elastic stiffness and its value is known by measurements. The numerical values of the other coefficients for sodium iodide and sodium chloride are given in Table I. These values are computed by the relations given in Ref. [39] with a minor correction.

Table 1. Material Coefficients.

	$a(10^{18})$	$-b_0(10^7)$	b_{11}	b_{12}	b_{44}	b_{77}	d_{11}	d_{12}	d_{44}
	dyn cm^2/C^2	dyn cm/C	10^4 dyn cm^4/C^2				10^7 dym cm/C		
Nal	87.19	1.26	4.55	7×10^{-7}	2.27	2.27	3.86	-1.51	-2.10
NaCl	110.37	1.44	4.44	-2×10^{-7}	2.18	2.18	4.71	-1.80	-2.44

Mindlin's theory, besides its simplicity and mathematical elegance, has many interesting properties. From Eqs. (41), one may note that coefficient d_{ijkl} provides for an additional electromechanical interaction even when $f_{ijk} = g_{ijk} = 0$ as in centrosymmetric materials. Therefore it is possible to excite thickness vibrations in a centrosymmetric, cubic crystal plate by applying an alternating voltage drop between electrodes on a pair of (100) faces of the plate. In Ref. [M105], electromechanical vibrations of a sodium chloride plate was studied. Frequencies and approximate mode shapes were calculated. It was found that the electromechanical coupling coefficient for sodium chloride can be as large as one four-hundredth of that of an X-cut quartz plate.

Surface Energy of Deformation and Polarization

In order to see how the surface energy of deformation and polarization is accommodated in Mindlin's theory, the total potential energy density W will be considered first. W is the sum of the energy density of deformation and polarization and the energy density of the Maxwell electric self-field,

$$W = W^L + \tfrac{1}{2}\epsilon_0 \phi_{,i} \phi_{,i} \tag{42}$$

Through the use of Eqs. (41), W^L, given in Eq. (40), can be rewritten as

$$W^L = \tfrac{1}{2} b_{ij}{}^0 P_{j,i} + \tfrac{1}{2} T_{ij} S_{ij} - \tfrac{1}{2} E_i P_i + \tfrac{1}{2} E_{ij} P_{j,i} \tag{43}$$

Substituting Eq. (43) into Eq. (42), integrating Eq. (42) over volume V'', and applying the chain rule, divergence theorem, the free boundary conditions and the equations of equilibrium, one obtains the total energy, in the case of equilibrium,

$$\int_{V''} W dV = \tfrac{1}{2} \int_S n_i b_{ij}{}^0 P_j dS + \tfrac{1}{2} \int_V (f_i u_i + E_i{}^0 P_i) dV \tag{44}$$

Hence, in the absence of external forces,

$$\int_{V''} W dV = \tfrac{1}{2} \int_S n_i b_{ij}{}^0 P_j dS \tag{45}$$

Accordingly, the surface energy of deformation and polarization, per unit area, is

$$W^S = \tfrac{1}{2}[n_i b_{ij}^0 P_j]_S \tag{46}$$

The total surface energy density due to the presence of a surface can be considered to be made up of two parts: the energy required to remove the part of the lattice on one side of the interface, while the rest of the lattice is held in its original equilibrium position, and the relaxation energy of the remaining part of the lattice due to the deformation and polarization of the lattice in going to a new equilibrium position [40]. W^S represents the long-wave, low-frequency limit of the latter portion of the surface energy density [39].

From Eq. (46), one sees that W^S depends on both the material constant b_{ij}^0 which depends, in turn, on the orientation of the surface, and the polarization at the surface, which has to be obtained as the solution of a boundary-value problem.

For the case of a free (100) surface of a semi-infinite, centrosymmetric, cubic crystal, the resulting fields are one-dimensional and the equations of equilibrium (37), for the half-space $x_1 \geq 0$, reduce to

$$\begin{aligned} c_{11}u_{1,11} + d_{11}P_{1,11} &= 0 \\ d_{11}u_{1,11} + b_{11}P_{1,11} - aP_1 - \phi_{,1} &= 0 \\ -\epsilon_0\phi_{,11} + P_{1,1} &= 0 \end{aligned} \tag{47}$$

and the free surface conditions, on $x_1 = 0$, become

$$\begin{aligned} c_{11}u_{1,1} + d_{11}P_{1,1} &= 0 \\ d_{11}u_{1,1} + b_{11}P_{1,1} + b_0 &= 0 \\ -\epsilon_0\phi_{,1} + P_1 &= 0 \end{aligned} \tag{48}$$

Mindlin's solution for this problem is [M90]

$$\begin{aligned} u_1 &= -\frac{b_0 d_{11}}{c_{11}\lambda(a+\epsilon_0^{-1})}\,e^{-x_1/\lambda} \\ P_1 &= \frac{b_0}{\lambda(a+\epsilon_0^{-1})}\,e^{-x_1/\lambda} \\ \phi &= -\frac{b_0}{a\epsilon_0 + 1}\,e^{-x_1/\lambda} \end{aligned} \tag{49}$$

and

$$W^S = -\frac{(b_0)^2}{2\lambda(a+\epsilon_0^{-1})} \tag{50}$$

where

$$\lambda^2 = \frac{b_{11}}{(a + \epsilon_0^{-1})}\left(1 - \frac{d_{11}^2}{c_{11}b_{11}}\right). \tag{51}$$

The parameter λ has unit of length and the expression for λ^2 in Eq. (51) is positive as required by the positive definiteness of W^L. Thus according to this theory, there exist displacement and polarization fields which decay exponentially into the interior, even when there are no external body and surface forces.

For sodium chloride, in additional to the values of material coefficients shown in Table I, the following ones are needed for some quantitative predictions:

$$c_{11} = 0.486 \times 10^{12} \text{ dyn/cm}^2$$

$$\epsilon_0^{-1} = 36\pi \times 10^{18} \text{ dyn cm}^2/\text{C}^2$$

$$r_0 = 2.81 \times 10^{-8} \text{ cm}$$

where r_0 is the interatomic distance between the nearest neighbors.

Inserting these values into above solutions,

$$\lambda = 1.32 \times 10^{-8} \text{ cm}, \qquad W^S = -35 \text{ erg/cm}^2$$

and the displacement at the free surface

$$u_0 = 0.047 \times 10^{-8} \text{ cm}$$

The surface energy density of deformation and polarization W^S predicted by this theory is of the same order of magnitude as compared to the results obtained by Benson [40] and Shuttleworth [41] based on discrete models and other experimental data [42]. The displacement of the particles at the surface was calculated to be of the order of 1–3% of the interparticle distance [40] while the present result shows that $u_0/r_0 = 1.69\%$. The exponential decaying of the displacement from the free surface had been found by Madelung [40] from a discrete analysis for sodium chloride. Furthermore, the surface effects are found to be confined to the first very few atomic layers of the surface. The parameter r_0/λ, which characterizes the rate of decay, is computed to be 2.13 from the present continuum theory. Consequently, the effects at the second and third layers are approximately 12 and 1%, respectively, of those at the first layer.

Functions analogous to the Papkovitch functions of classical elasticity are derived for Mindlin's theory by Schwartz [43]. Particular solutions were obtained for the concentrated force and for the surface energies of deformation and polarization at internal and external spherical surfaces. It was found that for concentrated force, the singularities of the displacement, polarization and Maxwell electric self-field are of order r^{-1}, as in Kelvin's solution in classical elasticity; and the surface energy density W^S for a plane surface cannot be a minimum compared to those of curved surfaces.

The effect of surface curvature and discontinuity on the surface energy density W^S were also studied by Askar, Lee, and Cakmak [44]. For cylindrical and spherical cavities in isotropic dielectrics, the surface energy densities on the free, curved surfaces are found

$$W^S = W_0{}^S \left(1 - n\alpha \frac{\lambda}{R}\right) \tag{52}$$

where $W_0{}^S$ is the surface energy density for a plane surface, λ is defined in Eq. (51), R is the radius of surface curvatures and

$$\alpha = 2\left(b_{44} - \frac{d_{44}{}^2}{c_{44}}\right) \bigg/ \left(b_{11} - \frac{d_{11}{}^2}{c_{11}}\right) \tag{53}$$

In Eq. (52), $n = 1$ is for cylindrical cavity and $n = 2$ is for spherical cavity. According to the numerical values obtained in Ref. [44], α is positive. Hence, the absolute value of the surface energy density for an interior surface is reduced by an amount directly proportional to λ, which is of order of interatomic distance, and inversely proportional to the radius of the curvature of the surface. As long as R is large compared to λ, the effect of surface curvature on the surface energy density W^S is small, as it was pointed out by Schwartz.

In the case of a plane, linear crack in an infinite body without external forces, it was found that stress singularity, induced by the surface energy, is of the type $\epsilon^{-1/2}$ at the crack tip as $\epsilon \to 0$; and that the surface energy density itself remains bounded.

Effects of gradient of polarization on stress-concentration at a cylindrical hole in an elastic dielectric subject to simple tension was investigated by Gou [45]. It was found that the stress-concentration factor at the surface of the hole is higher than the constant value 3 obtained from the classical theory of elasticity, even when the surface energy effect is neglected by setting $b_0 = 0$. It is interesting to note that the stress-concentration factors obtained from the continuum theories including strain gradients are lower than those from the classical theory of elasticity. (See article by Tiersten and Bleustein in this volume.)

Surface waves in elastic dielectrics with polarization gradient for both isotropic and central symmetric, cubic materials have been studied by Chao and Lee [46].

According to the classical theory of electrostatics, the capacitance of a metal-dielectric-metal sandwich is inversely proportional to the thickness of the dielectric, so that a graph of inverse capacitance versus thickness is a straight line through the origin. In a series of experiments with a variety of thin, dielectric films, Mead [47, 48] found that his experimental data fall on straight lines which, if extended to zero thickness, have positive intercepts of inverse capacitance. Mindlin showed that the augmented theory, rather than the classical theory of piezoelectricity, can account for this apparent anomaly [M97]. More detailed discussion of this study is given in the article by Gazis and Gong in this volume.

Diatomic, Elastic, Dielectric Continuum

The continuum theory of elastic dielectrics with polarization gradient corresponds to the long-wave limit of a monotomic lattice theory with shell model. It may be also obtained as the long-wave and low-frequency approximation from the dynamic theory of diatomic lattices, for which the displacements of the two different atoms within the same cell is disregarded and the atomic polarization is ignored. Hence, for the dynamic case, this theory can only accommodate long-wave acoustical vibrations.

Extending the polarization gradient theory to diatomic elastic dielectrics, two sets of mechanical displacements, $u_i^{(\kappa)}$, and two sets of electronic polarizations $P_i^{(\kappa)}$, $(\kappa = 1, 2)$ are introduced at each point of space which is occupied by the two interpenetrating deformable and polarizable continua [M103]. The stored energy density of deformation and polarization is assumed to be

$$W^L = W^L(S_{ij}^{(1)}, S_{ij}^{(2)}, P_i^{(1)}, P_i^{(2)}, P_{i,j}^{(1)}, P_{i,j}^{(2)}, u_i^*, \omega_{ij}^*) \tag{54}$$

where

$$S_{ij}^{(\kappa)} = \tfrac{1}{2}(u_{j,i}^{(\kappa)} + u_{i,j}^{(\kappa)})$$

$$u_i^* = u_i^{(2)} - u_i^{(1)}$$

$$\omega_{ij}^* = \tfrac{1}{2}(u_{j,i}^* - u_{i,j}^*) \tag{55}$$

It is seen that W^L depends not only on strains, polarizations, and polarization gradients, but also on the relative displacement and relative rotation of the two continua.

In a similar manner, an electric enthalpy density is defined by

$$H = W - (\epsilon_0 E_i + P_i^{(1)} + P_i^{(2)} + q_* u_i^*) E_i \tag{56}$$

where the total potential energy density

$$W = W^L + \tfrac{1}{2}\epsilon_0 \phi_{,i}\phi_{,i} \tag{57}$$

and $q_* u_i^*$ corresponds to the ionic, or atomic polarization in addition to the electronic polarizations $P_i^{(\kappa)}$.

The associated kinetic energy density is taken as

$$T = \sum_{\kappa} \tfrac{1}{2}\rho^{(\kappa)}\dot{u}_j^{(\kappa)}\dot{u}_j^{(\kappa)} \tag{58}$$

where $\kappa = 1, 2, j = 1, 2, 3$ and the summation convention applies only to repeated Latin indices.

A further extension of Toupin's variational principle takes the form

$$\delta \int_{t_0}^{t_1} dt \int_{V''} (T - H)dV + \sum_{\kappa} \int_{t_0}^{t_1} dt \int_{V} (f_i^{(\kappa)}\delta u_i^{(\kappa)} + E_i^0 \delta P_i^{(\kappa)} + E_i^0 q_* \delta u_i^*)dV$$

$$+ \sum_{\kappa} \int_{t_0}^{t_1} dt \int_{S} t_i^{(\kappa)} \delta v_i^{(\kappa)} dS = 0 \quad (59)$$

where $f_i^{(\kappa)}$ and $t_i^{(\kappa)}$ are the external body forces and surface-tractions on the two continua.

Inserting Eqs. (56) and (58) into Eq. (59) and applying the same procedure in obtaining Eqs. (37), (38), and (39), one obtains the Euler equations in V:

$$T_{ij,i}^{(\kappa)} + (-1)^{\kappa}(T_{ij,i}^* - T_j^* - q_*\phi_{,j}) + f_j^{(\kappa)} + (-1)^{\kappa} q_* E_j^0 = \rho^{(\kappa)} \ddot{u}_j^{(\kappa)}$$

$$E_j^{(\kappa)} + E_{ij,i}^{(\kappa)} - \phi_{,j} + E_j^0 = 0$$

$$-\epsilon_0 \phi_{,ii} + P_{i,i}^{(1)} + P_{i,i}^{(2)} + q_* u_{i,i}^* = 0 \quad (60)$$

and, in V':

$$\phi_{,ii} = 0 \quad (61)$$

and the boundary conditions on S:

$$n_i[T_{ij}^{(\kappa)} + (-1)^{\kappa} T_{ij}^*] = t_j^{(\kappa)} \quad \text{or specification of } u_i^{(\kappa)}$$

$$n_i E_{ij}^{(\kappa)} = 0 \quad \text{or specification of } P_i^{(\kappa)}$$

$$n_i(-\epsilon_0 [[\phi_{,i}]] + P_i^{(1)} + P_i^{(2)} + q_* u_i^*) = 0 \quad \text{or specification of } \phi \quad (62)$$

where

$$T_{ij}^{(\kappa)} \equiv \frac{\partial W^L}{\partial S_{ij}^{(\kappa)}}, \qquad E_i^{(\kappa)} \equiv \frac{\partial W^L}{\partial P_i^{(\kappa)}}$$

$$E_{ij}^{(\kappa)} \equiv \frac{\partial W^L}{\partial P_{j,i}^{(\kappa)}}$$

$$T_i^* \equiv \frac{\partial W^L}{\partial u_i^*}, \qquad T_{ij}^* \equiv \frac{\partial W^L}{\partial \omega_{ij}^*} \quad (63)$$

For centrosymmetric, cubic crystals, it can be shown that the surface energy density of deformation and polarization is

$$W^S = \tfrac{1}{2} n_i [b^{10} P_i^{(1)} + b^{20} P_i^{(2)} + c^{20} u_i^*]_S \quad (64)$$

where $b^{\kappa 0}$ and $c^{\kappa 0}$ are material coefficients associated to the terms linear in $P_{i,j}^{(\kappa)}$ and $S_{ij}^{(\kappa)}$, respectively, in the energy density function W^L. Comparing Eq. (64) with Eq. (46), one sees that the contribution to surface energy density now includes ionic or atomic polarization in addition to electronic polarizations.

In the same paper [M103], Mindlin showed that, for the one-dimensional, longitudinal case, the equations of motion of this theory are the long-wave limit of the equations of motion of a sodium chloride-type lattice with shell model. Also through the solutions of a plane-wave problem, it was demonstrated that this theory accommodates longitudinal and transverse optical branches as well as acoustical branches in dispersion relations. The simplified field equations of this theory for the case of long wavelengths and high frequencies are shown to be equivalent to the equations, given by Born and Huang [49], from which a dispersion formula for the dielectric constant is obtained.

The connection between the three-dimensional dynamical theory of sodium chloride-type lattice with polarizable atoms and Mindlin's continuum theory of diatomic crystals has been established by Askar and Lee [50].

Coupled Elastic, Electric, and Magnetic Fields

The field Eqs. (37) and (32) of elastic dielectrics, with polarization gradient taken into account, can be written as

$$\begin{aligned} T_{ij,i} + f_j &= \rho \ddot{u}_j \\ E_{ij,i} + \bar{E}_j + E_j &= 0 \\ \epsilon_{ijk} E_{k,j} &= 0 \\ \epsilon_0 E_{i,i} + P_{i,i} &= 0 \end{aligned} \tag{65}$$

where the external electric field E_j^0 is disregarded and ϵ_{ijk} is the unit alternating tensor.

To couple Eqs. (65), which are restricted to quasi-static electric field, to the equations of magnetic field, replace the third equation of Eqs. (65) by

$$\epsilon_{ijk} E_{k,j} + \dot{B}_i = 0 \tag{66}$$

and add the equations

$$\mu_0^{-1} \epsilon_{ijk} B_{k,j} - \epsilon_0 \dot{E}_i - \dot{P}_i = 0 \tag{67}$$

$$B_{i,i} = 0 \tag{68}$$

where B_i is the magnetic flux density and μ_0 is the magnetic permeability of a vacuum.

For convenience, B_i is eliminated from Eqs. (66) and (67) with the result

$$E_{j,ii} - E_{i,ij} = \epsilon_0 \mu_0 \ddot{E}_j + \mu_0 \ddot{P}_j \tag{69}$$

The last of Eqs. (65) and (68) are not independent of Eqs. (66) and (67), and therefore may be disregarded in considering mechanical and electromagnetic waves. Hence, the governing field equations are the first two equations of Eqs. (65) and (67). They are coupled through the constitutive equations (41) in which the b_{ij}^0 term may be disregarded, since it contributes only to the static portion of the solution [46].

These coupled equations have been employed by Mindlin and Toupin [M102] in the investigation of the acoustical and optical activity in alpha quartz. In the case of plane, transverse waves propagating along the X_3-axis, it is found that the presence of acoustical activity depends on the existence of the piezoelectric stress constant e_{123} and the interaction constant d_{3223} between strain and polarization gradient; whereas the presence of optical activity depends only on the existence of the interaction constant g_{132} between polarization and polarization gradient. The acoustical activity was previously accounted for by Toupin [51] on the basis of the first strain gradient theory (see article by Tiersten and Bleustein in this volume). Portigal and Burstein [52] found an equivalent result by assigning dependence of the elastic stiffness on the wave vector. In Ref. [M102], it is shown that both the acoustical and optical activities are accounted for in the polarization gradient theory.

In a similar manner, the governing Eqs. (60) of elastic and electric fields in the diatomic, dielectric continuum can be extended to include the magnetic field. The resulting equations for coupled mechanical and electromagnetic fields in diatomic, ionic crystals are [M104]

$$
\begin{aligned}
T_{ij,i}^{(\kappa)} + (-1)^{\kappa}(T_{ij,i}^* - T_j^* + q_* E_j) + f_j^{(\kappa)} + (-1)^{\kappa} q_* E_j^0 &= \rho^{(\kappa)} \ddot{u}_j^{(\kappa)} \\
E_j^{(\kappa)} + E_{ij,i}^{(\kappa)} + E_j + E_j^0 &= 0 \\
\epsilon_0 E_{i,i} + P_{i,i}^{(1)} + P_{i,i}^{(2)} + q_* u_{i,i}^* &= 0 \\
\epsilon_{ijk} E_{k,j} + \dot{B}_i &= 0 \\
\mu_0^{-1} \epsilon_{ijk} B_{k,j} - \epsilon_0 \dot{E}_i - P_i^{(1)} - P_i^{(2)} - q_* \dot{u}_i^* &= 0 \\
B_{i,i} &= 0 \qquad (70)
\end{aligned}
$$

whereas the constitutive relations obtained from Eqs. (63) remain unchanged.

Equations (70) are the extension of the ones given by Huang [53, 54]. They have been employed by Mindlin [M104] to obtain dispersion relations for longitudinal and transverse waves in the [100] direction. It is found that the longitudinal and transverse optical branches have the same long-wave limit as the electromagnetic field is taken into account. The long-wave limits of optical and electromagnetic branches are independent of constants b_{ijkl}, c_{ijkl}, and d_{ijkl}, i.e., independent of the polarization and polarization gradient. This result from the long-wave limit conforms with Huang's. In addition, the present theory accommodates also the acoustical branches. The long-wave behaviors of the acoustical branches depend only on the mass densities and the elastic stiffnesses and hence are independent of electrical and magnetic properties. However, when the wavelength diminishes from infinity, the displacement and polarization gradients affect the dispersion relation, as does the coupling with the elastic field. Then the results diverge from Huang's.

Conclusions

Although the differential equations of the classical theory of piezoelectricity do not correspond to the long-wave, low-frequency limit of the difference equations of the dynamical theory of crystal lattices with polarizable atoms, it still remains to be adequate and useful in most applications at macroscopic scales.

The mathematical difficulty and complexity encountered in solving problems of plate vibrations of three-dimensional theory often makes it necessary to resort to methods of approximation. Mindlin's general procedure for deducing approximate, two-dimensional plate equations from three-dimensional theory by expanding appropriate fields in series of functions in thickness coordinate in a variational principle is considered most fruitful. It has been applied to deriving two-dimensional equations of elastic, piezoelectric, thermoelastic, and thermopiezoelectric plate vibrations. Most existing closed-form solutions for finite plates were obtained from these two-dimensional theories.

By the addition of the gradient of electronic polarization to the energy density in the classical piezoelectricity, the resulting equations not only converge to the correct long-wave, low-frequency limit of the difference equations of the lattices of polarizable atoms, but also accommodate surface energy of polarization and deformation, anomalous capacitance of thin dielectric films, acoustical activity, and, with the inclusion of the magnetic field, optical activity.

The continuum theory of diatomic, dielectric crystals which is an extension of polarization gradient theory is the correct long-wave limit, not restricted to low frequencies, of the dynamical theory of sodium chloride-type lattices of both mass particles and polarizable atoms. The dispersion relation obtained from it provides both acoustical and optical branches with the correct long-wave behaviors. The surface energy of deformation and polarization includes the contributions of electronic as well as ionic polarizations.

These newly developed theories accommodate observed or observable physical phenomena not accounted for by the classical theory and extend the range of validity of continuum theories toward the domain of lattice theories in conformity. Hence, they represent a successful approach to bridging the gap between the continuum and lattice theories.

References

1. W. P. Mason, *Crystal Physics of Interaction Processes*, Academic Press, New York, 1966, p. 126.
2. H. F. Tiersten, *Linear Piezoelectric Plate Vibrations*, Plenum Press, New York, 1969, p. 37.
3. Reference [2], p. 30.
4. A. E. H. Love, *A Treatise on the Mathematical Theory of Elasticity*, Dover, 1944, p. 176.
5. J. L. Bleustein, A new surface wave in piezoelectric materials, *Appl. Phys. Letters* **13**, 412 (1968).
6. Reference [2], p. 45.
7. Reference [2], p. 50.
8. H. F. Tiersten, Natural Boundary and Initial Conditions from a Modification of Hamilton's Principle, *J. Math. Phys.* **9**, 1445 (1968).
9. E. P. Eer Nisse, Variational method for electroelastic vibration analysis, *IEEE Trans. on Sonics and Ultrasonics* **SU-14**, 153 (1967).
10. R. Holland and E. P. Eer Nisse, *Design of Resonant Piezoelectric Devices*, M.I.T. Press, 1968.
11. H. F. Tiersten, Wave propagation in an infinite piezoelectric plate, *J. Acoust. Soc. Amer.* **35**, 234 (1963).
12. H. F. Tiersten, Thickness Vibrations of piezoelectric plates, *J. Acoust. Soc. Amer.* **35**, 53 (1963).
13. G. A. Coquin and H. F. Tiersten, Analysis of the excitation and detection of piezoelectric surface waves in quartz by means of surface electrodes, *J. Acoust. Soc. Amer.* **41**, 921 (1967).
14. H. F. Tiersten, Elastic surface waves guided by thin films, *J. Appl. Phys.* **40**, 770 (1968).
15. R. M. White, Surface elastic waves, *Proc. IEEE* **58**, 1238 (1970).
16. J. Tasi and G. Herrmann, Thermoelastic dissipation in high frequency vibrations of crystal plates, *J. Acoust. Soc. Amer.* **36**, 100 (1964).
17. M. A. Biot, Thermoelasticity and irreversible thermodynamics, *J. Appl. Phys.* **27**, 240 (1956).
18. P. C. Y. Lee and Z. Nikodem, An approximate theory for high-frequency vibrations of elastic plates, *Int. J. Solids Struct.* **8**, 581 (1972).
19. Z. Nikodem and P. C. Y. Lee, Approximate theory of vibration of crystal plates at high frequencies, *Int. J. Solids Struct.* **10**, 177 (1974).
20. R. P. Jerrard, Vibration of quartz crystal plates, *Quart. Appl. Math.* **18**, 173 (1960).
21. P. C. Y. Lee and W. J. Spencer, Shear-flexure-twist vibrations in rectangular AT-cut quartz plates with partial electrodes, *J. Acoust. Soc. Amer.* **45**, 637 (1969).
22. P. C. Y. Lee and Shoei-sheng Chen, Vibrations of contoured and partially plated, contoured, rectangular, AT-cut quartz plates, *J. Acoust. Soc. Amer.* **46**, 1193 (1969).
23. W. S. Mortley, Circuit giving linear frequency modulation of a quartz crystal oscillator, *Wireless World* **57**, 399 (1951).
24. R. Bechmann, U.S. Patent No. 2,249,933 (22 July 1941).
25. D. W. Haines, Bechmann's number for harmonic overtones of thickness-shear vibrations in rotated-Y-cut quartz plates, *Int. J. Solids Struct.* **5**, 1 (1969).
26. J. L. Bleustein and H. F. Tiersten, Forced thickness-shear vibrations of discontinuously plated piezoelectric plates, *J. Acoust. Soc. Amer.* **43**, 1311 (1967).

27. R. J. Byrne, P. Lloyd and W. J. Spencer, Thickness-shear vibration in rectangular AT-cut quartz plates with partial electrodes, *J. Acoust. Soc. Amer.* **43,** 232 (1968).
28. W. D. Beaver, Analysis of elastically coupled piezoelectric resonators, *J. Acoust. Soc. Amer.* **43,** 972 (1968).
29. R. A. Toupin and B. Bernstein, Sound waves in deformed perfectly elastic materials, *J. Acoust. Soc. Amer.* **33,** 216 (1961).
30. R. N. Thurston and K. Brugger, Third-order elastic constants and the velocity of small amplitude elastic waves in homogeneously stressed media, *Phys. Rev.* **133,** A1604 (1964).
31. R. N. Thurston, H. J. McSkimin, and P. Andreatch, Jr., Third-order elastic coefficients of quartz, *J. Appl. Phys.* **37,** 267 (1966).
32. R. A. Graham, Strain dependence of longitudinal piezoelectric, elastic, and dielectric constants of X-cut quartz, *Phys. Rev.* **B6,** 4779 (1972).
33. A. E. Green and W. Zerna, *Theoretical Elasticity*, Oxford University Press, 1954, p. 114.
34. G. Herrmann and A. E. Armenàkas, Vibrations and stability of plates under initial stress, *J. Eng. Mech. Div., Proc. ASCE* **EM3,** 65 (1960).
35. P. C. Y. Lee, Y. S. Wang and X. Markenscoff, Elastic waves and vibrations in deformed crystal plates, *Proc. 27th Ann. Freq. Cont. Symp.*, U.S. Army Electronics Command, Fort Monmouth, N. J. 1, (1973).
36. R. A. Toupin, The elastic dielectrics, *J. Rat. Mech. Anal.* **5,** 849 (1956).
37. W. Cochran, Theory of lattice vibrations of germanium, *Proc. Roy. Soc.* **A253,** 260 (1959).
38. B. G. Dick, Jr. and A. W. Overhauser, Theory of the dielectric constants of alkali halide crystals, *Phys. Rev.* **112,** 90 (1958).
39. A. Askar, P. C. Y. Lee, and A. S. Cakmak, Lattice-dynamics approach to the theory of elastic dielectrics with polarization gradient, *Phys. Rev.* **B1,** 3525 (1970).
40. G. C. Benson and K. S. Yun, in *The Solid-Gas Interface,* edited by E. A. Flood, M. Dekker, New York, 1967, Vol. 1, p. 203.
41. R. Shuttleworth, The surface energies of inert-gas and ionic crystals, *Proc. Phys. Soc.* (London) **A62,** 167 (1949).
42. M. P. Tosi, IV. Surface energy, in *Solid State Phys.*, Academic Press, New York, 1964, Vol. 16, p. 92.
43. J. Schwartz, Solutions of the equations of equilibrium of elastic dielectrics: Stress functions, concentrated force, surface energy, *Int. J. Solids Struct.* **5,** 1209 (1969).
44. A. Askar, P. C. Y. Lee, and A. S. Cakmak, The effect of surface curvature and discontinuity on the surface energy density and other induced fields in elastic dielectrics with polarization gradient, *Int. J. Solids Struct.* **7,** 523 (1971).
45. P. F. Gou, Effects of gradient of polarization on stress-concentration at a cylindrical hole in an elastic dielectric, *Int. J. Solids Struct.* **7,** 1467 (1971).
46. T. Chao and P. C. Y. Lee, Surface waves in elastic dielectrics with polarization gradient, *Research Report No.* 70—*Mech.* **3,** Dept. of Civil and Geological Engineering, Princeton University, 1970.
47. C. A. Mean, Electron transport mechanisms in thin insulating films, *Phys. Rev.* **128,** 2088 (1962).
48. C. A. Mead, Electron transport in thin insulating films, *Proc. Int. Symp. on Basic Problems in Thin Film Physics*, Vandenhoeck and Ruprecht, Göttingen, 1966, p. 674. (Also M. McColl and C. A. Mead, Electron current through thin mica films, *Trans. Metall. Soc. AIME* **233,** 502 (1965).)

49. M. Born and K. Huang, *Dynamic Theory of Crystal Lattices*, Oxford University Press, 1956, p. 82.
50. A. Askar and P. C. Y. Lee, Lattice dynamics approach to the theory of diatomic dielectrics, to be published in *Phys. Rev.*, **B15**.
51. R. A. Toupin, Elastic materials with couple-stresses, *Arch. Rat. Mech. Anal.* **11,** 385 (1962).
52. D. L. Portigal and E. Burstein, Acoustical activity and other first order spatial dispersion effects in crystals, *Phys. Rev.* **170,** 673 (1968).
53. Reference [48], Chap. 11.
54. K. Huang, On the interaction between the radiation field and ionic crystals, *Proc. Roy. Soc.* **A208,** 352 (1951).

Lattice Theories and Continuum Mechanics

DENOS C. GAZIS

IBM Thomas J. Watson Research Center, Yorktown Heights, New York 10598

and

CHUNG GONG

Henry Krumb School of Mines, Columbia University, New York, New York 10027

1 Introduction

The early development of lattice and continuum theories paralleled the development of relevant mathematical methodologies. Lattice theory was developed first, starting with Newton's attempt [1] to evaluate the velocity of sound on the basis of a one-dimensional lattice model. Continuum models appeared much later, following the development of the theory of partial differential equations, and they supplanted lattice models because they were more powerful than lattice models in matching the observed phenomena of pre-atomic science. Lattice theory was revived in relatively recent years, getting impetus from the Born–von Karman [2] work on generalized lattice models, and continuing to this day. The revival was stimulated by the need to explain observations involving atomic interactions, which could not be conveniently described within the framework of a continuum theory. For example, the frequency density distribution (frequency spectrum) of a crystal lattice domain has been measured by neutron scattering experiments [3] and has been found to have the character of Fig. 1. A theoretical computation of this spectrum by Debye [4], based on a continuum theory, yields only the portion of the spectrum near the origin which is proportional to the second power of the frequency. However, recent computations using lattice models [5] reproduce many of the salient features of the entire frequency spectrum.

Mindlin's work over the past two decades included many contributions made with an eye toward bridging the gap between continuum and lattice theories, or, as he himself sometimes expressed it, "marrying the two theories."† His work on granular media and on micro-macroelasticity, described in other articles of this volume, brings continuum theory close to lattice theory. More recently, he undertook a methodical investigation of some basic physical prob-

† This expression led one of the authors of this article to say ten years ago, with Mindlin's cheerful concurrence, that this was likely to be a long engagement.

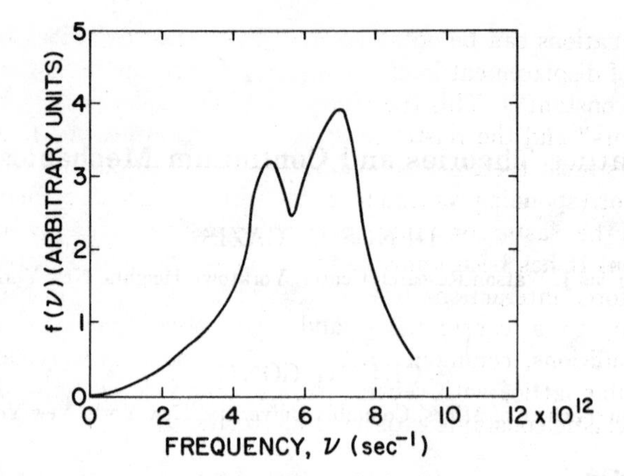

Fig. 1. Frequency spectrum of vanadium, obtained by neutron scattering experiments [3].

lems using both a lattice and a continuum theory. The juxtaposition of the two results, coupled with physical observations, gives insight into the limits of applicability and the missing ingredients of both theories.

In this article, we give a brief review of Mindlin's work which involves a comparison of lattice and continuum theories. A more detailed review of the entire area between lattice and continuum theories has been given recently by Mindlin himself [M105]†. The organization of the article is as follows: Section 2 gives the basic continuum and lattice equations for a cubic crystal. Section 3 discusses a problem in statics; the torsion of a rectangular bar. Section 4 contains a discussion of simple modes of vibration of cubic crystal plates and bars. Section 5 gives a comparison of diatomic lattice models with continuum models of compound media. Section 6 gives a discussion of shell models of dielectric media and a comparison of their results with those of continuum models of elastic dielectrics.

2 Basic Equations

The differential equations of harmonic motion of the atoms in a lattice domain can be obtained by assuming a quadratic potential function of the relative displacements of the atoms. The "force constants" multiplying each quadratic term of this function describe the interatomic interactions and must be chosen so as to insure that the function is positive definite and invariant under translation and rotation of the coordinate system. The corresponding

† Numbers preceded by the letter M in brackets refer to the Publications of R. D. Mindlin listed after the Preface to this volume.

continuum equations can be obtained from the lattice equations by assuming wavelengths of displacement long in comparison with the interatomic distance (the "lattice constant"). This transition establishes relationships between the "force constants" and the elastic constants of the continuum. The number of force constants must therefore be at least equal to the number of elastic constants in the corresponding continuum if one wishes to avoid restrictive relationships between the elastic constants. This, however, is a necessary but not sufficient condition. It has been known since Cauchy [6] that a lattice involving only central force interactions between atoms corresponds, in the long-wave approximation, to a continuum with elastic constants satisfying certain restrictive conditions, commonly referred to as the Cauchy conditions. For example, a cubic lattice with central force interactions corresponds to a continuum with elastic constants satisfying the relationship

$$C_{12} = C_{44} \tag{2.1}$$

Mindlin used in his work a non-central force lattice model such as that suggested by Gazis, Herman and Wallis [7] for cubic lattices. This model includes central force interactions between nearest neighbors and next nearest neighbors and "angular stiffness" interactions involving the change of angle of three non-collinear atoms. The simplest such model is a simple cubic monatomic model. Although there are no known elements which form a simple cubic lattice, this model is a useful one for investigating the effect of varying the ratio of the wavelength of deformation to the atomic distance, and has been used by Mindlin for this purpose. Other more complex lattices, such as a body-centered cubic (bcc) lattice and a face-centered cubic (fcc) lattice have been considered by his most recent students.

We shall discuss here in some detail the derivation of the lattice equations for a simple cubic lattice. Let $u_i^{l,m,n}$, $(i = 1, 2, 3)$ be the rectangular components of displacement of an atom located at a lattice site identified by the indices l, m, n. The equations of motion, given in Ref. [7] are

$$\begin{aligned} M\ddot{u}_1^{l,m,n} = {} & \alpha(u_1^{l+1,m,n} + u_1^{l-1,m,n} - 2u_1^{l,m,n}) \\ & + \beta(u_1^{l+1,m+1,n} + u_1^{l+1,m-1,n} + u_1^{l-1,m+1,n} + u_1^{l-1,m-1,n} \\ & + u_1^{l+1,m,n+1} + u_1^{l+1,m,n-1} + u_1^{l-1,m,n+1} + u_1^{l-1,m,n-1} - 8u_1^{l,m,n}) \\ & + (\beta + \gamma)(u_2^{l+1,m+1,n} + u_2^{l-1,m-1,n} - u_2^{l-1,m+1,n} - u_2^{l+1,m-1,n} \\ & + u_3^{l+1,m,n+1} + u_3^{l-1,m,n-1} - u_3^{l-1,m,n+1} - u_3^{l+1,m,n-1}) \\ & + 4\gamma(u_1^{l,m+1,n} + u_1^{l,m-1,n} + u_1^{l,m,n+1} + u_1^{l,m,n-1} - 4u_1^{l,m,n}) \end{aligned} \tag{2.2}$$

where M is the mass of an atom, α and β the nearest and next nearest neighbor central force constants, and γ a constant describing the resistance to the change of the angle of three atoms which form a right angle in their equilibrium position. Two other equations analogous to Eq. (2.2) are obtained by a cyclical permutation of the indices.

If we are interested in investigating problems associated with a bounded domain, we must also establish boundary conditions. This is done by assuming a fictitious extension of the lattice domain and imposing on the displacements of the fictitious atoms conditions such that the combined force exerted by these atoms on each atom of the real domain be zero. We will be discussing the motion of plates bounded between two planes normal to the x_1-axis, the planes defined by $l = \pm L$. The boundary conditions for such plates are

$$\begin{aligned}
&\pm\alpha(u_1^{\pm(L+1),m,n} - u_1^{\pm L,m,n})\\
&\quad \pm \beta(u_1^{\pm(L+1),m+1,n} + u_1^{\pm(L+1),m-1,n} + u_1^{\pm(L+1),m,n+1} + u_1^{\pm(L+1),m,n-1}\\
&\quad - 4u_1^{\pm L,m,n}) + \beta(u_2^{\pm(L+1),m+1,n} - u_2^{\pm(L+1),m-1,n} + u_3^{\pm(L+1),m,n+1}\\
&\quad - u_3^{\pm(L+1),m,n-1}) \pm 2\gamma(u_1^{\pm L,m+1,n} + u_1^{\pm L,m-1,n} + u_1^{\pm L,m,n+1} + u_1^{\pm L,m,n-1}\\
&\quad - 4u_1^{\pm L,m,n}) + \gamma(u_2^{\pm(L+1),m+1,n} - u_2^{\pm(L+1),m-1,n} - u_2^{\pm L,m+1,n} + u_2^{\pm L,m-1,n})\\
&\quad +\gamma(u_3^{\pm(L+1),m,n+1} - u_3^{\pm(L+1),m,n-1} - u_3^{\pm L,m,n+1} + u_3^{\pm L,m,n-1}) = 0\\
&\pm\beta(u_2^{\pm(L+1),m+1,n} + u_2^{\pm(L+1),m-1,n} - 2u_2^{\pm L,m,n})\\
&\quad + \beta(u_1^{\pm(L+1),m+1,n} - u_1^{\pm(L+1),m-1,n}) \pm 4\gamma(u_2^{\pm(L+1),m,n} - u_2^{\pm L,m,n})\\
&\quad + \gamma(u_1^{\pm L,m+1,n} - u_1^{\pm L,m-1,n} + u_1^{\pm(L+1),m+1,n} - u_1^{\pm(L+1),m-1,n}) = 0\\
&\pm\beta(u_3^{\pm(L+1),m,n+1} + u_3^{\pm(L+1),m,n-1} - 2u_3^{\pm L,m,n})\\
&\quad + \beta(u_1^{\pm(L+1),m,n+1} - u_1^{\pm(L+1),m,n-1}) \pm 4\gamma(u_3^{\pm(L+1),m,n} - u_3^{\pm L,m,n})\\
&\quad + \gamma(u_1^{\pm L,m,n+1} - u_1^{\pm L,m,n-1} + u_1^{\pm(L+1),m,n+1} - u_1^{\pm(L+1),m,n-1}) = 0 \qquad (2.3)
\end{aligned}$$

Mindlin pointed out [M6] that the relationship between continuum and lattice theories can be conveniently studied by first rewriting the lattice equations using the difference operators

$$\Delta_1^+ u_i^{l,m,n} = \frac{u_i^{l+1,m,n} - u_i^{l,m,n}}{a}$$

$$\Delta_1^- u_i^{l,m,n} = \frac{u_i^{l,m,n} - u_i^{l-1,m,n}}{a}$$

$$\Delta_1^2 u_i^{l,m,n} = \Delta_1^+\Delta_1^- u_i^{l,m,n} = \frac{u_i^{l+1,m,n} + u_i^{l-1,m,n} - 2u_i^{l,m,n}}{a^2}$$

$$\Delta_1\Delta_2 u_i^{l,m,n} = \tfrac{1}{4}(\Delta_1^+\Delta_2^+ + \Delta_1^-\Delta_2^- + \Delta_1^+\Delta_2^- + \Delta_1^-\Delta_2^+)u_i^{l,m,n} \qquad (2.4)$$

and analogous definitions for the forward, backward, second central, and cross differences in the other two coordinate directions. In Eq. (2.4), a is the lattice constant, i.e., the distance between two nearest neighbors.

Using the difference operators (2.4) we can write Eq. (2.2) in the form

$$\rho \ddot{u}_1{}^{l,m,n} = [C_{11}\Delta_1{}^2 + C_{44}(\Delta_2{}^2 + \Delta_3{}^2) + \tfrac{1}{2}a^2 C_{12}\Delta_1{}^2(\Delta_2{}^2 + \Delta_3{}^2)]u_1{}^{l,m,n} + (C_{12} + C_{44})(\Delta_1\Delta_2 u_2{}^{l,m,n} + \Delta_1\Delta_3 u_3{}^{l,m,n}) \quad (2.5)$$

where

$$\rho = \frac{M}{a^3}$$

$$C_{11} = \frac{\alpha + 4\beta}{a}$$

$$C_{12} = \frac{2\beta}{a}$$

$$C_{44} = \frac{2(\beta + 2\gamma)}{a} \quad (2.6)$$

If we now express Eq. (2.4) in Taylor series, then the difference operators on the $u^{l,m,n}$ can be expressed in terms of the partial derivatives of continuous displacement functions $u_i(x_i)$, namely

$$\Delta_i{}^2 u_j{}^{l,m,n} = \left(1 + \tfrac{1}{12}a^2 \frac{\partial^2}{\partial x_i{}^2} + \cdots\right)\frac{\partial^2 u_j}{\partial x_i{}^2}$$

$$\Delta_i\Delta_j u_k{}^{l,m,n} = \left(1 + \tfrac{1}{6}a^2 \frac{\partial^2}{\partial x_i{}^2} + \tfrac{1}{6}a^2 \frac{\partial^2}{\partial x_j{}^2} + \cdots\right)\frac{\partial^2 u_k}{\partial x_i \partial x_j}$$

$$\Delta_i{}^2\Delta_j{}^2 u_k{}^{l,m,n} = \left(1 + \tfrac{1}{12}a^2 \frac{\partial^2}{\partial x_i{}^2} + \tfrac{1}{12}a^2 \frac{\partial^2}{\partial x_j{}^2} + \cdots\right)\frac{\partial^4 u_k}{\partial x_i{}^2 \partial x_j{}^2} \quad (2.7)$$

Now, if we assume that the wavelengths of deformation are long in comparison with the lattice constant, we may neglect all but the second derivatives in Eq. (2.7). Then, combining these reduced expressions with Eq. (2.5) we obtain the equation

$$\rho \ddot{u}_1 = C_{11}\frac{\partial^2 u_1}{\partial x_1{}^2} + C_{44}\left(\frac{\partial^2 u_1}{\partial x_2{}^2} + \frac{\partial^2 u_1}{\partial x_3{}^2}\right) + (C_{12} + C_{44})\left(\frac{\partial^2 u_2}{\partial x_1 \partial x_2} + \frac{\partial^2 u_3}{\partial x_1 \partial x_3}\right) \quad (2.8)$$

which is the equation of dynamic equilibrium in the x_1-direction for a cubic continuum with elastic constants C_{11}, C_{12}, and C_{44}. Similarly, the lowest order of a Taylor series expansion of the boundary conditions is

$$\left.\begin{aligned} C_{11}\frac{\partial u_1}{\partial x_1} + C_{12}\left(\frac{\partial u_2}{\partial x_2} + \frac{\partial u_3}{\partial x_3}\right) &= 0 \\ \frac{\partial u_1}{\partial x_2} + \frac{\partial u_2}{\partial x_1} &= 0 \\ \frac{\partial u_1}{\partial x_3} + \frac{\partial u_3}{\partial x_1} &= 0 \end{aligned}\right\} x_1 = \pm La \tag{2.9}$$

which are the traction-free boundary conditions for a cubic crystal plate bounded by the planes $\pm La$.

Equations (2.6) show how the force constants α, β, γ may be chosen to match the elastic constants C_{ij}, which are obtained, for example, by means of ultrasonic experiments. A reasonable choice of α, β, γ is the one that yields the correct C_{ij} if used in Eqs. (2.6), i.e.,

$$\begin{aligned} \alpha &= a(C_{11} - 2C_{12}) \\ \beta &= \frac{aC_{12}}{2} \\ \gamma &= \frac{a(C_{44} - C_{12})}{4} \end{aligned} \tag{2.10}$$

It may be pointed out that this is not the only possible way of selecting the force constants. One may, for example, select them in order to match an appropriate set of points of a frequency spectrum (frequency density function) or dispersion curves obtained by neutron diffraction experiments. Matching the C_{ij} according to Eqs. (2.10) corresponds to matching the initial slopes of the dispersion curves for plane waves in an infinite domain, and it yields a unique set of force constants because three relationships are available for the determination of three force constants. If we introduce more than three interactions, associated with more than three independent force constants, we can still obtain three relationships such as Eqs. (2.10) by the long-wave approximation (Taylor series expansion). Thus there will be an indeterminacy in the force constants, which may be removed either by choosing all but three of them arbitrarily, or by obtaining additional relationships through matching of additional experimental data, e.g., frequencies at the Brillouin zone boundaries, as was done by two of Mindlin's most recent students, Brady [8] for fcc lattices, and Gong [9] for bcc lattices.

3 Torsional Equilibrium of a Rectangular Bar

Consider a bar bounded by free faces at $l = \pm L$ and $m = \pm M$, and at equilibrium under the action of a torsional moment about the axis x_3 which is assumed to produce a twist τ per unit length. Mindlin [M96] obtained the displacements and warping function for this problem by using a procedure analogous to the Saint Vernant procedure [10] for the analogous continuum problem. Let

$$
\begin{aligned}
u_1^{l,m,n} &= -\tau a^2 mn \\
u_2^{l,m,n} &= \tau a^2 ln \\
u_3^{l,m,n} &= \tau a^2(lm + A \sin l\theta \sinh m\phi), \qquad 0 < \theta < \pi
\end{aligned} \tag{3.1}
$$

The equations of equilibrium are of the type (2.2) or (2.5) with the left-hand side set equal to zero. The equations expressing equilibrium in the x_1- and x_2-directions are identically satisfied by the displacement (3.1). Equilibrium in the x_3-direction is satisfied if

$$\cosh \phi = 2 - \cos \theta \tag{3.2}$$

The boundary conditions (2.3), and analogous conditions for the boundaries $m = \pm M$ reduce, when used in conjunction with Eqs. (3.1), to

$$
\begin{aligned}
&\pm 2(u_3^{\pm(l+1),m,n} - u_3^{\pm l,m,n}) + u_1^{l,m,n+1} - u_1^{l,m,n-1} = 0 \quad \text{on } l = \pm L \\
&\pm 2(u_3^{l,\pm(m+1),n} - u_3^{l,\pm m,n}) + u_2^{l,m,n+1} - u_2^{l,m,n-1} = 0 \quad \text{on } m = \pm M
\end{aligned} \tag{3.3}
$$

Substituting Eqs. (3.1) in the first of Eqs. (3.3) we find

$$\theta = \theta_p = \frac{(2p-1)\pi}{2L+1}, \qquad p = 1, 2, \ldots, L \tag{3.4}$$

Hence, the most general form for the third of Eqs. (3.1) is

$$u_3^{l,m,n} = \tau a^2\left(lm + \sum_{p=1}^{L} A_p \sin l\theta_p \sinh m\phi_p\right) \tag{3.5}$$

which, when substituted into the second of Eqs. (3.3) yields a system of L linear algebraic equations on the coefficients A_p, namely

$$\sum_{p=1}^{L} A_p \sin l\theta_p \sinh \tfrac{1}{2}\phi_p \cosh (M + \tfrac{1}{2})\phi_p = -l, \qquad l = 1, 2, \ldots, L \tag{3.6}$$

where θ_p and ϕ_p satisfy Eq. (3.2).

The system (3.6) may be solved for A_p by a procedure analogous to that used for determining the coefficients of a Fourier series. Multiplying both sides of Eq. (3.6) by $\sin l\theta_q$ and summing over l from $l = 1$ to $l = L$, we obtain a relationship which is reducible to a single equation for each A_p by employing the properties of sums of trigonometric functions and Eq. (3.4). Specifically, we use the identity [11]

$$\sum_{l=1}^{L} l \sin l\theta_p = \frac{\sin L\theta_p}{4 \sin^2 (\frac{1}{2}\theta_p)} - \frac{L \cos (L + \frac{1}{2})\theta_p}{2 \sin \frac{1}{2}\theta_p} \tag{3.7}$$

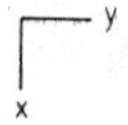

0	0	0	
0	0	0	L = 1, M = 1
0	0	0	

y, x

$\frac{4}{5}$	$\frac{3}{5}$	0	$-\frac{3}{5}$	$-\frac{4}{5}$	
0	0	0	0	0	L = 1, M = 2
$-\frac{4}{5}$	$-\frac{3}{5}$	0	$\frac{3}{5}$	$\frac{4}{5}$	

$\frac{23}{13}$	$\frac{20}{13}$	$\frac{11}{13}$	0	$-\frac{11}{13}$	$-\frac{20}{13}$	$-\frac{23}{13}$	
0	0	0	0	0	0	0	L = 1, M = 3
$-\frac{23}{13}$	$-\frac{20}{13}$	$-\frac{11}{13}$	0	$\frac{11}{13}$	$\frac{20}{13}$	$\frac{23}{13}$	

$\frac{47}{17}$	$\frac{43}{17}$	$\frac{31}{17}$	$\frac{16}{17}$	0	$-\frac{16}{17}$	$-\frac{31}{17}$	$-\frac{43}{17}$	$-\frac{47}{17}$	
0	0	0	0	0	0	0	0	0	L = 1, M = 4
$-\frac{47}{17}$	$-\frac{43}{17}$	$-\frac{31}{17}$	$-\frac{16}{17}$	0	$\frac{16}{17}$	$\frac{31}{17}$	$\frac{43}{17}$	$\frac{47}{17}$	

Fig. 2. Values of the warping function at the lattice points, for

and Eq. (3.4) to obtain

$$A_p = \frac{2L \sin \frac{1}{2}\theta_p \cos (L + \frac{1}{2})\theta_p - \sin L\theta_p}{(2L + 1) \sinh^3 \frac{1}{2}\phi_p \cosh (M + \frac{1}{2})\phi_p} \tag{3.8}$$

which, when substituted in Eq. (3.5) defines the warping function $u_3{}^{l,m,n}$ and completes the solution to the problem.

The corresponding solution for a continuum domain in the form of a rectangular bar [10] is given by the displacements

$$u_1(x_i) = -\tau x_2 x_3$$

$$u_2(x_i) = \tau x_1 x_3$$

$$u_3(x_i) = \tau \left[x_1 x_2 + \frac{1}{2}\left(\frac{4}{\pi}\right)^3 b^2 \sum_{p=1}^{\infty} A_p \sin \frac{(2p-1)\pi x_1}{2b} \sinh \frac{(2p-1)\pi x_2}{2b} \right] \tag{3.9}$$

$$
\begin{array}{ccccc}
0 & \frac{1}{3} & 0 & -\frac{1}{3} & 0 \\
-\frac{1}{3} & 0 & 0 & 0 & \frac{1}{3} \\
0 & 0 & 0 & 0 & 0 \\
\frac{1}{3} & 0 & 0 & 0 & -\frac{1}{3} \\
0 & -\frac{1}{3} & 0 & \frac{1}{3} & 0
\end{array}
\qquad L = 2,\ M = 2
$$

y

x

$$
\begin{array}{ccccccc}
\frac{448}{281} & \frac{498}{281} & \frac{300}{281} & 0 & -\frac{300}{281} & -\frac{498}{281} & -\frac{448}{281} \\
\frac{117}{281} & \frac{184}{281} & \frac{121}{281} & 0 & -\frac{121}{281} & -\frac{184}{281} & -\frac{117}{281} \\
0 & 0 & 0 & 0 & 0 & 0 & 0 \\
-\frac{117}{281} & -\frac{184}{281} & -\frac{121}{281} & 0 & \frac{121}{281} & \frac{184}{281} & \frac{117}{281} \\
-\frac{448}{281} & -\frac{498}{281} & -\frac{300}{281} & 0 & \frac{300}{281} & \frac{498}{281} & \frac{448}{281}
\end{array}
\qquad L = 2,\ M = 3
$$

$$
\begin{array}{ccccccccc}
\frac{7788}{2245} & \frac{8080}{2245} & \frac{6242}{2245} & \frac{3342}{2245} & 0 & -\frac{3342}{2245} & -\frac{6242}{2245} & -\frac{8080}{2245} & -\frac{7788}{2245} \\
\frac{3006}{2245} & \frac{3475}{2245} & \frac{2814}{2245} & \frac{1539}{2245} & 0 & -\frac{1539}{2245} & -\frac{2814}{2245} & -\frac{3475}{2245} & -\frac{3006}{2245} \\
0 & 0 & 0 & 0 & 0 & 0 & 0 & 0 & 0 \\
-\frac{3006}{2245} & -\frac{3475}{2245} & -\frac{2814}{2245} & -\frac{1539}{2245} & 0 & \frac{1539}{2245} & \frac{2814}{2245} & \frac{3475}{2245} & \frac{3006}{2245} \\
-\frac{7788}{2245} & -\frac{8080}{2245} & -\frac{6242}{2245} & -\frac{3342}{2245} & 0 & \frac{3342}{2245} & \frac{6242}{2245} & \frac{8080}{2245} & \frac{7788}{2245}
\end{array}
\qquad L = 2,\ M = 4
$$

various cross-sectional dimensions of a lattice crystal bar under torsion.

where

$$A_p = \frac{(-1)^p}{(2p-1)^3 \cosh[(2p-1)\pi c/2b]} \tag{3.10}$$

and

$$b = La, \qquad c = Ma \tag{3.11}$$

It will be observed that the lattice equations are equivalent to a finite difference approximation of the continuum equations. Therefore, the lattice solution may be expected to match the continuum one when there is a reasonably large number of lattice points across both dimensions of the cross section. A detailed comparison of the lattice and continuum solutions is shown in Figs. 2 and 3. Figure 2 gives the numerical value of the warping function, $u_3^{l,m,n}$, for various combinations of L and M, while Fig. 3 shows the form of contours of constant

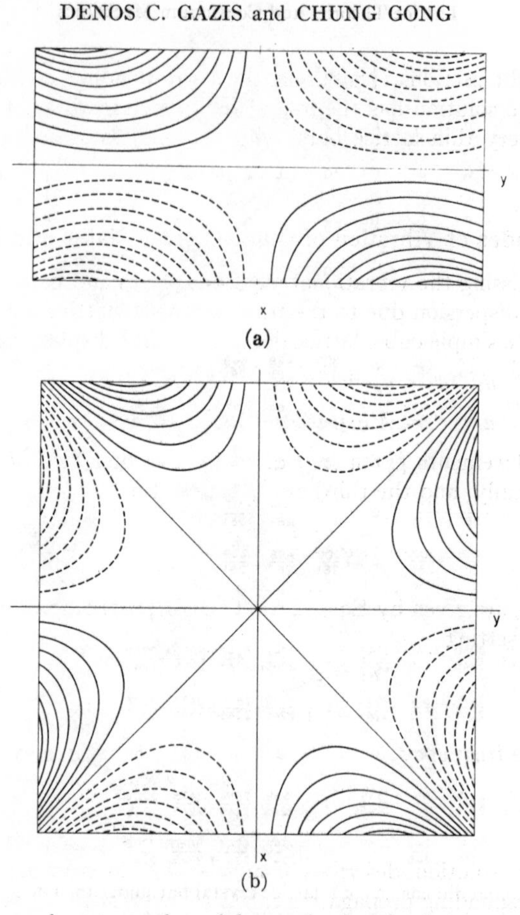

Fig. 3. Contours of constant values of the warping function for a rectangular and a square lattice crystal bar under torsion.

values of the warping function $u_3(x_i)$ for two distinct cases, a rectangular bar with substantially unequal cross-sectional dimensions, and a bar with a square cross section. It is seen that for $L = M = 1$ the lattice solution has no resemblance to the cross section. As L and M increase, all the qualitative features of the continuum (Saint Vernant) solution begin to appear. For example, for $L = 2$, $M = 3$ and $L = 2$, $M = 4$, the character of the warping function $u_3{}^{l,m,n}$ is qualitatively identical to that of the continuum solution; there is a maximum and a minimum of the warping function along the long side of the rectangular cross section, both in the lattice and in the continuum solution. Also, for $L = 2$, $M = 2$ the square cross section is divided into eight sections (instead of four for the rectangular cross section), with $u_3{}^{l,m,n}$ alternating sign around the eight sections, just as in the corresponding continuum solution. Thus the lattice solution has the major qualitative features of the continuum one even for very thin lattice bars, with a thickness as small as four lattice constants.

4 Simple Modes of Vibration of Cubic Crystal Plates and Bars

Before discussing the vibrational modes of plates and bars, let us introduce the concept of dispersion due to the periodicity of a lattice domain.

In an infinite simple cubic lattice domain, assume displacements of the form

$$u_1{}^{l,m,n} = u_2{}^{l,m,n} = 0$$
$$u_3{}^{l,m,n} = A \exp i(\omega t + \xi na), \qquad 0 \leq \xi a \leq \pi \tag{4.1}$$

With the displacements given in Eq. (4.1), the first two of Eqs. (2.2) are satisfied identically, and the third one is satisfied if

$$\omega = \frac{2}{a}\sqrt{\frac{C_{11}}{\rho}} \sin\frac{\xi a}{2} \tag{4.2}$$

where ρ and C_{11} are given by Eqs. (2.6). The corresponding continuum solution is obtained by setting

$$u_1 = u_2 = 0$$
$$u_3 = A \exp i(\omega t + \xi x_3) \tag{4.3}$$

which yield the frequency

$$\omega = \xi\sqrt{\frac{C_{11}}{\rho}} \tag{4.4}$$

The continuum solution describes nondispersive extentional modes of vibration. The corresponding propagating motion travels with phase velocity

$$V_c = \frac{\omega}{\xi} = \sqrt{\frac{C_{11}}{\rho}} \tag{4.5}$$

regardless of the frequency ω. On the other hand, the lattice solution yields a phase velocity

$$V_e = \frac{\omega}{\xi} = \frac{2}{\xi a}\sqrt{\frac{C_{11}}{\rho}} \sin\frac{\xi a}{2} \tag{4.6}$$

which varies with the frequency ω (or the wave number ξ). However, in the long-wave approximation, i.e., when $\xi a \ll 1$, the velocity V_e tends to the value V_c. As the wavelength decreases and becomes comparable to the lattice constant a, the phase velocity decreases, tending to the value $(2/\pi)V_c$ as $\xi a \to \pi$ (when the wavelength $\lambda = 2\pi/\xi$ is equal to twice the lattice constant). The group velocity $d\omega/d\xi$ tends to zero as $\xi a \to \pi$.

Similarly, shear waves correspond to displacements and frequency given by

$$u_1^{l,m,n} = u_2^{l,m,n} = 0$$

$$u_3^{l,m,n} = A \exp i(\omega\tau + \xi a l)$$

$$\omega = \frac{2}{a}\sqrt{\frac{C_{44}}{\rho}} \sin\frac{\xi a}{2} \tag{4.7}$$

for an infinite lattice domain, or

$$u_1 = u_2 = 0$$

$$u_3 = A \exp i(\omega t + \xi x_1)$$

$$\omega = \xi\sqrt{\frac{C_{11}}{\rho}} \tag{4.8}$$

for a continuum domain. The lattice solution matches the continuum one in the long-wave approximation, provided that C_{44} and ρ satisfy Eqs. (2.6), but has a dispersion character which manifests itself by a decrease in group velocity from the shear velocity $\sqrt{C_{44}/\rho}$ to zero, as the wavelength decreases from infinity to $2a$.

Let us now consider some more complex modes, such as those arising in plates and bars of a cubic crystal, following Mindlin [M94].

4.1 *Thickness-Shear Vibrations of a Plate*

Consider a plate bounded by $x_1 = \pm La$ and assume displacements of the form

$$u_1^{l,m,n} = u_2^{l,m,n} = 0$$

$$u_3^{l,m,n} = (A_1 \cos \xi l a + A_2 \sin \xi l a) e^{i\omega\tau}, \qquad 0 \leq \xi a \leq \pi \tag{4.9}$$

The displacements (4.9) satisfy the first two of Eqs. (2.2) identically, and they also satisfy the third one if

$$\rho a^2 \omega^2 = 4C_{44} \sin^2 \frac{\xi a}{2} \tag{4.10}$$

Furthermore, substituting Eqs. (4.9) in the boundary conditions (2.3) we find that the first two are satisfied identically and the third one is satisfied if

$$\xi a = \frac{p\pi}{2L+1}, \qquad p = 0, 1, 2 \ldots < 2L + 1 \tag{4.11}$$

when even and odd values of p correspond to symmetric modes ($A_2 = 0$) and antisymmetric modes ($A_1 = 0$), respectively. From Eqs. (4.10) and (4.11) we find that the frequencies are given by

$$\omega = \frac{2}{a}\sqrt{\frac{C_{44}}{\rho}} \sin \frac{p\pi}{2(2L+1)}, \qquad p = 0, 1, 2 \ldots < 2L + 1 \qquad (4.12)$$

and hence, the displacements are given by

$$u_3{}^{l,m,n} = \left[A_1\mu \cos \frac{p\pi l}{2L+1} + A_2(1-\mu) \sin \frac{p\pi l}{2L+1}\right] e^{i\omega t} \qquad (4.13)$$

where the parameter μ is

$$\mu = \begin{cases} 1 & \text{if } p \text{ is even} \\ 0 & \text{if } p \text{ is odd} \end{cases} \qquad (4.14)$$

To compare the lattice solution with the analogous continuum solution, consider a plate bounded by the planes $x_1 = \pm h$, where

$$2h = (2L+1)a \qquad (4.15)$$

i.e., each lattice layer is replaced by a continuum layer of thickness a. Solving the corresponding continuum Eqs. (2.8) subject to boundary conditions (2.9), for thickness-shear motion, we obtain

$$u_1 = u_2 = 0$$

$$u_3 = \left[A_1\mu \cos \frac{p\pi x_1}{2h} + A_2(1-\mu) \sin \frac{p\pi x_1}{2h}\right] e^{i\omega t}$$

$$p = 0, 1, 2 \ldots \infty \qquad (4.16)$$

where the frequencies are given by

$$\omega = \left(\frac{p\pi}{2h}\right)\sqrt{\frac{C_{44}}{\rho}}, \qquad p = 0, 1, 2, \ldots \infty \qquad (4.17)$$

We first observe that the continuum solution yields an infinity of thickness-shear modes, whereas the lattice solution yields no more than $2L + 1$ of them. This is because we may have no more than one node between lattice points, for a mode within the first Brillouin zone. We also see that in the long-wave, low-frequency approximation, i.e., when $p \ll 2L + 1$, the lattice solution becomes identical with the continuum one.

If the frequency for $p = 1$ is taken as a reference frequency, the normalized frequencies of the lattice plate are

$$\Omega = \frac{\omega}{\omega_1} = \frac{2(2L+1)}{\pi} \sin \frac{p\pi}{2(2L+1)}, \qquad p = 0, 1, 2 \ldots < 2L + 1 \qquad (4.18)$$

where

$$\omega_1 = \frac{\pi}{a(2L+1)}\sqrt{\frac{C_{44}}{\rho}} = \frac{\pi}{2h}\sqrt{\frac{C_{44}}{\rho}} \tag{4.19}$$

For the continuum solution, the normalized frequencies are simply the integers from 0 to infinity.

Table 1, from Ref. [M94], gives the normalized frequencies for lattice plates of various thicknesses, from 0 to 15 atomic layers thick (i.e., for L varying from 0 to 7). It may be seen that the lattice frequencies are always smaller than the corresponding continuum frequencies, but they get closer and closer to those frequencies as the number of layers increases. This behavior may be understood if we observe that the thickness-shear frequencies correspond to a set of points on the dispersion curve for shear waves described by Eq. (4.10). For example, Fig. 4 shows this set of points for $L = 7$. It is seen that they are evenly spaced along the abscissa (wave number) axis, just as they are for the continuum modes which lie on the tangent to zero wave number to the dispersion curve. Thus all the lattice frequencies are below the corresponding continuum ones, but the ones near the origin are approximately equal; also, as the number of layers increases, more and more lattice frequencies are close to the origin, and hence close to those of the continuum solution. Another comparison of the lattice and continuum solutions is shown in Fig. 5, which depicts the displacements corresponding to the fifteen thickness-shear modes of

Table 1. Normalized Frequencies, Ω, of Thickness-Shear Vibrations in Plates $2L + 1$ Atomic Layers Thick.

P/L	0	1	2	3	4	5	6	7
0	0	0	0	0	0	0	0	0
1		0.955	0.984	0.990	0.995	0.997	0.998	0.998
2		1.654	1.871	1.934	1.960	1.972	1.978	1.985
3			2.575	2.778	2.865	2.909	2.935	2.951
4			3.027	3.484	3.683	3.786	3.845	3.884
5				4.015	4.389	4.586	4.701	4.775
6				4.344	4.962	5.292	5.488	5.613
7					5.385	5.891	6.196	6.390
8					5.643	6.370	6.811	7.097
9						6.719	7.328	7.726
10						6.932	7.738	8.270
11							8.038	8.724
12							8.216	9.082
13								9.340
14								9.497

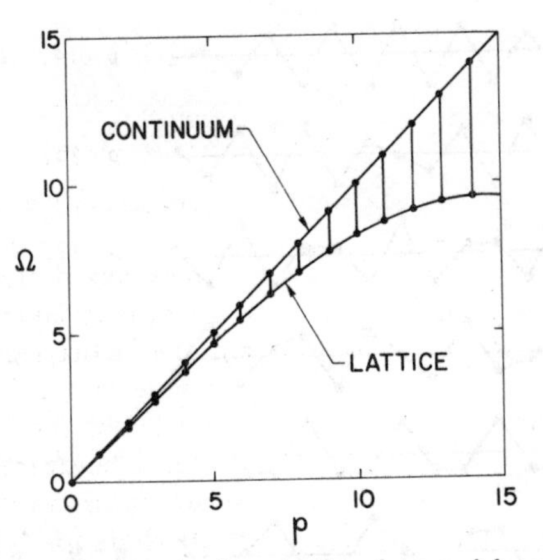

Fig. 4. Thickness-shear frequencies for a lattice plate, and for a continuum plate with the same long-wave properties.

a lattice plate fifteen atomic layers thick. The lowest modes, which are close to the origin of the dispersion curve of Fig. 4, are associated with mode shapes which are almost like the sinusoidal ones of the continuum solution. As we go to higher and higher modes, the wavelength becomes more and more comparable to the atomic constant, and the mode shapes deviate more and more from the sinusoidal form.

4.2 *Face-Shear and Thickness-Twist Waves in a Plate*

A face-shear wave in a plate is one in which motion is normal to the sagittal plane (the plane defined by the direction of propagation and the normal to the plate boundaries), and uniform across the thickness of the plate. A thickness-twist wave also involves motion normal to the sagittal plane, but with a sinusoidal variation across the thickness. It will be seen that the thickness-shear vibrations just described in Section 4.1 are the limits, for infinite wavelength, of the face-shear and thickness-twist waves, which correspond to displacements of the form

$$u_1{}^{l,m,n} = u_2{}^{l,m,n} = 0$$

$$u_3{}^{l,m,n} = (A_1 \cos \xi la + A_2 \sin \xi la) \exp [i(\eta ma - \omega t)] \qquad (4.20)$$

where $0 \leq \xi a \leq \pi$, $0 \leq \eta a \leq \pi$. The equations of motion are satisfied by these displacements, provided that

$$\rho a^2 \omega^2 = 4C_{44}(\sin^2 \tfrac{1}{2}\xi a + \sin^2 \tfrac{1}{2}\eta a) \qquad (4.21)$$

The boundary conditions again require that Eq. (4.11) be satisfied. As a result,

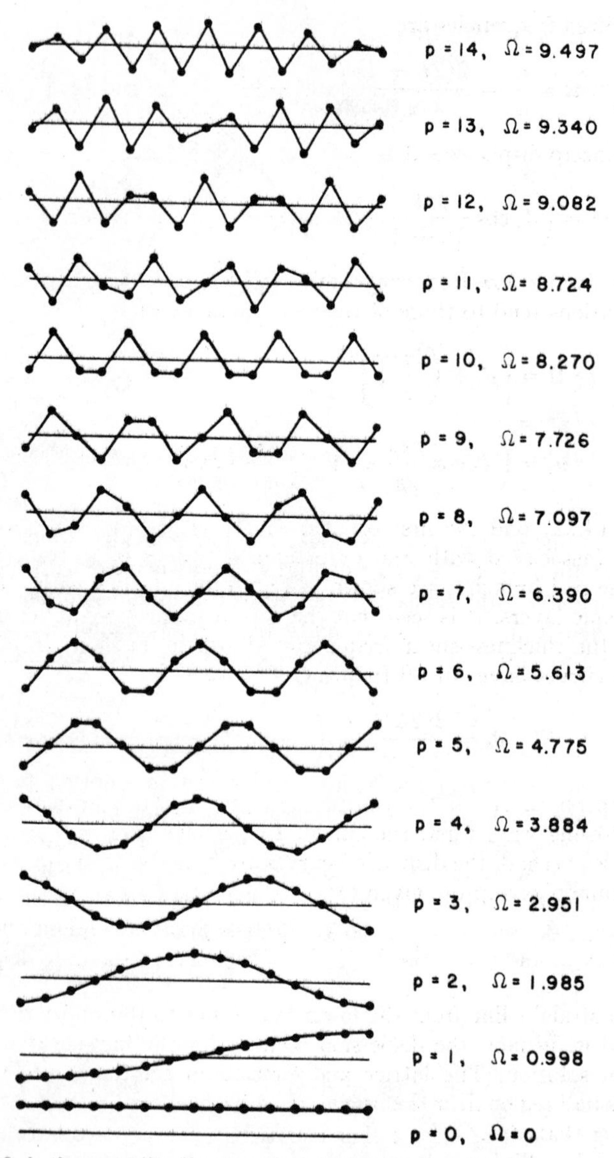

Fig. 5. Mode shapes of the fifteen modes of thickness-shear vibration of a simple cubic lattice plate composed of fifteen atomic layers.

the normalized frequencies are

$$\Omega = \frac{\omega}{\omega_1} = \frac{2(2L+1)}{\pi}\left(\sin^2 \frac{p\pi}{2(2L+1)} + \sin^2 \tfrac{1}{2}\eta a\right)^{1/2} \tag{4.22}$$

and the nonzero displacement is

$$u_3{}^{l,m,n} = \left(A_1 \cos \frac{p\pi l}{2L+1} + A_2 \sin \frac{p\pi l}{2L+1}\right) \exp\left[i(\eta m a - \omega t)\right] \tag{4.23}$$

In the long-wave, low-frequency limit ($la \to x_1$, $ma \to x_2$, $p \ll 2L+1$), the above equations tend to those of the continuum solution

$$\Omega = \left[p^2 + \left(\frac{2\eta h}{\pi}\right)^2\right]^{1/2}$$

$$u_3 = \left(A_1 \cos \frac{p\pi x}{2h} + A_2 \sin \frac{p\pi x}{2h}\right) \exp\left[i(\eta x_2 - \omega t)\right] \tag{4.24}$$

Equation (4.22) and the first of Eqs. (4.24) yield both real and imaginary branches, (associated with real or imaginary values of η, respectively), for real Ω. The real branches are shown in Fig. 6 for a lattice with a thickness of fifteen atomic layers. It is seen that the cut-off frequencies at zero wave number η are the thickness-shear frequencies shown in Table 1. In addition, the dispersion curves have cut-off frequencies for $\eta a = \pi$,

$$\Omega_b = \frac{2(2L+1)}{\pi}\left(1 + \sin^2 \frac{p\pi}{2(2L+1)}\right) \tag{4.25}$$

The dispersion curves of the continuum solution differ from those of Fig. 6 in the following way. First, the cut-off frequencies at $\eta = 0$ are the integers from 0 to ∞. Second, the dispersion curves are hyperbolas in the $(\Omega - \eta)$-plane with a common asymptote given by

$$\Omega_\infty = \frac{2\eta h}{\pi} = \frac{(2L+1)\eta a}{\pi} \tag{4.26}$$

which is a straight line from the lower left corner to the upper right corner of Fig. 6 and is, in fact, the dispersion relation for the face-shear waves in the continuum solution. The lattice and continuum solutions are in agreement only in a small region near the origin, when both ηa and ξa are relatively small, say, smaller than $\pi/4$.

Two of Mindlin's students have carried out investigations of face-shear and thickness-twist waves in plates of more complex lattice structures, Brady [8] for fcc lattices, and Gong [9] for bcc lattices.

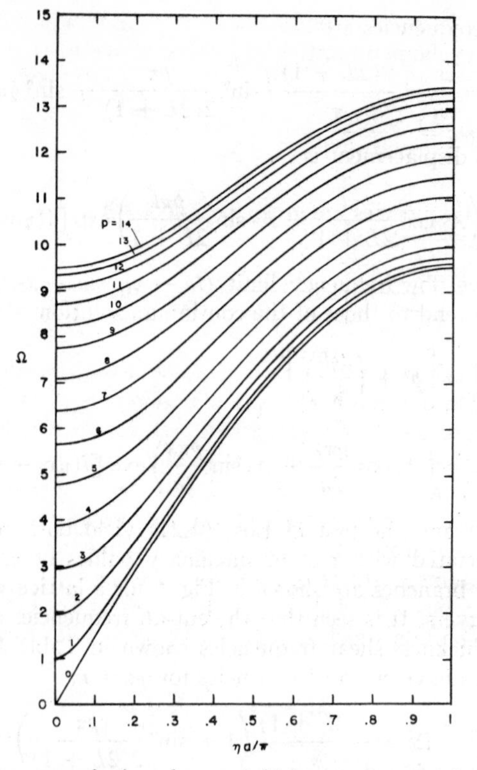

Fig. 6. Dispersion curves for face-shear and thickness-twist waves in a simple cubic lattice plate composed of fifteen atomic layers.

4.3 *Axial Shear Vibrations of a Rectangular Bar* [*M94*]

We consider a rectangular bar composed of a lattice medium and bounded by the planes $l = \pm L$ and $m = \pm M$, and assume displacements of the form

$$u_1{}^{l,m,n} = u_2{}^{l,m,n} = 0$$

$$u_3{}^{l,m,n} = (A_1 \cos \xi la \sin \eta ma + A_2 \sin \xi la \sin \eta ma + A_3 \cos \xi la \cos \eta ma + A_4 \sin \xi la \cos \eta ma) e^{i\omega t} \qquad (4.27)$$

These displacements satisfy the equations of motion (2.2) and boundary conditions (2.3) provided that

$$\xi a = \frac{p\pi}{2L+1}, \qquad p = 0, 1, 2 \ldots < 2L+1$$

$$\rho a^2 \omega^2 = 4C_{44}(\sin^2 \tfrac{1}{2}\xi a + \sin^2 \tfrac{1}{2}\eta a) \qquad (4.28)$$

They also satisfy boundary conditions at $m = \pm M$ which are obtained from Eqs. (2.3) by a cyclic permutation of indices, provided that

$$ma = \frac{q\pi}{2M+1}, \qquad q = 0, 1, 2 \ldots < 2M+1 \tag{4.29}$$

The normalized frequencies are given by

$$\Omega = \frac{2(2L+1)}{\pi}\left[\sin^2 \frac{p\pi}{2(2L+1)} + \sin^2 \frac{q\pi}{2(2M+1)}\right]^{1/2} \tag{4.30}$$

which are exactly those given in Eq. (4.22), with ηa equal to its values defined by Eq. (4.29). This means that we can use a figure such as Fig. 6 to obtain the normalized frequencies of axial shear for a rectangular bar. For example, Fig. 6 can be used for a bar with a thickness of fifteen atomic layers in the x_1-direction and any thickness in the x_2-direction. Given this lattice thickness (i.e., given M) we can obtain fifteen eigen frequencies for each value of $\eta a/\pi = q/(2M+1)$, with q taking values from 0 to $2M$. The total number of frequencies is equal to the number of atoms in the cross section of the bar.

If both the ratios $q/(2M+1)$ and $p/(2L+1)$ are very small, we obtain the continuum solution which is associated with normalized frequencies

$$\Omega = \left(p^2 + \frac{q^2 h_1^2}{h_2^2}\right)^{1/2} \tag{4.31}$$

where h_1 and h_2 are the half-thicknesses in the x_1- and x_2-directions, respectively. As in the case of face-shear and thickness-twist modes, the lattice and continuum solutions are in agreement if the wavelengths of deformation along both directions x_1 and x_2 are moderately large in comparison with the interatomic distance.

5 Diatomic Lattices and Compound Continuum

In the preceding section we discussed simple lattices of the Bravais-type, in which there is only one atom at each lattice site. The long-wave approximation for such lattices yields only branches of low frequency which are identical to those of the continuum case, after appropriate matching of the force constants and elastic constants. There are, however, more complex, non-Bravais lattices which have a "basis," i.e., a group of two or more distinct atoms corresponding to each lattice site, and which can be formed by two or more interpenetrating Bravais lattices. For such lattices the long-wave approximation yields not only low-frequency acoustic branches, but also high-frequency optical branches.

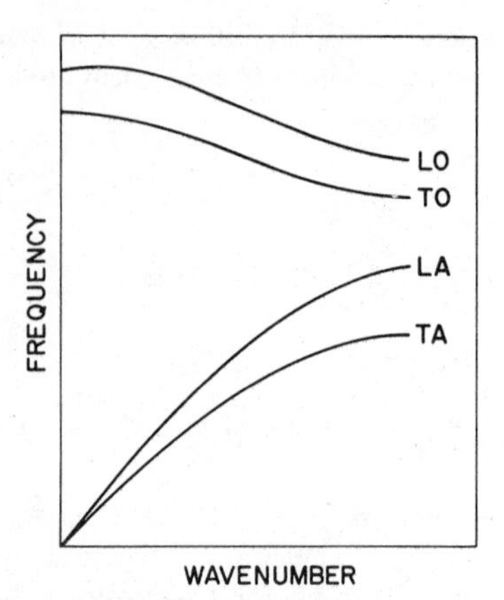

Fig. 7. Acoustical and optical branches for a diatomic lattice (LA = Longitudinal Acoustical, LO = Longitudinal Optical, TA = Transverse Acoustical, TO = Transverse Optical).

(Both, the acoustical and the optical branches are continued over the entire first Brillouin zone as shown in Fig. 7). Mindlin showed [M93] that a continuum model could be developed which would also have an optical branch similar to the long-wave optical branch of a lattice with a base. In analogy with the lattice, he assumed that the continuum was formed by two or more interpenetrating continua and the potential energy of this compound medium depended not only on the strains of the component continua, but also on their relative displacement. The lattice and continuum treatments of compound media will be illustrated by examining the simplest such medium, a sodium chloride-type diatomic cubic crystal. The sodium chloride lattice has atoms at each point of a simple cubic lattice but the sodium and chlorine particles alternate along each row of atoms. The lattice can be formed by two interpenetrating face-centered cubic Bravais lattices, as shown in Fig. 8, where dots and circles represent sodium and chlorine atoms, respectively.

Lattice equations analogous to Eqs. (2.2) can be obtained for such a lattice as was done for the Bravais lattice. We shall first label the sodium and chlorine atoms by the superscripts 1 and 2, and assume that their masses are $\overset{1}{M}$ and $\overset{2}{M}$, respectively. We then consider nearest neighbor central force interactions associated with force constants α, between sodium and chlorine atoms; and next nearest neighbor interaction associated with force constants $\overset{i}{\beta}$, for atoms of the ith type. The noncentral, angular-stiffness interactions are assumed similar to those in Section 2, and associated with a force constant γ regardless of the

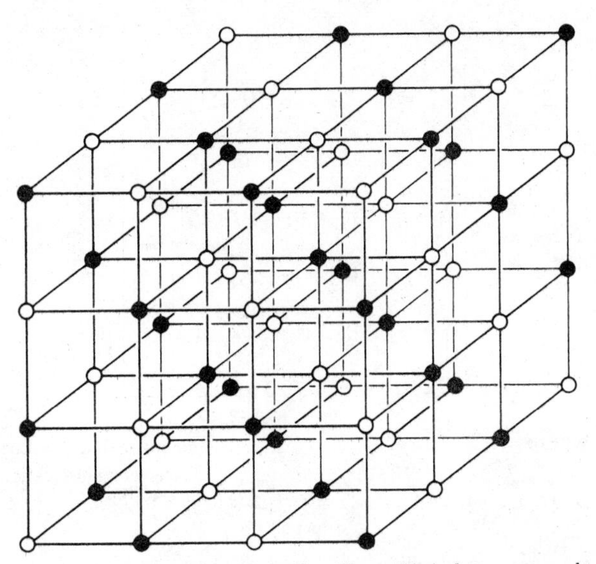

Fig. 8. Structure of a sodium chloride lattice. The solid circles correspond to sodium atoms and the open circles to chlorine atoms.

type of atom at the apex of the deformed right angle. Following the procedure outlined in Section 2, and using the difference operators given in Eq. (2.4), we obtain three equations of motion of the type

$$\begin{aligned}\overset{1}{M}\frac{\partial^2 \overset{1}{u}_1{}^{l,m,n}}{\partial t^2} &= 2(\alpha+8\gamma)(\overset{2}{u}_1{}^{l,m,n}-\overset{1}{u}_1{}^{l,m,n}) + 2a^2\beta(2\Delta_1^2+\Delta_2^2+\Delta_3^2)\overset{1}{u}_1{}^{l,m,n}\\ &\quad + 4(\beta+\gamma)a^2(\Delta_1\Delta_2\overset{1}{u}_2{}^{l,m,n}+\Delta_1\Delta_3\overset{1}{u}_1{}^{l,m,n})\\ &\quad + \beta a^4(\Delta_1^2\Delta_2^2+\Delta_1^2\Delta_3^2)\overset{1}{u}_1{}^{l,m,n}\\ &\quad + \alpha a^2\Delta_1^2\overset{2}{u}_1{}^{l,m,n} + 4\gamma a^2(\Delta_2^2+\Delta_3^2)\overset{2}{u}_1{}^{l,m,n} \qquad (5.1)\end{aligned}$$

and three more equations obtained by interchange of superscripts 1 and 2. Assuming displacements of the form

$$\overset{\nu}{u}_1{}^{l,m,n} = \overset{\nu}{A}_j \exp\left[i(l\theta_1+m\theta_2+n\theta_3-\omega t)\right], \qquad (\nu = 1, 2) \qquad (5.2)$$

we find that the equations of motion are satisfied provided that the dispersion relation

$$\begin{vmatrix} \overset{1}{d}_{11} & \overset{1}{d}_{12} & \overset{1}{d}_{13} & d_{11} & 0 & 0 \\ \overset{1}{d}_{21} & \overset{1}{d}_{22} & \overset{1}{d}_{23} & 0 & d_{22} & 0 \\ \overset{1}{d}_{31} & \overset{1}{d}_{32} & \overset{1}{d}_{33} & 0 & 0 & d_{33} \\ d_{11} & 0 & 0 & \overset{2}{d}_{11} & \overset{2}{d}_{12} & \overset{2}{d}_{13} \\ 0 & d_{22} & 0 & \overset{2}{d}_{21} & \overset{2}{d}_{22} & \overset{2}{d}_{23} \\ 0 & 0 & d_{33} & \overset{2}{d}_{31} & \overset{2}{d}_{32} & \overset{2}{d}_{33} \end{vmatrix} = 0 \tag{5.3}$$

is satisfied, where

$$\overset{\nu}{d}_{ij} = \overset{\nu}{M}\omega^2 - 2(\alpha + 8\gamma) - 4\overset{\nu}{\beta}(2 - \cos\theta_i \sum_{k\neq i} \cos\theta_k), \qquad i = j$$

$$\overset{\nu}{d}_{ij} = -4(\overset{\nu}{\beta} + \gamma)\sin\theta_i \sin\theta_j, \qquad i \neq j$$

$$d_{ii} = 2\alpha\cos\theta_i + 8\gamma \sum_{k\neq i} \cos\theta_k \tag{5.4}$$

At long wavelength and low frequency, we obtain the equation of the classical theory of elasticity, after expanding the difference operators in Taylor series according to Eqs. (2.7) and neglecting all derivatives higher than the second. The resulting approximation is described by equations such as Eq. (4.1), with the displacement components $\overset{1}{u}_i = \overset{2}{u}_i = u_i$, density

$$\rho = \frac{\overset{1}{M} + \overset{2}{M}}{2a^3} \tag{5.5}$$

and the elastic constants given by Eqs. (2.6) with

$$\beta = \frac{\overset{1}{\beta} + \overset{2}{\beta}}{2} \tag{5.6}$$

If we assume long wavelength but not necessarily low frequency, we can obtain the long-wave approximation of both the acoustical and the optical branches.

Setting

$$\overset{1}{\rho} = \frac{\overset{1}{M}}{2a^3} \tag{5.7}$$

we obtain three equations of the type

$$a\overset{1}{\rho}\frac{\partial^2 \overset{1}{u}_1}{\partial t^2} = a^{-2}(\alpha + 8\gamma)(\overset{2}{u}_1 - \overset{1}{u}_1) + \beta(2\partial_1{}^2 + \partial_2{}^2 + \partial_3{}^2)\overset{1}{u}_1$$

$$+ 2(\beta + \gamma)(\partial_1\partial_2\overset{1}{u}_2 + \partial_1\partial_3\overset{1}{u}_3)$$

$$+ \tfrac{1}{2}\alpha\partial_1{}^2\overset{2}{u}_1 + 2\gamma(\partial_2{}^2 + \partial_3{}^2)\overset{2}{u}_1 \tag{5.8}$$

and three more obtained by interchanging the superscripts 1 and 2. In Eq. (5.8) we have used the notation

$$\partial_i = \frac{\partial}{\partial x_i} \tag{5.9}$$

Mindlin obtained [M93] a continuum theory with the properties of a diatomic lattice by assuming two interpenetrating continua, identified by superscripts 1 and 2, which are associated with a potential energy that depends on the strains of the individual continua and their relative displacements. Specifically, the potential energy was taken as

$$W = C_0{}^*(\overset{2}{S}_{ii} - \overset{1}{S}_{ii}) + \tfrac{1}{2}a^{**}u_i{}^*u_i{}^* + C^{**}\omega_{ij}{}^*\omega_{ij}{}^* + \tfrac{1}{2}\sum_{\kappa,\lambda} C_{ijkl}{}^{\kappa\lambda}\overset{\kappa}{S}_{ij}\overset{\lambda}{S}_{kl} \tag{5.10}$$

where

$$u_i{}^* = (\overset{2}{u}_i - \overset{1}{u}_i), \quad \overset{\kappa}{S}_{ij} = \tfrac{1}{2}(\partial_i\overset{\kappa}{u}_j + \partial_j\overset{\kappa}{u}_i), \quad \omega_{ij}{}^* = \tfrac{1}{2}(\partial_i u_j{}^* - \partial_j u_i{}^*)$$

$$C_{ijkl}{}^{\kappa\lambda} = C_{ijkl}{}^{\lambda\kappa} = (C_{11}{}^{\kappa\lambda} - C_{12}{}^{\kappa\lambda} - 2C_{44}{}^{\kappa\lambda})\delta_{ijkl} + C_{12}{}^{\kappa\lambda}\delta_{ij}\delta_{kl}$$

$$+ C_{44}{}^{\kappa\lambda}(\delta_{ik}\delta_{jl} + \delta_{il}\delta_{jk}) \tag{5.11}$$

and the $\delta_{ij}\ldots$ are equal to unity if all subscripts are the same, and zero otherwise. The a^{**} and C^{**} are additional stiffness coefficients. Now, assuming kinetic energy given by

$$T = \tfrac{1}{2}\sum_{i,\kappa}\overset{\kappa}{\rho}\left(\frac{\partial \overset{\kappa}{u}_i}{\partial t}\right)^2 \tag{5.12}$$

and using the usual Hamilton's variational principle, for independent variations $\delta \overset{1}{u}_i$ and $\delta \overset{2}{u}_i$, we obtain six Euler equations of the type

$$
{}^{\kappa}\rho \frac{\partial \overset{\kappa}{u}_j}{\partial t^2} = -(-1)^{\kappa}[a^{**}(\overset{2}{u}_j - \overset{1}{u}_j) + C^{**} \sum_{\lambda} \partial_i(\partial_i \overset{\lambda}{u}_j - \partial_j \overset{\lambda}{u}_i)]
$$

$$
+ \sum_{\lambda} [(C_{11}{}^{\kappa\lambda} - C_{12}{}^{\kappa\lambda} - 2C_{44}{}^{\kappa\lambda}) \delta_{ijkl} \partial_i \partial_k \overset{\lambda}{u}_l]
$$

$$
+ \sum_{\lambda} [C_{12}{}^{\kappa\lambda} \partial_i \partial_j \overset{\lambda}{u}_i + C_{44}{}^{\kappa\lambda} \partial_i(\partial_i \overset{\lambda}{u}_j + \partial_j \overset{\lambda}{u}_i)] \tag{5.13}
$$

where we have employed the summation convention of the indices (repetition of indices implies summation). Equations (5.13) are identical to Eqs. (5.8) if we set

$$
aC_{11}{}^{\kappa\kappa} = 2aC_{44}{}^{\kappa\kappa} = 2\overset{\kappa}{\beta}, \quad aC_{12}{}^{\kappa\kappa} = \beta + 2\gamma, \quad C^{**} = 0
$$

$$
aC_{44}{}^{12} = -aC_{12}{}^{12} = 2\gamma, \quad aC_{11}{}^{12} = \tfrac{1}{2}\alpha, \quad a^{**} = \frac{\alpha + 8\gamma}{a^3} \tag{5.14}
$$

Thus Mindlin's theory of a compound continuum produces equations of motion which are the long-wave approximation of the diatomic lattice equations for a sodium chloride-type lattice, just as the classical elasticity equations of motion are the long-wave limit of the monatomic lattice equations.

6 Shell Models of Dielectric Media

The discussion given so far in this article shows where continuum and lattice theories may be expected to agree or diverge in their results. The lattice theories show a dispersion associated with a characteristic length, the interatomic distance. When the wavelength of deformation is comparable to this distance, the results of the lattice theory deviate substantially from those of the corresponding continuum theory. If we expand the difference operators of the lattice theory in Taylor series and retain only derivatives up to the second, we obtain the "long-wave approximation" which is identical to the continuum theory. If we retain even higher derivatives, we obtain equations analogous to those of a continuum with a potential function dependent on the gradients of strain as well as the strain itself (see also article by Tiersten and Bleustein in this volume). This observation gives a procedure for developing a continuum theory which accounts for specific material properties, guided by a lattice model which contains an explicit description of the origin of these properties.

We already gave an example of an application of this procedure by Mindlin in developing a continuum theory of compound media. Mindlin also developed an improved continuum theory of an elastic dielectric, guided by the appropriate lattice theory for such a medium, the shell-model of a lattice. The continuum theory, discussed in the article by Lee and Haines in the volume, includes the polarization gradient in the potential energy function. In what follows we shall discuss the shell-model of a lattice, and show once more that its results in the long-wave limits match those of the Mindlin continuum theory.

Mindlin developed [M97] his improved continuum theory, and the corresponding shell-model of a lattice, in order to explain an anomaly observed by Mead [12] in measurements of the capacitance of thin dielectric films. According to the classical theory of electrostatics, the capacitance of a metal-dielectric-metal sandwich is inversely proportional to the thickness of the dielectric. Mead measured capacitances which, when plotted in a diagram of inverse capacitance versus thickness (Fig. 9), lie on a straight line that does not go through the origin, as it should according to classical theory. No satisfactory explanation of this anomaly existed until Mindlin showed [M97] that electromechanical coupling could account for it, and gave both a continuum and a lattice description of this coupling.

The lattice considered is a one-dimensional, monatomic one of the Cochran-type [13] based on the Dick–Overhauser [14] shell-model of the atom: a core comprising the nucleus and the inner electrons plus a shell of outer electrons

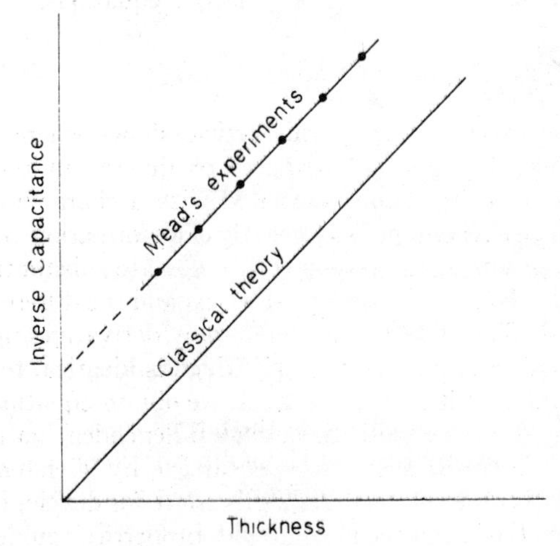

Fig. 9. Experimental results of Mead [12] for the capacitance of thin dielectric films.

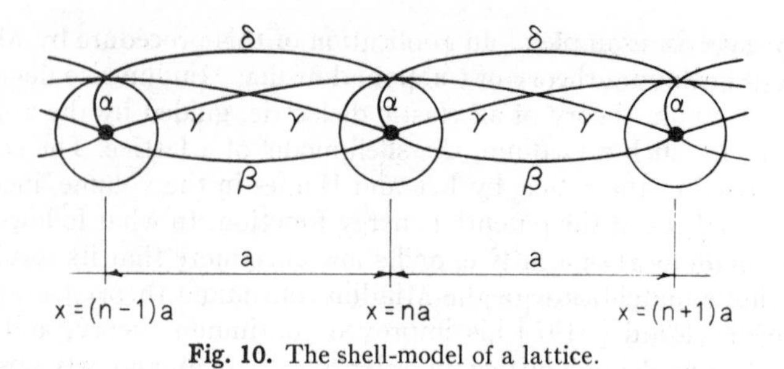

Fig. 10. The shell-model of a lattice.

(Fig. 10). The one-dimensional lattice is assumed extracted from a three-dimensional, monatomic lattice of shell-model atoms. The atoms are situated at positions $x_1 = na$ when in equilibrium, where n is a positive or negative integer. During motion, we assume core displacements u_n and shell displacements S_n. We assume nearest neighbor interaction only, and a harmonic potential which depends not only on the usual core-core interaction but also on core-shell interaction within an atom, core-shell interaction between atoms and shell-shell interactions, which are associated with the force constants α, β, γ, δ, shown in Fig. 10.

The potential energy of a line of atoms, per unit volume a^3, is

$$W = \sum_n \left[\tfrac{1}{2}\epsilon_0 E_n{}^2 + \tfrac{1}{2}\epsilon_0{}^{-1}\eta^{-1}P_n{}^2 + \tfrac{1}{2}b_{11}(P_{n+1} - P_n)^2 + b_0(P_{n+1} - P_n)\right.$$
$$\left. + \tfrac{1}{2}c_{11}(u_{n+1} - u_n)^2 + d_{11}(P_{n+1} - P_n)(u_{n+1} - u_n)\right] \quad (6.1)$$

where P_n is the polarization of the nth atom, per unit area of the three-dimensional lattice, defined by

$$P_n = \frac{(S_n - u_n)q}{a^3} \quad (6.2)$$

q is the electric charge of the atom, E_n is the value of the Maxwell self-field at the lattice points $x_1 = na$, and where we have used the notation

$$c_{11} = (\beta + 2\gamma + \delta)a^{-1}, \qquad b_{11} = \delta a^5 q^{-2}$$
$$d_{11} = (\gamma + \delta)a^2 q^{-1}, \qquad a_{11} = (\alpha + 2\gamma)a^3 q^{-2} = \epsilon_0{}^{-1}\eta^{-1} \quad (6.3)$$

in order to bring out the relationship of the lattice theory with the continuum theory given in the article by Lee and Haines in this volume. The constant c_{11} is an elastic stiffness, ϵ_0 is the permittivity of a vacuum and $\epsilon_0 a_{11}$ is the reciprocal dielectric susceptibility $(= \eta^{-1})$. The remaining three constants b_{11}, d_{11}, and b_0 are associated with terms introduced by Mindlin in the energy density to account for the effect of the polarization gradient, b_{11} being associated with the square of the polarization gradient, d_{11} with the product of the polarization and displacement gradients, and b_0 with the first power of the polarization gradient. The Maxwell field is now expressed in terms of a potential function ϕ, according to

$$E_n = -\left[\frac{\partial\phi}{\partial x_1}\right]_{x_1=na} \tag{6.4}$$

Furthermore, the gradient of ϕ may be expressed in terms of the values ϕ_n at the lattice points, by expanding $\partial\phi/\partial x_1$ in an infinite series of forward differences [15]. Thus

$$E_n = \partial^+\phi_n = \sum_{m=1}^{\infty} (-1)^{m-1} m^{-1} a^{m-1} \Delta_+{}^m \phi_n \tag{6.5}$$

where $\Delta_+{}^m$ are forward difference operators defined by

$$\Delta_+\phi_n = \frac{\phi_{n+1} - \phi_n}{a}$$

$$\Delta_+{}^2\phi_n = \frac{\phi_{n+2} - 2\phi_{n+1} + \phi_n}{a^2}$$

$$\Delta_+{}^3\phi_n = \frac{\phi_{n+3} - 3\phi_{n+2} + 3\phi_{n+1} - \phi_n}{a^3} \tag{6.6}$$

Now, following Toupin [16], we define an electric enthalpy function by

$$\begin{aligned}\mathcal{H} &= W - \sum_n E_n(\epsilon_0 E_n + P_n) \\ &= \sum_n [b_0\Delta_+P_n + \tfrac{1}{2}\epsilon_0^{-1}\eta^{-1}P_n{}^2 + \tfrac{1}{2}b_{11}(\Delta_+P_n)^2 + \tfrac{1}{2}c_{11}(\Delta_+u_n)^2 \\ &\quad + d_{11}(\Delta_+P_n)(\Delta_+u_n) - \tfrac{1}{2}\epsilon_0(\partial^+\phi_n)^2 + P_n(\partial^+\phi_n)]\end{aligned} \tag{6.7}$$

Then the equations of equilibrium (or motion) are obtained by differentiating $\mathcal{H}$ with respect to u_n, P_n, and ϕ_n. Thus we obtain

$$-\frac{\partial\mathcal{H}}{\partial u_n} = c_{11}\Delta_1{}^2u_n + d_{11}\Delta_1{}^2P_n = \rho\ddot{u}_n$$

$$-\frac{\partial\mathcal{H}}{\partial P_n} = d_{11}\Delta_1{}^2u_n + b_{11}\Delta_1{}^2P_n - \epsilon_0^{-1}\eta^{-1}P_n - \partial^+\phi_n = 0$$

$$-\frac{\partial\mathcal{H}}{\partial\phi_n} = -\epsilon_0\partial^-\partial^+\phi_n + \partial^-P_n = 0 \tag{6.8}$$

where $\Delta_1{}^2$ is the central difference operator defined in Eq. (2.4). The inertia of the shell is assumed negligible, and ∂^- is a Taylor series expansion of $\partial/\partial x_1$

in terms of backword differences, analogous to the ∂^+ defined by Eqs. (6.5) and (6.6). In deriving the third of Eqs. (6.8) we have used the identity

$$\sum_m P_m \frac{\partial(\partial^+\phi_m)}{\partial\phi_n} = -\partial^- P_n \tag{6.9}$$

If the lattice is of finite thickness spanning an odd number of atoms with the end ones at $n = \pm N$, the conditions for free boundaries are

$$-\frac{\partial\mathcal{H}}{\partial(\Delta_+ u_{\pm N})} = c_{11}\Delta_+ u_{\pm N} + d_{11}\Delta_+ P_{\pm N} = 0$$

$$-\frac{\partial\mathcal{H}}{\partial(\Delta_+ P_{\pm N})} = d_{11}\Delta_+ u_{\pm N} + b_{11}\Delta_+ P_{\pm N} + b_0 = 0$$

$$-\frac{\partial\mathcal{H}}{\partial(\partial^+\phi_{\pm N})} = \epsilon_0\partial^+\phi_{\pm N} - P_{\pm N} = 0 \tag{6.10}$$

where

$$\Delta_+ f_{\pm N} = \frac{f_{\pm(N+1)} - f_{\pm N}}{a}$$

$$\partial^+\phi_{\pm N} = (\partial^+\phi_n)_{n=\pm N} \tag{6.11}$$

In general, admissible boundary conditions are the specification at each end of the lattice of one member of each of the three products

$$u_{\pm N}(c_{11}\Delta_+ u_{\pm N} + d_{11}\Delta_+ P_{\pm N})$$

$$P_{\pm N}(d_{11}\Delta_+ u_{\pm N} + b_{11}\Delta_+ P_{\pm N} + b_0)$$

$$\phi_{\pm N}(\epsilon_0\partial^+\phi_{\pm N} - P_{\pm N})$$

giving a total of eight possibilities at each end.

Let us now consider the capacitance of a metal-dielectric-metal sandwich whose middle plane, $x_1 = 0$ is a plane of geometric and material symmetry. We consider the case of equilibrium, $\ddot{u}_1 = 0$, and neglect the effects of plate ends, thus reducing the problem to the one-dimensional case. Let us assume a voltage $\pm V$ at the boundaries $x_1 = \pm h = \pm Na$ which are traction-free. The appropriate boundary conditions are

$$P_{\pm N} = \frac{-k\epsilon_0\eta V}{h}, \qquad 0 \leq k \leq 1$$

$$c_{11}\Delta_+ u_{\pm N} + d_{11}\Delta_+ P_{\pm N} = 0$$

$$\phi_{\pm N} = \pm V \tag{6.12}$$

In the first of Eqs. (6.12), the constant k is introduced to account for the physical properties of the adjacent electrode and metal-dielectric interface in a metal-dielectric-metal sandwich. The polarization of the metal is zero, so if we assumed continuity of the polarization function at the interface, we should set $k = 0$. On the other hand, the classical theory of elastic dielectrics would yield, for a dielectric layer of thickness $2h$ with traction-free boundaries and a voltage $\pm V$ at those boundaries, a polarization at those boundaries corresponding to $k = 1$. It is therefore reasonable to assume that the surface polarization of the dielectric, if not actually zero, lies between zero and the value corresponding to $k = 1$ in the first of Eqs. (6.12). We now assume

$$u_n = B_1 \cosh \frac{na}{\lambda}$$

$$P_n = A_2 + B_2 \cosh \frac{na}{\lambda}$$

$$\phi_n = A_3 na + B_3 \sinh \frac{na}{\lambda} \tag{6.13}$$

which, when substituted in Eqs. (6.8), with $\ddot{u}_n = 0$, yield the relationships

$$A_2 = \epsilon_0 \eta A_3, \qquad B_3 = \frac{\lambda B_2}{\epsilon_0} = \frac{-\lambda c_{11} B_1}{\epsilon_0 d_{11}} \tag{6.14}$$

where

$$\sinh \frac{a}{2\lambda} = \frac{a}{2l}$$

$$l = \left[\frac{b_{11} c_{11} - d_{11}^2}{c_{11}(a_{11} + \epsilon_0^{-1})} \right]^{1/2} \tag{6.15}$$

Applying the boundary conditions (6.12) we obtain the additional conditions

$$A_3 = \frac{B_3}{\eta\lambda} \cosh \frac{h}{\lambda} + \frac{kV}{h}, \qquad B_3 = \frac{(1-k)\eta V}{\eta \sinh (h/\lambda) + (h/\lambda) \cosh (h/\lambda)} \tag{6.16}$$

The capacitance can now be computed according to the formula

$$C = \frac{\epsilon_0 \partial^+ \phi_{\pm N} - P_{\pm N}}{2V} \tag{6.17}$$

if we ignore any voltage drop which may occur in the electrodes. Using Eqs. (6.13) through (6.17) we finally obtain

$$C = \frac{\epsilon}{2h} \frac{1 + (k\eta\lambda/h) \tanh (h/\lambda)}{1 + (\eta\lambda/h) \tanh (h/\lambda)} \tag{6.18}$$

where ϵ is the permittivity of the dielectric given by

$$\epsilon = \epsilon_0(1 + \eta) \tag{6.19}$$

The corresponding continuum solution, using Mindlin's augmented theory of an elastic dielectric which is discussed in the article by Lee and Haines in this volume, is as follows:

The equations of equilibrium are

$$\begin{aligned} c_{11}\partial_1^2 u_1 + d_{11}\partial_1^2 P_1 &= 0 \\ d_{11}\partial_1^2 u_1 + b_{11}\partial_1^2 P_1 - a_{11}P_1 - \partial_1\phi &= 0 \\ -\epsilon_0\partial_1^2\phi + \partial_1 P_1 &= 0 \end{aligned} \tag{6.20}$$

where $\partial_1 = \partial/\partial x_1$, and the boundary conditions are

$$\begin{aligned} [P_1]_{x_1=\pm h} &= -\frac{k\epsilon_0\eta V}{h}, \qquad 0 \leq k \leq 1 \\ [c_{11}\partial_1 u_1 + d_{11}\partial_1 P_1]_{x_1=\pm h} &= 0 \\ [\phi]_{x_1=\pm h} &= \pm V \end{aligned} \tag{6.21}$$

The displacements

$$\begin{aligned} u_1 &= B_1' \cosh\frac{x_1}{l} \\ P_1 &= A_2' + B_2' \cosh\frac{x_1}{l} \\ \phi &= A_3' x_1 + B_3' \sinh\frac{x_1}{l} \end{aligned} \tag{6.22}$$

together with Eqs. (6.20) yield

$$A_2' = \epsilon_0\eta A_3', \qquad B_3' = \frac{lB_2'}{\epsilon_0} = \frac{-lc_{11}B_1'}{\epsilon_0 d_{11}} \tag{6.23}$$

with l given by Eqs. (6.15). Finally, application of the boundary condition (6.21) yields the relationships

$$\begin{aligned} A_2' + B_2' \cosh\frac{h}{l} &= -\frac{k\epsilon_0\eta V}{h} \\ A_3' h + B_3' \sinh\frac{h}{l} &= V \end{aligned} \tag{6.24}$$

which, combined with Eqs. (6.22) and (6.23) lead to the formula for the capacitance

$$C = \frac{[\epsilon_0 \partial_1 \phi - P_1]_{x_1=\pm h}}{2V} = \frac{\epsilon}{2h} \frac{1 + (k\eta l/h) \tanh (h/l)}{1 + (\eta l/h) \tanh (h/l)} \tag{6.25}$$

We observe that the lattice solution is identical to the continuum one, except that in the former, the displacements and polarization are defined only at the atom sites, and l is replaced by λ. The length λ is a material property which is approximately equal to l, which is also a material property. The lattice and continuum solutions yield the same capacitance provided that both λ and l be large compared to the lattice constant a. (Actually, if $l/a \gg 1$, then also $\lambda/a \gg 1$ according to the first of Eqs. (6.15).)

The relationship between continuum and lattice theory may be further elucidated by considering the dispersion curves obtained by assuming wave motion of the form

$$u_1 = Ae^{i(\xi x_1 - \omega t)}, \quad P_1 = Pe^{i(\xi x_1 - \omega t)}, \quad \phi = Ce^{i(\xi x_1 - \omega t)} \tag{6.26}$$

for the continuum, or

$$u_n = Ae^{i(\xi na - \omega t)}, \quad P_n = Pe^{i(\xi na - \omega t)}, \quad \phi_n = Ce^{i(\xi na - \omega t)} \tag{6.27}$$

for the lattice. Then the equations of motion yield [M97] the dispersion relationship

$$l^2 \xi^4 + \left[1 - \frac{b_{11} \epsilon_0 \rho \omega^2}{c_{11}(1 + \eta^{-1})}\right] \xi^2 - \frac{\rho \omega^2}{c_{11}} = 0 \tag{6.28}$$

for the continuum, and

$$\frac{16 l^2}{a^4} \sin^4 \left(\frac{\xi a}{2}\right) + \left[1 + \frac{b_{11} \epsilon_0 \rho \omega^2}{c_{11}(1 + \eta^{-1})}\right] \frac{4}{a^2} \sin^2 \left(\frac{\xi a}{2}\right) - \frac{\rho \omega^2}{c_{11}} = 0 \tag{6.29}$$

for the lattice. Both the continuum and the lattice dispersion curves are shown in Fig. 11, where it is seen that they both have a real and an imaginary branch in the frequency range of the "passing band" (i.e., the range of the acoustic, real branch). The initial slope of the real branch is the same for both the continuum and the lattice case, hence the two match in the long-wave approximation. However, the imaginary branch, at $\omega \approx 0$, approaches i/l for the continuum case and i/λ for the lattice case. Since at equilibrium this imaginary

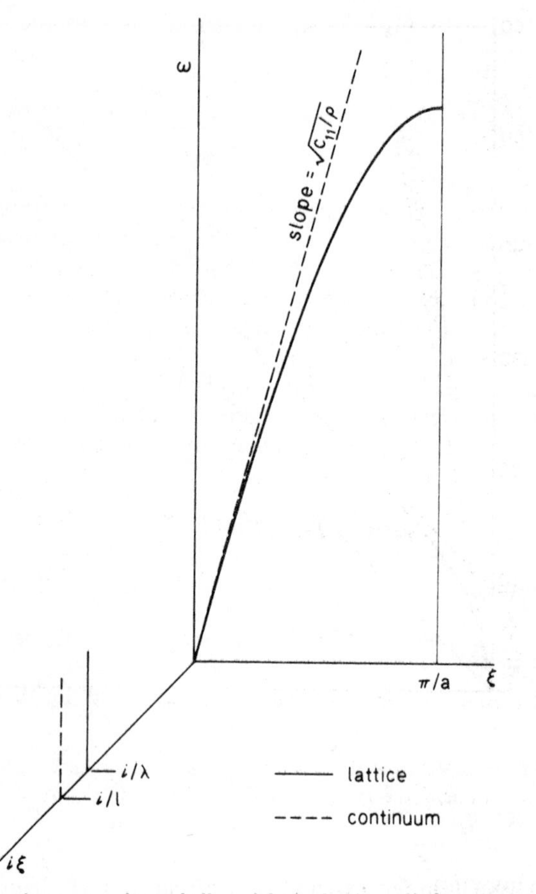

Fig. 11. Dispersion curves for a shell-model of a lattice and the corresponding dielectric continuum proposed by Mindlin.

branch manifests itself as a boundary effect, the amplitude and spatial decay of the displacement will be different in the continuum and lattice solutions if l and λ are appreciably different.

Let us now return to the Mead experiments. Figure 12 shows the normalized capacitance versus normalized thickness, according to the continuum solution (or also the lattice solution for $\lambda \approx l$), and for the case $\eta = 10$, $k = 0.1$. Calculations made by Mindlin using Mead's data indicate that the material property l for all his dielectrics is of the order of a few angstroms. Hence,

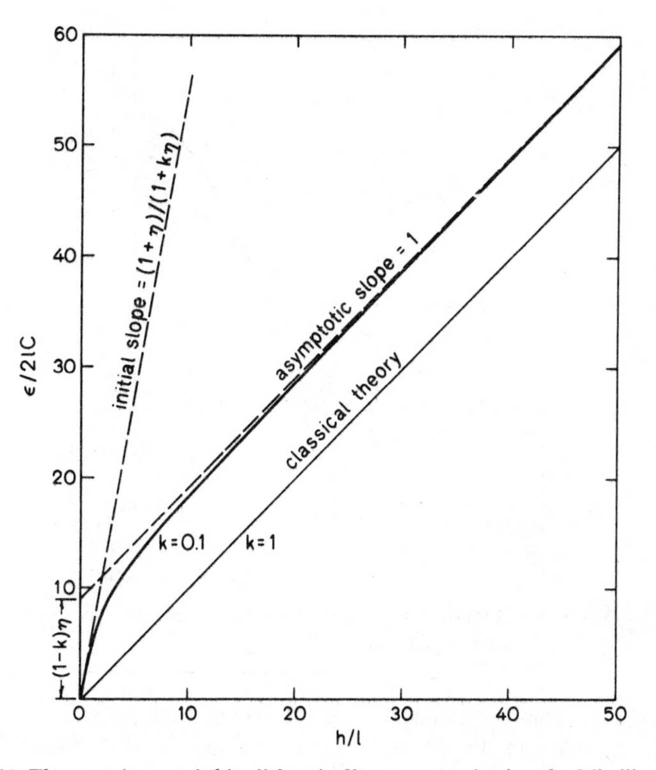

Fig. 12. The capacitance of thin dielectric films computed using the Mindlin augmented continuum theory of elastic dielectrics, or the shell-model of lattice.

Mead's data, which do not extend below a thickness of about 30 Å, would be well into the almost linear portion of the curve of Fig. 12, and hence would tend to indicate a nonzero intercept for zero thickness, as observed by Mead in disagreement with the classical theory.

Another comparison of the classical theory and the lattice theory, or the augmented Mindlin continuum theory, is shown in Fig. 13 where the variation of the polarization P and the potential ϕ are plotted versus the thickness dimension. The absolute value of the polarization is almost uniform across the thickness, and slightly less than the uniform value given by the classical theory; it falls sharply to a smaller value near the boundaries of the dielectric, to a value specified by the first of Eqs. (6.12). The value of the potential varies almost linearly across the thickness, with a slope slightly smaller than that of the

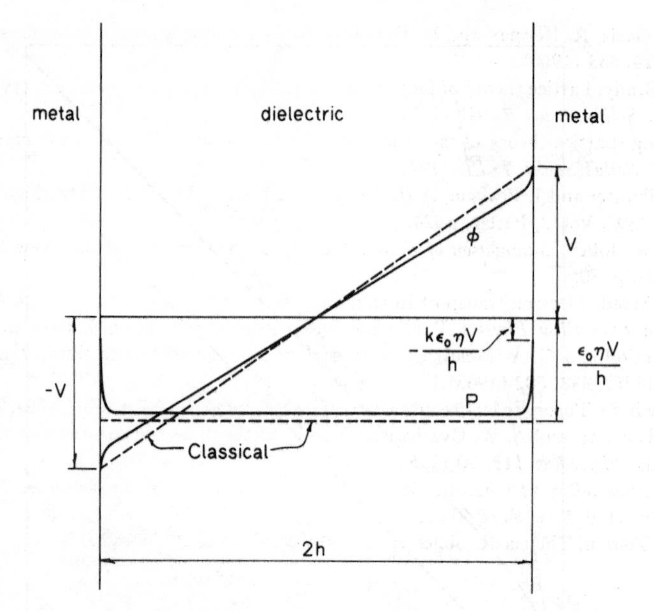

Fig. 13. Variation of polarization, P, and potential, ϕ, across the thickness of a thin dielectric film, according to Mindlin [M97].

classical theory, and tends sharply to $\pm V$ near the boundaries. These boundary effects are mathematically attributable to the imaginary dispersion curve which is absent in the classical theory. They are clearly due to the effect of the polarization gradient which is present explicitly in the augmented Mindlin continuum theory and implicitly in his shell-model lattice theory.

References

1. Sir Isaac Newton, *Principia,* translation by A. Motte, revised by F. Cajori, University of California Press, Berkeley, 1962, Vol. I, Book II, p. 372.
2. M. Born and K. Huang, *Dynamical Theory of Crystal Lattices*, Clarendon Press, Oxford, 1954.
3. C. M. Eisenhauser, I. Pelah, D. J. Hughes and H. Palevsky, Measurements of lattice vibrations in vanadium by neutron scattering, *Phys. Rev.* **109,** 1046 (1958).
4. P. Debye, Zur Theorie der spezifischen Wärmen, *Annalen der Physik* **39,** 789 (1912).
5. B. C. Clark, D. C. Gazis and R. F. Wallis, Frequency Spectra of Body-Centered Cubic Lattices, *Phys. Rev.* **134,** A1486 (1964).
6. A. L. Cauchy, *Exercises de mathêmatique,* De la pression ou tension dans un système de points matériels (1828).

7. D. C. Gazis, R. Herman and R. F. Wallis, Surface elastic waves in cubic crystals, *Phys. Rev.* **119,** 533 (1960).
8. K. J. Brady, Lattice theory of face-shear and thickness-twist waves in f.c.c. crystal plates, *Int. J. Solids Struct.* **7,** 941 (1971).
9. C. Gong, Lattice theory of face-shear and thickness-twist waves in b.c.c. crystal plates, *Int. J. Solids Struct.* **7,** 751 (1971).
10. J. Todhunter and J. Pearson, *A History of the Theory of Elasticity*, Cambridge University Press, 1893, Vol. 2, Part 1, p. 24.
11. L. B. W. Jolley, *Summation of Series*, 2nd edition, Dover Publication, New York, 1968, No. 427, p. 80.
12. C. A. Mead, Electron transport in thin insulating films, *Proc. Int. Symp. on Basic Problems in Thin Film Physics,* Vanderhoeck and Ruprecht, Göttingen, 1966, p. 674. (Also M. McColl and C. A. Mead, Electron current through thin mica films, *Trans. Metall. Soc. AIME* **233,** 502 (1965).)
13. W. Cochran, Theory of lattice vibrations of germanium, *Proc. Roy. Soc.* **A253,** 260 (1959).
14. B. S. Dick, Jr. and A. W. Overhauser, Theory of the dielectric constants of alkali halide crystals, *Phys. Rev.* **112,** 90 (1958).
15. M. G. Salvadori and M. L. Baron, *Numerical Methods in Engineering*, 2nd edition, Prentice-Hall, New York, 1961.
16. R. A. Toupin, The elastic dielectric, *J. Rat. Mech. Anal.* **5,** 849 (1956).